Springer Series in Materials Science

Volume 210

The Springer Series in Materials Science covers the complete spectrum of materials physics, including fundamental principles, physical properties, materials theory and design. Recognizing the increasing importance of materials science in future device technologies, the book titles in this series reflect the state-of-the-art in understanding and controlling the structure and properties of all important classes of materials.

More information about this series at http://www.springer.com/series/856

Soo-Jin Park

Carbon Fibers

Second Edition

Springer

Soo-Jin Park
Department of Chemistry
Inha University
Incheon
Korea (Republic of)

ISSN 0933-033X ISSN 2196-2812 (electronic)
Springer Series in Materials Science
ISBN 978-981-13-0537-5 ISBN 978-981-13-0538-2 (eBook)
https://doi.org/10.1007/978-981-13-0538-2

Library of Congress Control Number: 2018942158

Printed on acid-free paper

This Springer imprint is published by the registered company Springer Nature Singapore Pte Ltd.
part of Springer Nature
The registered company address is: 152 Beach Road, #21-01/04 Gateway East, Singapore 189721, Singapore

Preface

Recently, carbon-based materials have received much attention for their many potential applications. The carbon fibers are very strong, stiff, and lightweight, enabling the carbon materials to deliver improved performance in several applications such as aerospace, sports, automotive, wind energy, oil and gas, infrastructure, defense, and semiconductors. However, the use of carbon fibers in cost-sensitive, high-volume industrial applications is limited because of their relatively high costs. However, its production is expected to increase because of its widespread use in high-volume industrial applications; therefore, the methods used for manufacturing carbon fibers and carbon fiber-reinforced composites and their structures and characteristics need to be investigated.

This second edition book contains ten chapters that discuss the manufacturing methods, surface treatment, composite interfaces, microstructure–property relationships with underlying fundamental physical and mechanical principles, and applications of carbon fibers and their composites.

Chapter 1 provides a brief overview of carbons and carbon fibers, including their origin, history, manufacturing technologies, performance, and global market trends. Chapter 2 introduces the precursors and manufacturing processes of carbon fibers. Chapter 3 lists the various matrices for carbon fiber composites, and focuses on the thermosetting resins and thermoplastic resins. Chapter 4 reviews the effect of surface treatment on the properties of carbon fiber composites. The characteristics of carbon fibers were measured using XPS, X-ray diffraction (XRD), EA, Raman spectroscopy, scanning tunneling microscopy (SEM), and atomic force microscopy (AFM), and are presented in Chap. 5. The selection of manufacturing processes, matrix type, and molding processes for carbon fiber composites are presented in Chap. 6. Chapter 7 describes the recent use of carbon fibers for applications such as adsorbents, energy storage, molecular sieves, catalysts, carbon fiber-reinforced composites. Chapter 8 discusses the carbon/carbon composites, how to manufacturing them, and their properties such as thermal and mechanical properties.

Chapter 9 introduces low-cost techniques for general industries, thin carbon fibers for extreme industries, and smart carbon/carbon composites. Lastly, Chap. 10 describes prospectives and outlook of carbon fibers.

The authors are grateful to everyone who has contributed to this book.

Incheon, Korea (Republic of) Soo-Jin Park, Ph.D.

Contents

Chapter 1
History and Structure of Carbon Fibers

Abstract In this chapter, we will introduce the brief overview of carbon fibers showing superior performance and comprising expensive materials. Further, the origin and history, definition and classification, manufacturing technology, and world market and industry trends in carbon fibers will be presented. Recently, carbon fibers have attracted attention for their potential use as structural materials in aerospace, sports, cars, and bridges. Also, carbon fibers have been considered as the next-generation materials in the aerospace/aviation industry, which has presented a huge opportunity to the entire supply chain of carbon fibers.

1.1 Introduction

Carbon has a number of distinct molecular or crystalline forms termed as allotropes. These carbon allotropes have distinct properties, which are derived from their unique structures. The two most notable allotropes of carbon, each containing an infinite network, are diamond and graphite. These two solids share the same chemistry (carbon) but possess different structures and properties. For example, diamond is the hardest known material, while graphite is extremely soft.

Diamond is the perfect atomic crystal of a giant molecule. It is comprised of sp^3 hybridized carbon atoms joined by four strong covalent bonds (each 0.154-nm long) with complete electron pairs, as typically found in covalent molecules. As shown in Fig. 1.1, the other four carbon atoms point toward the corners of a regular tetrahedron to form a strongly bonded covalent structure. Diamond has three major crystal forms: cubic (100 plane), dodecahedral (110 plane), and octahedral forms (111 plane) (Fig. 1.2) [1].

Graphite is a layered structure with hexagonally arranged carbon atoms in a planar condensed ring system. It has three 0.1415-nm-long strong bonds within the layer that are sp^2 hybridized with one electron capable of taking on a dual role, i.e., formation of coplanar and interplanar bondings, as shown in Fig. 1.3. These distributions are called "π" and "σ" distributions, respectively. The graphite layers are weakly bonded through van der Waal forces. In 1985, a new true carbon form having both sp^2 and sp^3 hybridizations was discovered and was named "fullerenes" [2].

S.-J. Park, *Carbon Fibers*, Springer Series in Materials Science 210,
https://doi.org/10.1007/978-981-13-0538-2_1

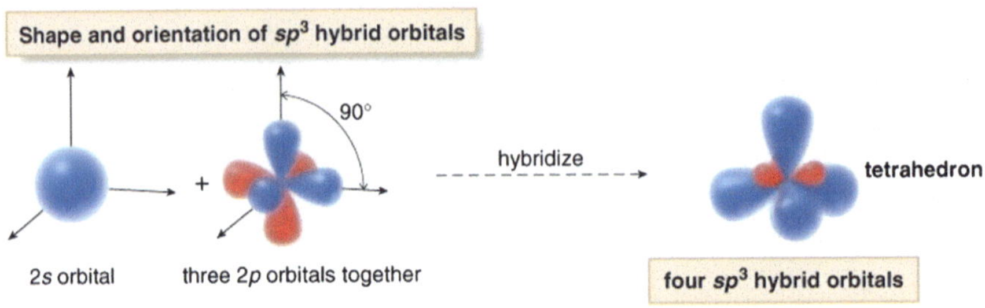

Fig. 1.1 Shape and orientation of sp^3 hybridized orbitals

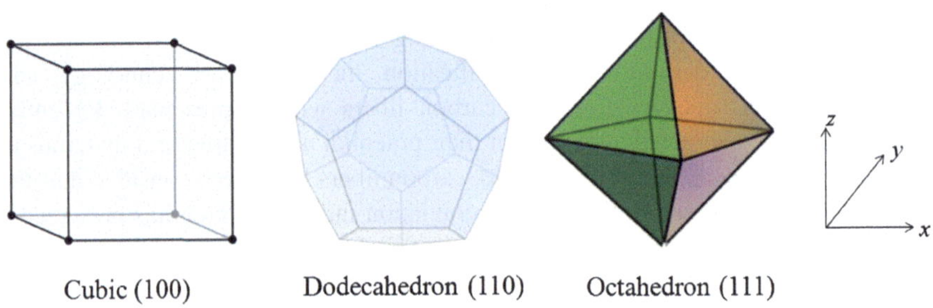

Fig. 1.2 Major crystal forms of diamond

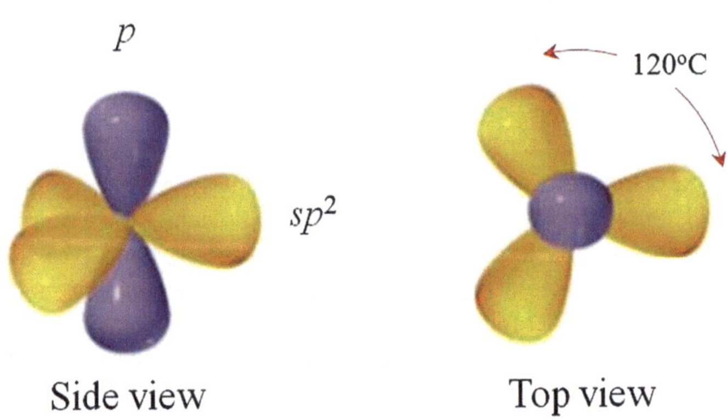

Fig. 1.3 Shape and orientation of sp^2 hybridized orbitals

The relationship among the different allotropes of carbon is presented in Fig. 1.4 (Figs. 1.5 and 1.6).

In addition, there are fibrous forms of carbon such as pyrolytic graphite [3], glass-like carbon [4], graphite whiskers [5], vapor-grown carbon fibers (VGCFs) and catalytic chemical vapor-deposited filaments [6–9], and carbon fibers [10]. Some other forms of carbon are carbon black, charcoal, coal, coke, and soot [11].

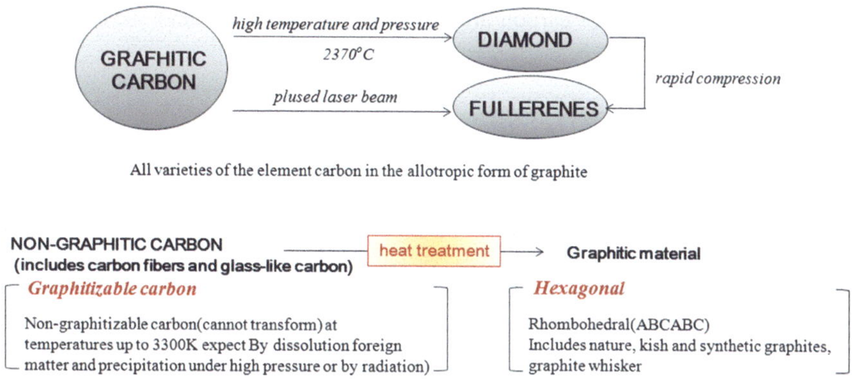

Fig. 1.4 Relationship among the allotropes of carbon

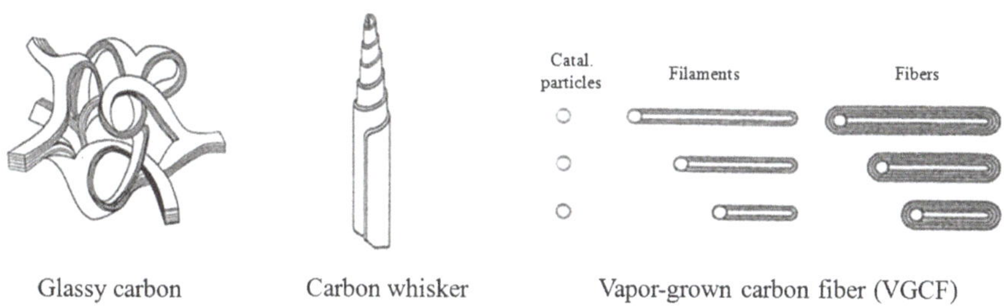

Fig. 1.5 Fibrous forms of carbon

Fig. 1.6 Various types of carbonCarbon

Finally, the newer forms of carbon are fullerenes [12], carbon nanotubes [13], and graphene [14].

1.2 Origin and History of Carbon Fibers

The earliest known commercial use of carbon fibers was in the carbonization of cotton and bamboo fibers for incandescent lamp filaments. In 1879, Thomas Alva Edison, for the first time, used carbon fiber filaments (testing materials) in his early incandescent light bulb experiments, which used electricity to heat a thin strip of material, called a filament, until it glowed. Later, carbon fibers were discovered in 1880, and he patented the use of carbon fibers as filament materials for his electric lamp [15]. Also, he might have created the first commercial carbon fibers. The carbon fibers that Edison made using the natural bamboo fibers or cotton threads were fire resistant, making them ideal for his early incandescent filaments. Cotton and bamboo mostly consist of cellulose, a natural linear polymer composed of repeating units of glucose. When heated, the filament carbonized in the absence of air in a heated gas furnace and then formed a true carbon copy of the starting materials with exact sizes or shapes. Meanwhile, the use of carbon filaments for electric lamps was relatively short-lived; tungsten wire soon displaced these carbon filaments. However, the carbon filaments were in use on US Navy ships as late as 1960 because they withstood ship vibrations better than tungsten (Fig. 1.7).

The carbon fibers were first employed by Edison as a filament material in the initial development of the electric light, and that paved the way for the use of more efficient and durable materials within a few years. In particular, the fibrous nature turned out to be of great importance for structural developments in the aerospace and automotive fields. It was expected that the carbon fibers would substitute glass fibers in several fields, including transportation, building construction, marine, electrical and thermal insulation, and industrial products.

The practical commercial uses of carbon fibers, such as reinforcement, came into being since the 1960s when it became apparent that the carbon fibers, which contributed significantly to the strength and stiffness of structural products, could be developed.

The Union Carbide Corporation (UCC) during World War II investigated the fabricated carbon fibers analogously to manufacture rayon and polyacrylonitrile (PAN). Two manufacturing processes of high-strength and high-modulus carbon fibers from rayon and PAN precursor fibers were simultaneously developed in 1959 and 1962, respectively. In 1963, the pitch-derived carbon fibers, with high-modulus properties, were invented. Until now, many precursor materials such as polyesters, polyamides, polyvinyl alcohol, polyvinylidene, poly-p-phenylene, and phenolic resins have been investigated for manufacturing carbon fibers [16–32]. The invention of carbon filaments by Edison paved the way for the oriented process, which was developed nearly a century later to produce the carbon fibers from synthetic fibers.

Fig. 1.7 Replica of Edison's first electric lamp containing a carbon fiber filament [15]

1.3 Definition of Carbon Fibers

Historically, a distinction was made between the fibers, which were heat-treated over the range 1000–1500 °C, called carbon fibers, and those heat-treated above 2000 °C, called graphite fibers. The proportion of graphite in carbon fibers could range from 0 to 100%. When the proportion was above 99%, the fibers were called graphite fibers.

Generally, carbon fibers are widely defined as fibers containing at least 92 and up to 100 wt% carbon. Moreover, carbon fibers are polycrystalline, and usually in the nongraphitic stage. They possess a two-dimensional long-range order of carbon atoms in planar hexagonal networks (Fig. 1.8), but without any measurable crystallographic order in the third direction (z-direction) apart from more or less parallel stacking [33, 34].

On the other hand, a graphite whisker (or carbon whisker) is a single crystal with the carbon layer rolled up like a scroll. The TEM observations of graphite whiskers are presented in Fig. 1.9 [35]. Because of the single crystal nature, graphite whiskers are virtually flaw-free and have exceptionally high strength. However, the production yield of graphite whiskers is too low to be commercially significant [35, 36].

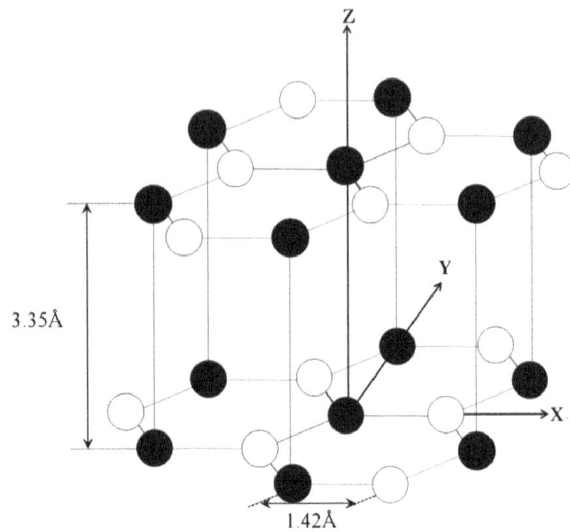

Fig. 1.8 Crystal structure of graphite

1.4 Classification of Carbon Fibers

The carbon fibers play a crucial role in a variety of specialized applications such as aerospace, automobiles, chemical industry, general engineering, missiles, nuclear energy, reinforcement in composite materials, and textiles, owing to their inherent properties, including high strength and stiffness, dimensional stability, low coefficient of thermal expansion, biological compatibility, and fatigue resistance. In this chapter, we have classified the carbon fibers based on their performance, precursors, and commercial availability. Their unique characteristics have also been described.

1.4.1 Performance

Carbon fibers have been classified on the basis of the fiber structure and degree of crystallite orientation: ultrahigh-modulus (UHM), high-modulus (HM), intermediate-modulus (IM), high tensile strength (HT), and isotropic carbon fibers. Broadly speaking, these types have been categorized into three or four groups as listed in Table 1.1.

The UHM and HM carbon fibers are highly graphitized and characterized by a high modulus. The UHM carbon fibers are characterized with a modulus greater than 500 GPa, while the HM carbon fibers are characterized with a modulus greater than 300 GPa and a strength-to-modulus ratio of less than 1%.

The IM and HT carbon fibers have high strength and low modulus owing to the heat treatment at lower temperatures. The IM carbon fibers have a modulus of up to 300 GPa and a strength-to-modulus ratio of $>1 \times 10^{-2}$. The HT carbon fibers are characterized with strength greater than 3 GPa and a strength-to-modulus ratio lying between 1.5 and 2×10^{-2}.

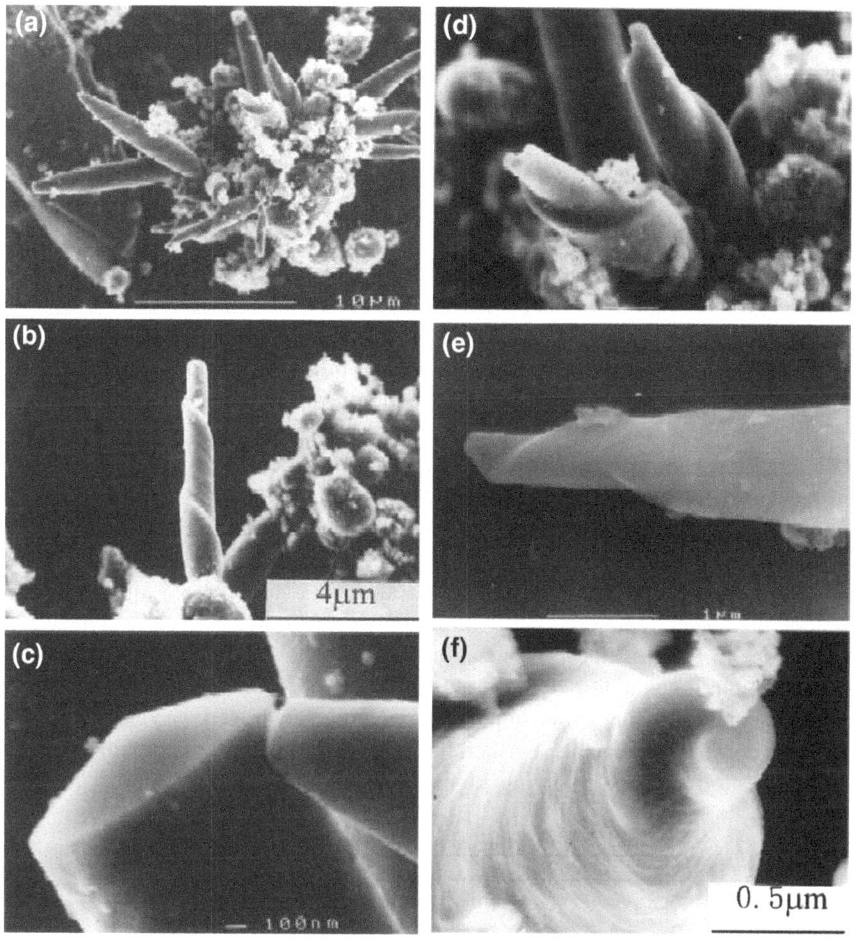

Fig. 1.9 Field emission scanning electron micrographs of whiskers growing at 2100 °C and 100 Pa: **a** aggregation of several whiskers with different morphologies on ground graphite particle; **b** a whisker with long spiral; **c** fracture surface of a whisker with conical shape proves that the whisker is composed of conical carbon layers; **d** a couple of whiskers with different spirals; **e** a whisker with two different spirals; and **f** top view of tip of a whisker with spiral [35]

Table 1.1 Classification of carbon fibers

Carbon fiber type	Heat-treatment temperature (°C)	Crystallite orientation	Long-distance order	
Type I high modulus	>2000	Mainly parallel to the fiber axis	High	UHM
				HM
				IM
Type II high strength	≈1500	Mainly parallel to the fiber axis	Low	HT
Type III isotropic	<1000	Random	Very low	Isotropic

Finally, the isotropic carbon fibers show a random orientation of the crystallites and possess a modulus as low as 100 GPa combined with low strength. Their main advantage is low cost (Fig. 1.10).

1.4.2 Precursor

Carbon fibers are manufactured from synthetic fibers (precursor fibers) through heating and stretching treatments. The processing of carbon fibers from different precursors requires different conditions to obtain the satisfactory quality end products. The essential features are similar. The processing paths for various precursors are similar at the macrolevel. In addition, the precursor materials of the carbon fibers are important because the combination of various properties and behaviors (mechanical, physical, chemical) on the carbon fibers depends strongly on the starting precursor materials [38–42].

Generally, the carbon fibers are synthesized through a controlled pyrolysis of the stabilized precursor fibers (Fig. 1.11) [43]. The precursor fibers were first stabilized and stretched over the temperature range 200–400 °C in air using an oxidization process (thermoset treatment), which strongly depended on the natural chemistry of the precursor. Then, in the carbonization process, the stabilized fibers were subjected to high temperatures over the range 800–1600 °C in an oxygen-free environment to remove noncarbon impurities, including hydrogen, oxygen, nitrogen, and other noncarbon elements. Further, the carbonized fibers were graphitized using a graphitization process involving higher temperatures of up to 3000 °C, which

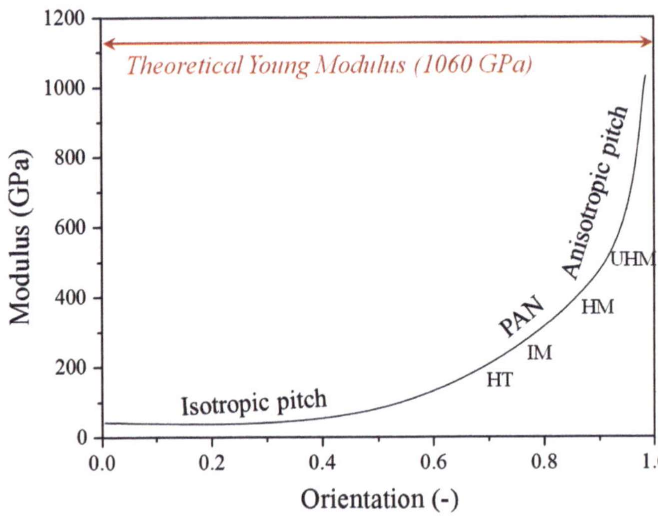

Fig. 1.10 Correlation of orientation and Young's modulus for carbon fibers made from different precursor systems; PAN, isotropic pitch, and anisotropic pitch (mesophase pitch) [37]

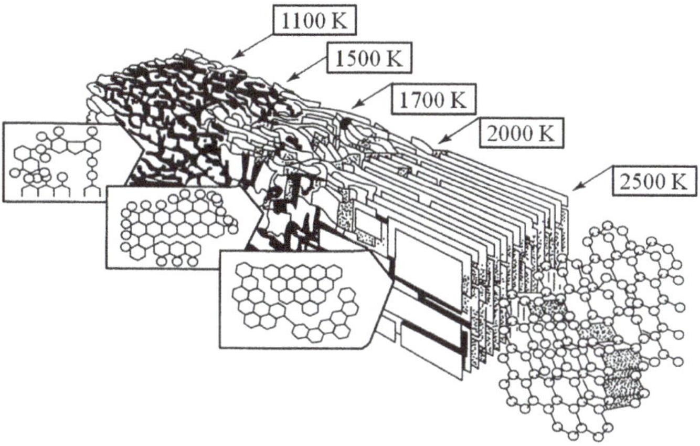

Fig. 1.11 Structural model for carbon fibers during graphitization process [51]

stretched the fibers resulting in 50–100% elongation. The stretching of fibers ensured a preferred crystalline orientation, which resulted in the desired modulus being higher than those of carbonized fibers. The properties of the resultant carbon fibers were relevant to their crystallinity, crystalline distribution, carbon content, molecular orientation, and the proportion of defects. Finally, the surface-treatment processes and epoxy sizing of carbon fibers, i.e., posttreatment processes, were performed to enhance their adhesion to the composite matrices [44–50].

It is known that the polyacrylonitrile (PAN), pitch, and rayon are the three most common precursors for the carbon fibers. A simplified schematic of the process for the preparation of precursor and typical carbon fibers is presented in Fig. 1.12. The process for the manufacturing of PAN-based carbon fibers and pitch-based carbon fibers is shown in Figs. 1.13 and 1.14, respectively.

Figure 1.15 explains in detail the basic chemistry and physics of different processes for fabrication. In case of the PAN fibers, the stretching of the coiled macromolecule in the direction of the polymer chain was parallel to the fiber axis and the subsequent transformation of the linear molecule into a ladder polymer was

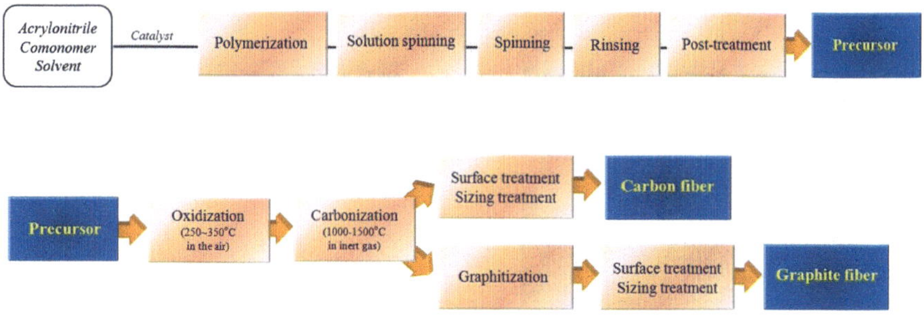

Fig. 1.12 Typical schematic of the process for preparation of precursor and typical carbon fibers

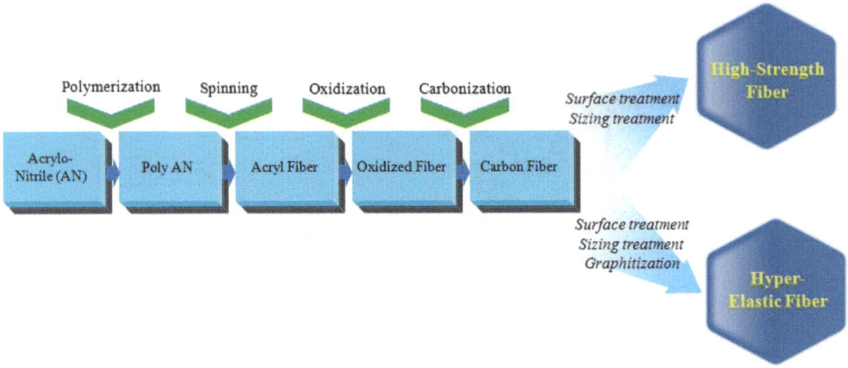

Fig. 1.13 Manufacturing process for PAN-based carbon fibers

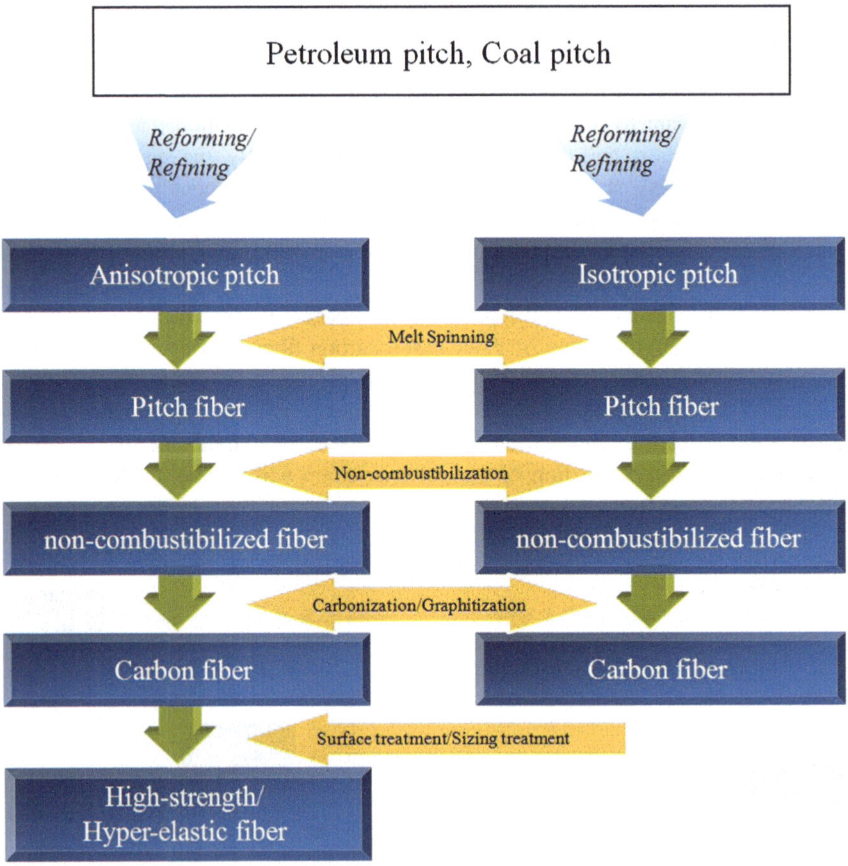

Fig. 1.14 Manufacturing process for pitch-based carbon fibers

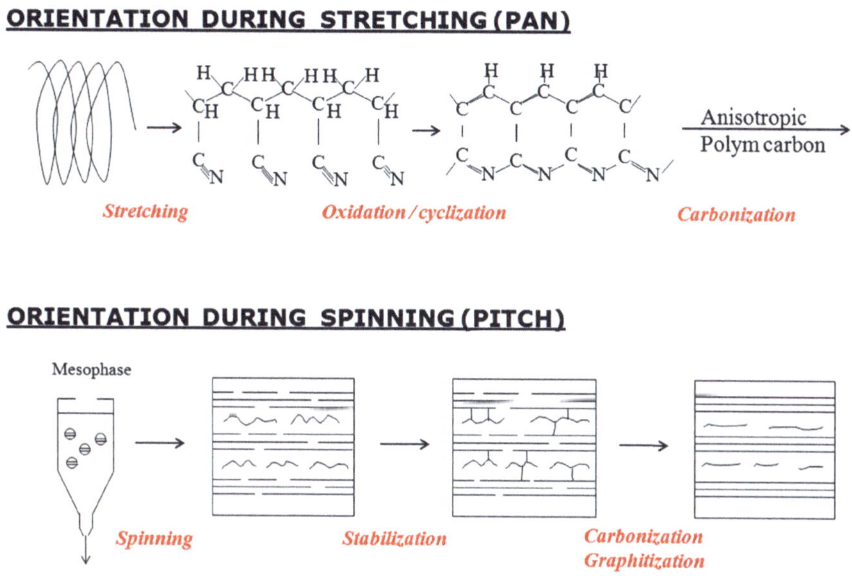

Fig. 1.15 Two types of fabrication processes

observed. The PAN-based precursor fibers are usually fabricated using the conventional spinning technique such as wet spinning. They must be converted to a form which is flameproof and stable at the high temperatures involved in carbonization. Consequently, before carbonization, they are stabilized in the case of the PAN precursor in an oxidizing atmosphere.

In the case of the pitch fibers, the preferred orientation was realized by the spinning process of the liquid crystals in the pitch (melt-spinning and melt-blowing methods) [52–59], which consisted of polyaromatic molecules, in a direction parallel to that of the fibers [55]. The step in the pitch process, involving the introduction of the structurally preferable orientation of the planar molecules, has to be combined with chemical treatment to conserve the resulting carbonaceous defect structure during all the subsequent heat-treatment processes.

Over the last 100 years, it is well known in the graphite industry what type of thermally stable defect structures form by the carbonization of thermosetting polymers. On the other hand, nonstable defect structures, i.e., well-graphitizing cokes, could be achieved by the carbonization of carbon–hydrogen compounds in a fluid state. From these general rules, it is believed that the PAN-based precursor fibers prepared by the cyclization of PAN polymers must yield either nongraphitizing carbon or carbon fibers with a thermally stable, defective structure. In contrast, the anisotropic pitch-based carbon fibers possess the ideal preconditions for good graphitization. This capability of the anisotropic pitch must be destroyed before carbonization by forced oxidation and cross-linking reactions.

Meanwhile, the pitch precursors are divided into the isotropic and anisotropic pitches (mesophase pitch) based on their optical characteristics. On the basis of their mechanical properties, pitch-based carbon fibers can be further classified into high-performance carbon fibers (HP-carbon fibers) and general-purpose carbon fibers (GP-carbon fibers). The isotropic pitch forms an isotropic carbon fiber, which belongs to the category of GP-carbon fibers, whereas the anisotropic pitch forms HP-carbon fibers, wherein the carbon layers are preferentially parallel to the fiber axis.

The pitch is usually derived as a by-product from either coal- or petroleum-based chemistry. The pitch from petroleum is preferred because the coal pitch is difficult to clean owing to natural inclusions as defect-sensitive particles. In contrast to the PAN-based precursors, the pitch is a low molecular weight material.

In addition, the synthetic pitch could be produced from the synthetic polymers such as polyvinyl chloride (PVC) and naphthalene [56]. Ohtani [57] reported the molten pyrolysis product of PVC for manufacturing carbon fibers. This method suggested the possibility of utilizing tar pitches from oil refinery and coal pyrolysis processes as new precursors for carbon fibers. Today, pitch-based carbon fibers have been introduced in the market [58].

Wazir et al. reported that when a petroleum pitch was heated at 420 °C for 7 h in a nitrogen atmosphere, a carbon fiber precursor is formed with a softening point of 295 °C. The precursor was successfully melt-spun into fibers through a circular nozzle of a monofilament spinning apparatus. Wazir et al. focused on optimizing the operating conditions for melt spinning of the precursor pitch [59] (Fig. 1.16).

Rayon-based carbon fibers are obtained from the rayon precursor fibers using chemical pretreatments and carbonization. These are isotropic carbon fibers and can be transformed into anisotropic carbon fibers with high strength and stiffness by stretching at temperatures above 2500 °C. However, as shown in Fig. 1.17, all glycosidic bonds generate volatile carbon dioxide during pyrolysis. It is difficult to avoid the fracture of the bimolecular unit of the polymer in the volatile ether glucosan. Mass loss from the polymer structure at different stages of pyrolysis causes an enormous amount of shrinkage, resulting in a low carbon yield of $<30\%$. Consequently, an extremely porous carbon residue with low density is formed.

Figure 1.18 shows the carbon yields of aromatic polymers such as poly(p-phenyleneacetylene) and others, compared to those of rayon, PAN, and pitch [60, 61].

It is well known that there are primarily three types of precursor materials of commercial significance, as previously mentioned, PAN, pitch, and rayon. Although the worldwide production of carbon fibers has increased rapidly, the prohibitive cost of carbon fiber production and high demand limit the widespread use of carbon fibers. The overall challenge is the combination of precursor cost, yield, processing cost, and similar factors.

In addition, carbon fibers have been previously manufactured from lignin. Lignin-based carbon fibers such as Kayacarbon were first developed and made commercially available by Nippon Kayaku Co. on a pilot scale. The process involved carbonization of dry-spun fibers from lignin dissolved in an alkali solution containing poly(vinyl alcohol) as a plasticizer [62–64].

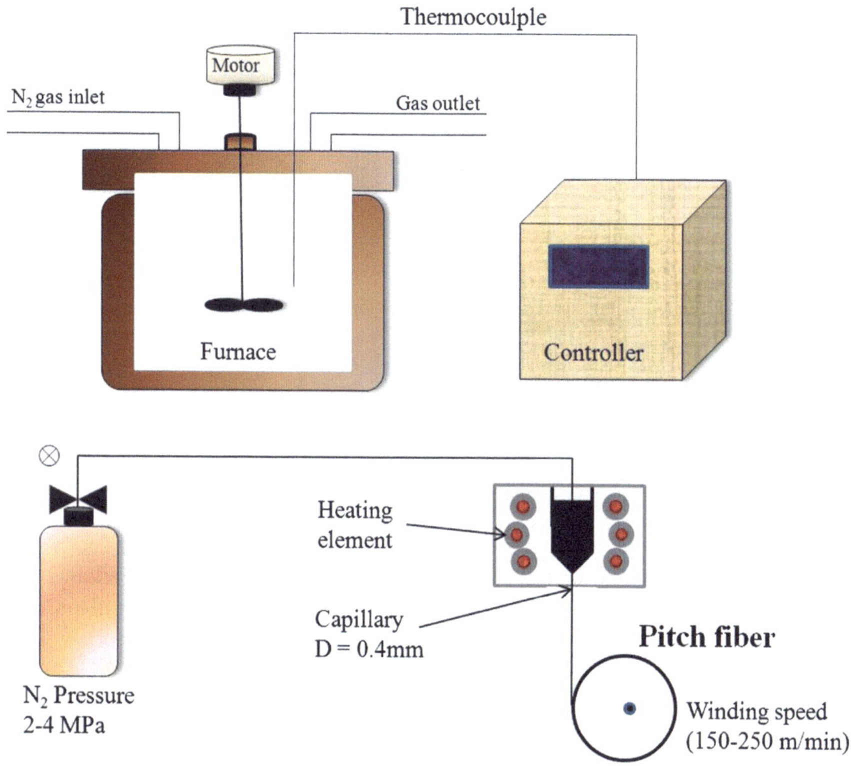

Fig. 1.16 Schematic of reactor and melt-spinning apparatus [59]

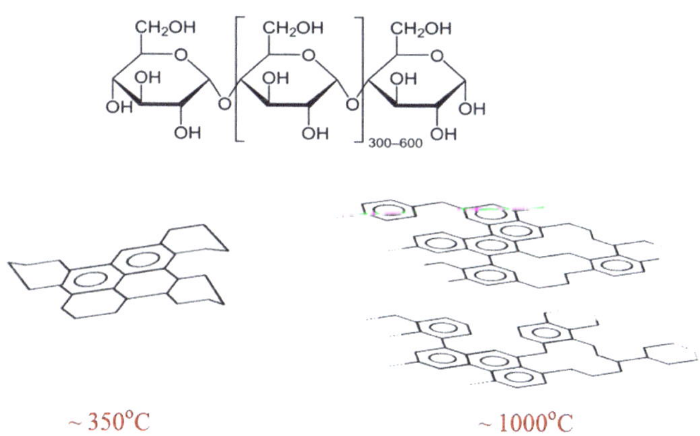

Fig. 1.17 Schematic of thermal degradation of cellulose to carbon

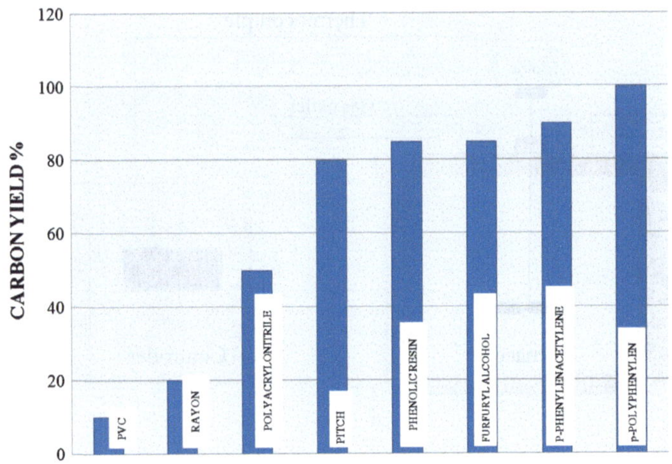

Fig. 1.18 Comparison of carbon yield of polymers after pyrolysis

Lignin, second only to cellulose in natural abundance, is readily available, relatively inexpensive, and structurally appealing as a precursor for carbon fibers. It is a high molecular weight polyaromatic macromolecule with a reported total worldwide production of approximately 26 million ton/year [65].

However, lignin-based carbon fibers have a low tensile strength of approximately 0.3 GPa, in comparison with typical commercial PAN-based carbon fibers (above 3.5 GPa). Thus, a few researchers have reported the lignin-based carbon fibers with mechanical properties, which are suitable for either general purpose (GP) or above grades, through the various pretreatment processes of lignin [66–68]. Kadra et al. have reported the lignin-based carbon fibers modified by the kraft process (a process for converting wood into wood pulp comprising almost pure cellulose fibers) and blends, thereby reaching a tensile strength of approximately 0.5 GPa and a modulus of up to 60 GPa [69]. Kubo et al. have investigated lignin-based carbon fibers by blending them with PET, which resulted in a tensile strength of 0.7 GPa and a modulus of 94 GPa [70]. A more recent investigation by the Oak Ridge National Laboratory (ORNL) on lignin-based carbon fibers revealed a tensile strength of approximately 0.51 GPa and a modulus of 28.6 GPa (Fig. 1.19) [71].

Despite the weak mechanical properties, the development of processes utilizing lignin, as a low-cost and renewable resource precursor material for low-cost carbon fiber production, has been relentlessly pursued. A few studies reported the possibility of producing carbon fibers from lignin. Also, the modification of lignin for producing high-performance carbon fibers has been studied using purification, chemical modification, and plasticization [72–76].

Short carbon fibers are being rapidly developed through the preparation in the vapor of hydrocarbon atmosphere with low molecular weight compounds, including acetylene and ethylene. This process involves catalytic growth using solid catalyst particles (Fe, Co, and other transition metals) to form carbon filaments,

Fig. 1.19 Fiber spools produced from organic-purified hardwood lignin utilizing pilot scale unit with 12-hole spinneret [71]

which could be as small as 0.1 μm in diameter. They consist of graphitizable carbon and could be transformed into graphite fibers by heat treatments above 2500 °C. Then, subsequent chemical vapor deposition from the carbonaceous gas in the same chamber caused the filaments to grow in diameter, resulting in vapor-grown carbon fibers (VGCFs) or gas-phase-grown carbon fibers [77–80].

Unlike the complex PAN-based processes, the manufacturing process for "vapor-grown" production technology is simpler, faster, and cheaper. As indicated in Fig. 1.20, VGCFs are produced in the vapor phase by decomposing methane, ethane, or coal gas in the presence of a metal catalyst, hydrogen sulfide, and ammonia. The catalyzed gas decomposes into carbon and hydrogen and gets conveyed to a reactor furnace. Carbon remains in the reactor for only a few milliseconds, growing into a fiber of 60–200 nm in diameter and approximately 100 μm in length. The fiber is debulked for either sale or additional processing (such as heat treatment). In particular, the process of preparing VGCFs overcame many costly production steps and processing compared to those of the PAN-based process; it also provided an innovative approach for fabricating high-performance fibers at lower costs.

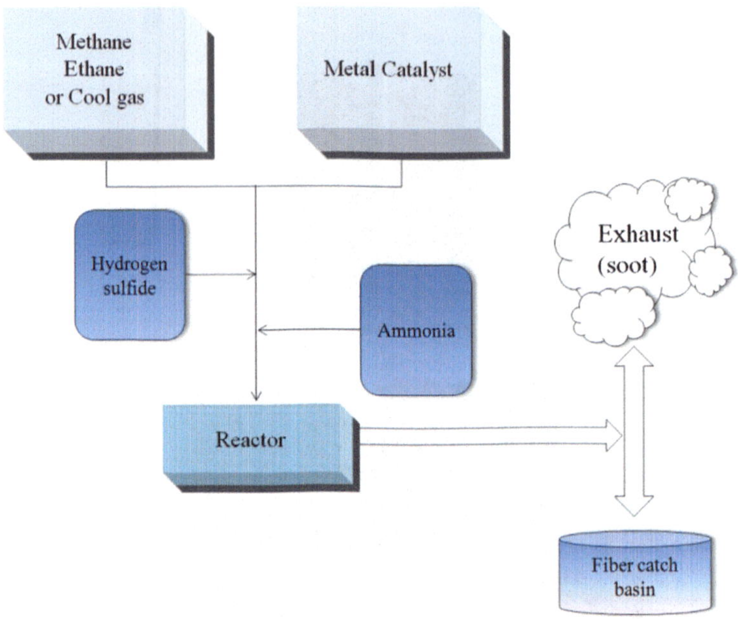

Fig. 1.20 Vapor-grown carbon fiber manufacturing process

1.4.3 Commercial Availability

Carbon fibers could alternatively be classified on the basis of their commercial availability, namely general-purpose (GP), high-performance (HP), and activated carbon fibers (ACF).

The GP-carbon fibers have low tensile strength, low tensile modulus, and low cost because of their amorphous and isotropic structures. The greatest advantage of these carbon fibers is their low cost. The short GP-pitch-based carbon fibers are used in the reinforcement of concrete because it is easy to use in large quantities. The GP-carbon fibers are also used for thermal insulation, sealing materials, electrically conducting materials, antistatic materials, heating elements, filters, fraction materials, sorbents, and catalysts [81].

The HP-carbon fibers are characterized by high strength and modulus in comparison with the GP-carbon fibers. Among the HP-carbon fibers, a higher modulus is associated with a higher proportion of graphite and more anisotropy. The HP-carbon fibers are widely used in polymer–matrix composites for aircraft, which are lightweight for the purpose of saving fuel. Nearly 90% of the structure of the aircraft, Voyager, is made of such composites. In 1986, the Voyager successfully performed a nonstop, unfueled, round-the-world flight. The use of such composites in passenger aircraft is rapidly rising. The HP-carbon fibers are also used in carbon–matrix composites for high-temperature aerospace applications such as the space shuttle, as the carbon matrix is more temperature resistant compared to a polymer matrix. These fibers are now used in metal matrices such as aluminum for aerospace applications as aluminum is more temperature resistant compared to polymers.

Finally, activated carbon fibers (ACFs) have a large specific surface area and micropore volume for the adsorption of either gas or liquid. Compared to the conventional granular or powder-activated carbons, ACFs have been widely used in the separation, purification, and catalytic processes owing to their extended specific surface area, high adsorption capacity, highly porous structure, and special surface reactivity. ACFs have been used not only as reducing agents by themselves but also as catalyst supports for removing pollutants such as NO_x, SO_x, NH_3, and even HCl [82–106]. Also, the microstructure of ACFs is created during the activation, and is affected by many factors such as the degree of activation and carbonization conditions [107]. The adsorption/desorption rate of carbonaceous adsorbents is greatly dependent on not only microporous structures but also surface properties [108–111].

Most recently, ACFs have been drawing attention as adsorbents for hydrogen storage and capture of CO_2 from a thermal power plant. ACFs offer several advantages in terms of cost with rapid adsorption kinetics, availability, large surface area, easy-to-design pore structure, easy surface functionalization, hydrophobicity, reversibility, and low energy requirements for regeneration. With regard to gas adsorption, ACFs are also considered promising materials because of their nanostructure, abundant micrometer porosity, and properties such as high specific surface area and narrow pore size distribution. The fibrous shape of ACFs has the advantage of easy handling compared to granular and powdered adsorbents [112–115].

In particular, the amount of CO_2 adsorbed by the ACF samples (modified by thermally chemical activation) was substantially higher than that obtained for most other well-known adsorbent materials designed for CO_2 capture, including MOF-5, zeolite (Fig. 1.21), and activated carbons [116]. This is because of their well-developed pore structures and inherent CO2 affinity of ACF surfaces for high-performance CO_2 capture (Fig. 1.22).

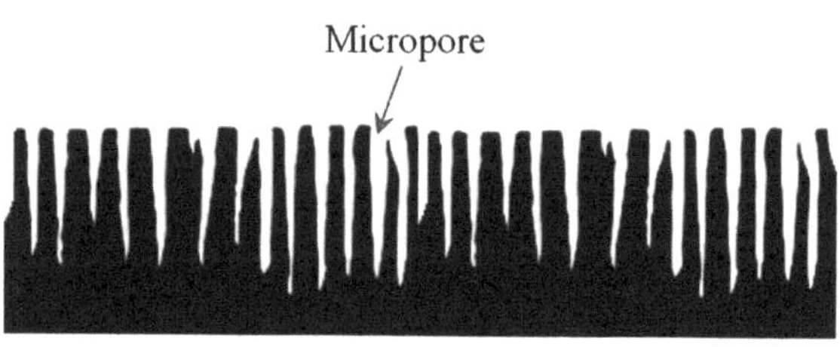

Fig. 1.21 Pore shape and structure of activated carbon fibers

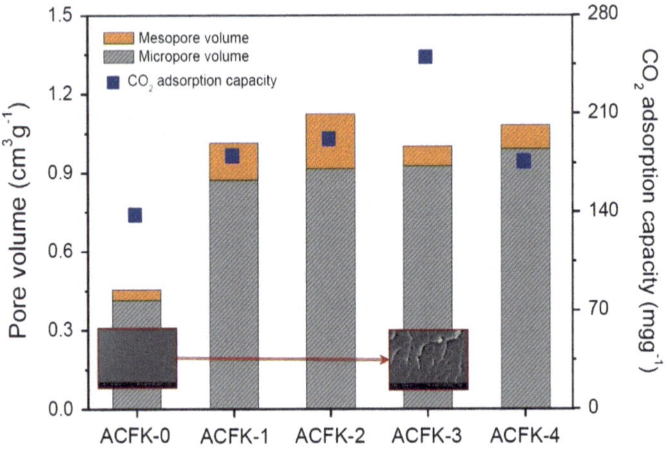

Fig. 1.22 Relationship between micro-/mesopore volume ratios and CO2-adsorption capacity of ACFK samples at various KOH ratios (inset: SEM images before and after KOH heat treatment)

1.5 Structure of Carbon Fibers

Carbon fibers generally have excellent tensile properties, low densities, and high thermal and chemical stabilities in the absence of oxidizing agents, good thermal and electrical conductivities, and excellent creep resistance. They have been extensively used in composites in the form of woven textiles, prepregs, continuous fibers/rovings, and chopped fibers. The composite parts could be produced through filament winding, tape winding, pultrusion, compression molding, vacuum bagging, liquid molding, and injection molding.

Several techniques, including wide- and small-angle X-ray diffraction, electron diffraction, transmission, scamming and surface replica electron microscopy , and optical microscopy, have been investigated for the structure of carbon fibers [117–121]. From these techniques, valuable information about the various micro- and macroaspects of the fiber structure has been provided continually hitherto such as qualitative and quantitative information about several structural parameters, including crystallite size and orientation, stack height and width, shape and size of pores, and presence of three-dimensional order.

Here, we will briefly summarize the unique structures of carbon fibers as discovered using X-ray diffraction and electron microscopy, which are the simplest and most effective techniques for the elucidation and foundational understanding of the basic science of structure in carbon fibers.

Carbon fibers could be either short or continuous. In addition, the structure of carbon fibers could be crystalline, amorphous, or partly crystalline. The atomic structure of a carbon fiber is similar to that of graphite wherein the *c*-direction distance between the layer planes is 3.35 Å (d-spacing, $d_{(002)}$), and consists of carbon atom layers arranged in a regular hexagonal pattern (as previously shown in

Fig. 1.1). It has a high modulus, 0.18–0.35 GPa; however, in humid atmosphere, its shear strength is weak along the a-axis.

In graphitic crystalline regions (microdomains), the crystal structure of carbon fibers consists of sp^2 hybridized carbon atoms arranged two-dimensionally in a honeycomb structure in the x–y plane with van der Waals forces, which are stacked parallel to one another in a regular fashion. In the graphite layer, the carbon atoms are bonded covalently and through metallic bonds owing to the overlap of the sp^2 hybridized orbital and delocalization of the p_z orbitals (the π electrons), respectively. The good electrical conductivity and thermal conductivity of graphite are attributable to the delocalization in the x–y plane.

However, the basic structural unit of most of the carbon fibers consists of a stack of turbostratic layers. In a turbostratic structure, the distance between the parallel graphene sheets is greater than that in graphite (Fig. 1.23). The basic structural unit could either irregularly or haphazardly split, tilt, twist, fold, and join other basic structural units to form microdomains. As shown in Fig. 1.24, these microdomains could also split, tilt, twist, fold, and join each other in carbon fibers [121]. Thus, the structure of carbon fibers is not homogeneous. It has been reported that the irregular stacking and presence of sp^3 bonding could increase the d-spacing to 3.44 Å [122–124].

Johnson and Watt [125] have investigated the crystallite structure of PAN-based carbon fibers treated at 2500 °C and found that the turbostratic crystallites had L_c (crystallite height) of at least 12 layer planes and L_a (crystallite width) of 6–12 nm. Both L_c and L_a tended to increase with the heat-treatment temperature.

The basic structural unit and some of the microdomains were initially established by the fiber-formation process, including the precursors and processing conditions.

Generally, it is known that a well-stacked graphitic crystalline structure could only be observed in the mesophase pitch and vapor-grown carbon fibers, while the turbostratic structure could be observed in carbon fibers with other precursors such as PAN. In the graphitization of stabilized PAN-based fibers, the crystalline domain is developed by either amalgamating with adjacent crystallites or incorporating the surrounding disorganized carbons. Furthermore, the layer planes within the crystalline domain relocated through rotating and shifting. However, these

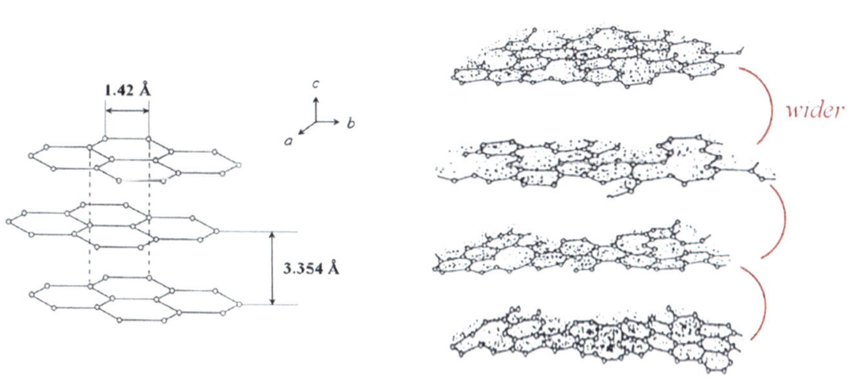

Fig. 1.23 Crystal structure of a graphite crystal (L) and structure of turbostratic carbon (R)

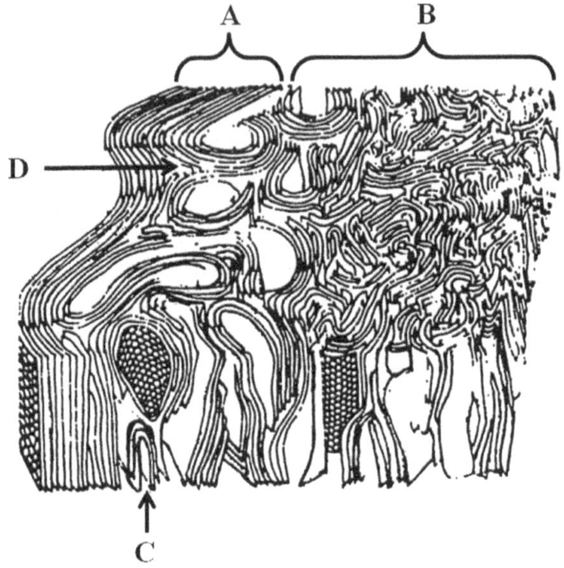

Fig. 1.24 Combination of basic structural units into microdomains within a carbon fiber: (A) skin region; (B) core region; (C) a hairpin defect; and (D) a wedge disclination [121]

arrangements had appeared locally and the graphite fibers still possessed significant turbostratic domains with the existence of extensive rotational disordering of layer planes.

Bourrat et al. [126] reported that noncarbonized mesophase pitch-based fibers exhibited three levels of structural organization. The same three levels of organization, with slight modification, could be applied to all types of carbon fibers. These are as follows:

1. At the molecular range level, molecules are stacked together creating small coherent domains, as in the starting mesophase pitch, known as "basic structural units."
2. At the next level of organization, i.e., from one to a few hundred nm, a microtexture is observed. It is defined by microdomains which consist of assemblies of molecules with a long-range orientation order. These microdomains are elongated along the fiber-axis direction, up to several μm, and limited laterally by pores and wedge disclinations. The descriptive "micro-" is used here to specify this level of organization. The characteristics of the basic structural unit and microtexture are determined using a combination of wide-angle X-ray scattering (WAXS), small-angle X-ray scattering (SAXS), selected area diffraction (SAD) by transmission electron microscopy (TEM), and use of bright- and dark-field imaging in the TEM.
3. The last level of organization defines the fibers' texture. It reflects the changing statistical orientation of the molecules at a very long range. The texture is readily observed in the SEM and could be deduced through the optical microscope studies with the acid of reflected polarized light (detailed information and optical images are shown in [125]).

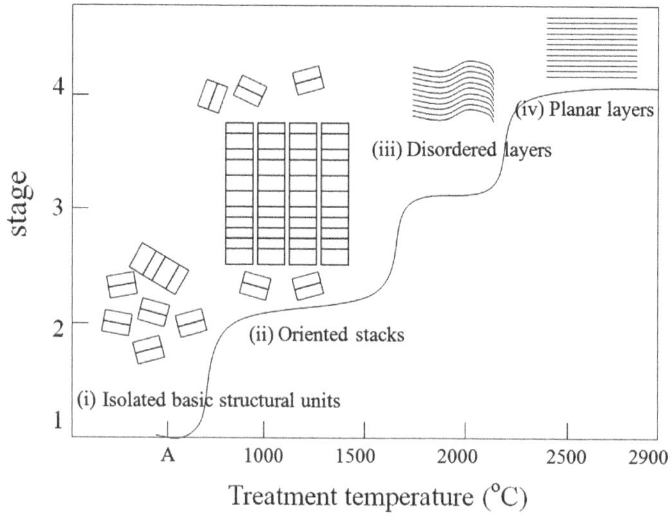

Fig. 1.25 Various stages of graphitization and sketches of structures (At point A: nonheat-treated carbonaceous materials) [125]

Figure 1.25 shows the basic structural unit at the several stages during the conversion of a fiber from the stabilized material to graphite. At low temperatures, after stabilization, the oriented basic structural units were isolated and the heteroatoms were eliminated at the higher temperatures. Then, some stacks were formed by assembling the oriented basic structural units, which led to the creation of pores and increase the densification. At higher temperatures, some microdomains were formed as the dense and orderly structures. At further higher temperatures, disordered layers of turbostratic carbon appeared. Finally, at extreme temperatures, planar layers of graphite resulted. The change in the perfection and ordering of carbon in fibers could be readily observed using high-resolution electron microscopes.

1.6 State of Carbon Fiber Industry

1.6.1 Technology Development Trends

Carbon fibers have remarkable properties such as tensile strength, stiffness, low density, electrical conductivity, and chemical inertness. Carbon fibers have been used to manufacture sports cars as well as sports equipment such as fishing rods, golf shafts, and tennis racquets. Currently, a great deal of attention is being paid to reduce the weight of passenger vehicles to increase vehicle fuel economy and lower the greenhouse gas emissions. Thus, carbon fibers are now used in aircraft and industrial applications such as pressure vessels, windmills, civil engineering/construction-related uses, and cars and yachts.

However, owing to the tricky and multistage manufacturing processes of carbon fibers, only eight companies, including Toray, Toho Tenax, Mitsubishi Chemical, Hexcel, and Zoltek, have succeeded in the mass production of carbon fibers. In each country, the export of carbon fibers is strictly prohibited because carbon fibers are widely used in the state-of-the-art weapons. In addition, historically, the carbon fiber industry has been cyclical with a period of limited supply resulting in high prices to periods of oversupply resulting in falling prices.

Toray leads in the development of high-performance carbon fibers. Figure 1.26 shows the classification of high-performance carbon fibers made by Toray. In particular, T (Torayca)-1000 contains high-performance carbon fibers with a tensile strength of 7 GPa and a tensile modulus 300 GPa, which enabled the diversification of the application fields. Figure 1.27 presents the application based on the mechanical characteristics of regular carbon fibers [127].

1.6.2 Utility Development Trends

Figure 1.28 shows the trends and compound annual growth rates (CAGRs) of carbon fiber applications until 2020 [127]. After the storm of the financial crisis in 2009, the market for carbon fibers gradually increased. The demand for carbon fibers is classified into industrial, sporting goods, and aerospace applications. The sporting goods is one of the oldest applications for carbon fibers, but it is experiencing only modest growth, as there are no new significant applications to be expected.

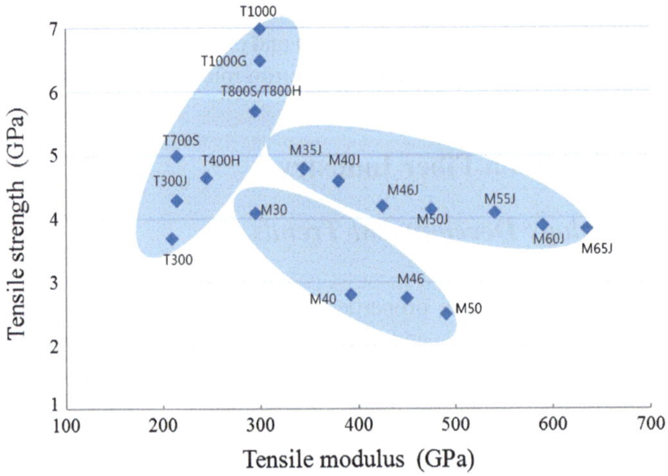

Fig. 1.26 Classification of high-performance carbon fibers (Toray) [127]

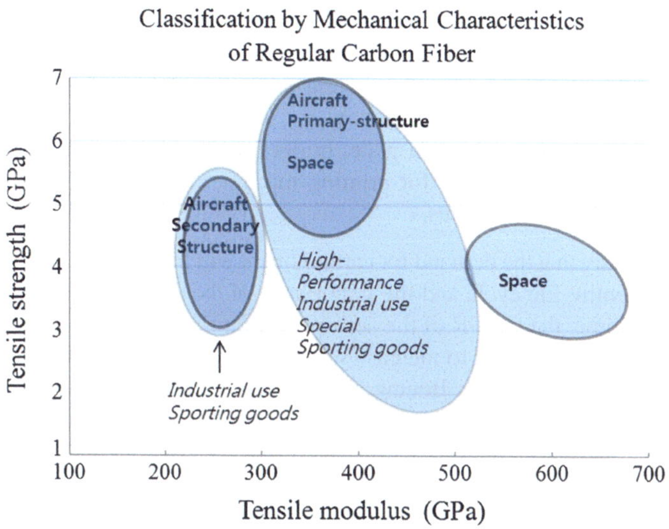

Fig. 1.27 Applications based on mechanical characteristics of regular carbon fibers (Torayca) [127]

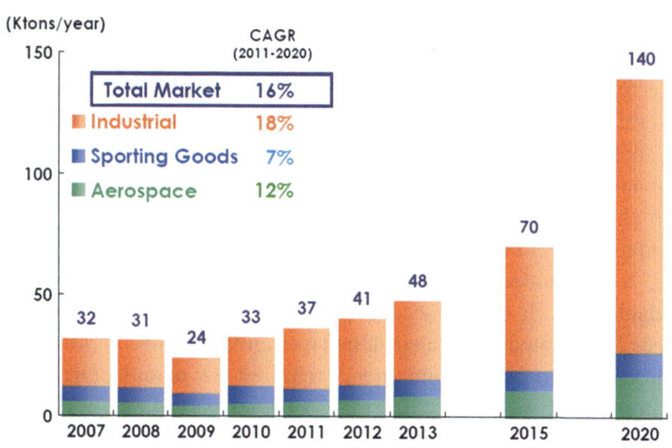

Fig. 1.28 Carbon fibers market trends and compound annual growth rate (CAGR) of carbon fiber applications (Toray) [127]

The industrial sector of the market breadth and CAGR has led to the expanded market for carbon fibers. The industrial sector is also growing in importance with a wide range of applications. The industrial sector overall, excluding aerospace and sporting goods, is forecast to represent more than 75% of the total usage of carbon fibers by 2020 (see below).

- Energy: blades for wind power, pressure vessels, electrode gas diffusion layers for fuel cells, offshore oil industry, high risers;
- Transportation: vehicle parts (for weight reduction, safety, and environment), marine parts (decks, hulls, and masts);
- Civil engineering (antivibration plate, bridges, lightweight buildings);
- Industrial equipment: rollers for printing machines, PC frames, machine components, medical components.

It is noteworthy that the demand for carbon fibers will increase with the increasing quality of the entire life cycle and the importance of the environmental issues.

In the meantime, the CAGR of the aerospace sector is somewhat sluggish. In the next few years, it is expected to increase substantially owing to the demand of large aerospace companies (such as Boeing and Airbus) for the development of carbon fiber composites, which could result in the weight reduction and fuel savings.

1.6.3 Market Trends

Since 1971, Toray Co. [127] has been manufacturing PAN-based carbon fibers using a proprietary technology and became the top carbon fiber manufacturer in the world, capturing 34% (regular tow) of the global market share with a production capacity of 7600 ton/year. Toray has shifted its focus from the aviation, space, sports, and industrial sectors and directed attention toward the carbon fiber composites for the Boeing Aircraft. In addition, they have developed the carbon fiber manufacturing techniques such as precursor emulsion, stretching, mixed spinning, carbonization/graphitization techniques for carbon fibers, and molding of carbon fiber composites to obtain the high-performance carbon fibers. The steady development of a variety of purposes and usages for the carbon fibers has been observed. Toho Tenax Co. [128], since 1975, has manufactured high-performance PAN-based carbon fibers and flame-retardant fibers. It ranked second in the world with a production capacity of 5600 ton/year. The company has also gained recognition in the aircraft industry. They supplied the entire structural components required for manufacturing vertical tail wings and flooring of Airbus A380, which occupies part of the 70% of the aircraft industry. Toho Tenax Co [129]. has also developed industrial components such as spring, rotor, and brake, and reinforced technology for concrete structure. Since 1976, Mitsubishi Chemical Co. [130] has produced prepregs using the high-performance carbon fibers obtained from Courtaulds Co. (England). Later, carbon fibers were produced through a technical partnership with HITCO Co. (USA) since 1983. With a production capacity of approximately 3400 ton/year, Mitsubishi was ranked third in the world. Then, Hexcel Co. [131], Cytec Co. [132], and Formosa Co. [133] accelerated the manufacture and development of carbon fibers. Table 1.2 lists the PAN-based carbon fiber manufacturers and their markets in the world.

Table 1.3 lists the pitch-based carbon fiber manufacturers and their respective market shares in the world. As listed in Table 1.3, the pitch-based carbon fibers

Table 1.2 PAN-based carbon fiber manufacturers and their respective global market shares

Division		2007	2008	2009	2010	2011	2012	2013	Product name	Main fields
Small tow	Toray	1350	1390	1390	1390	1390	1390	1390	Torayca	Aircraft (Boeing)
	Toho Tenax	780	1050	1050	1050	1050	1050	1050	Besfight	Aircraft (Airbus)
	Mitsubishi chemical	740	740	740	740	740	740	740	Pyrofil	Sporting goods
	Hexcel	380	380	380	380	380	380	380	–	Aircraft
	Cytec	200	200	200	200	200	200	200	Thornel	Advanced composites materials
	Formosa	300	300	300	300	300	300	300	–	Sporting goods
	Quantity	3750	4060	4060	4060	4060	4060	4060	–	–
	Cost	1350	1462	1462	1462	1462	1462	1462	–	–
Large tow	Toho Tenax	130	130	130	130	130	130	130	–	–
	SGL	200	250	250	250	250	250	250	Panox	Brake (aircraft)
	CFT (Aldila)	100	100	100	100	100	100	100	–	Golf shafts
	Zoltek	450	600	600	600	600	600	600	Panex	Carbon/carbon composites (aircraft)
	Quantity	880	1080	1080	1080	1080	1080	1080	–	–
	Cost	270	331	331	331	331	331	331	–	–
Total amount		4630	5140	5140	5140	5140	5140	5140	–	–
	Total cost	1620	1793	1793	1793	1793	1793	1793		

A basis of calculated tax amount: Materials Technology Publications 2008 "The Carbon Fiber Industry"—The Carbon Fiber price small tow USD 30/kg, large tow USD 25.5/kg
(Unit: ton, billion)

Table 1.3 Pitch-based carbon fiber manufacturers and their global market shares

		2007	2008	2009	2010	2011	2012	2013	Product name	Main fields
Anisotropy	Mitsubishi Chemical	75	100	100	125	125	125	125	Dialead	Aircraft, aerospace, electric components, cement reinforcement
	Nippon Graphite	12	12	18	18	40	40	40	Granoc	Aircraft, aerospace, civil construction, sports
	Cytec	36	36	36	36	36	36	36	Pitch Thornel	High thermal conductivity/structure/friction
	Quantity	123	148	154	179	201	201	201	–	–
	Cost	89	107	111	129	145	145	145	–	–
Isotropy	Petoca Materials	120	120	120	120	120	120	120	Melblown	Electrode, capacitors, heating
	Kureha Chemicals	90	90	90	90	90	90	90	Kreca	Curtain wall, insulation, electromagnetic shielding
	Donac	30	30	30	30	30	30	30	–	Electromagnetic sealer
	Quantity	240	240	240	240	240	240	240	–	–
	Cost	173	173	173	173	173	173	173	–	–
Total amount		363	388	394	419	441	441	441	–	–
Total cost		262	280	284	302	318	318	318	–	–

A basis of calculated tax amount: Materials Technology Publications 2008 "The Carbon Fiber Industry"—The Carbon Fiber price USD 60/kg
(Unit: ton, billion)

have a relatively smaller market size compared to the PAN-based carbon fibers. While the universal type of pitch-based carbon fibers was used as fillers and insulating materials, the high-performance pitch-based carbon fibers were competing with the PAN-based carbon fibers in the field of advanced composites materials, which resulted in the creation of a new market owing to their unique characteristics. In particular, it is expected that the mesophase pitch-based carbon fiber market will grow significantly in the aerospace sector for applications in extreme environments such as the structure of the antenna and exterior materials of electronic devices.

1.7 Summary

In this chapter, we have provided a brief overview of carbon fibers, showing superior performance and comprising expensive materials, in that the origin, history, manufacturing technology, performance, and world market trends.

Despite the high performance, the carbon fibers were only used for military purposes and aerospace/aviation owing to the higher price. Recently, carbon fibers have attracted attention for their potential use as structural materials in aerospace, sports, cars, and bridges. Also, carbon fibers have been considered as the next-generation materials in the aerospace/aviation industry, which has presented a huge opportunity to the entire supply chain of carbon fibers.

References

1. Handbook of Carbon, Graphite, Diamond and Fullerenes. (Noyes Publications, Park Ridge, New Jersey, 1993)
2. R.F. Furl, R.E. Smalley, Sci. Am. **265**, 54 (1991)
3. A.R.G. Brown, W. Watt, R.W. Powell, R.P. Tye, J. Appl. Phys. **7**, 73 (1956)
4. G.M. Jenkins, K. Kawamura, Nature **231**, 175 (1971)
5. R. Growth, J. Appl. Phys. **31**, 283 (1960)
6. T.V. Hughes, C.R. Chambers, Manufacture of carbon filaments, US Patent 405, 480 (1889)
7. R.T.K. Baker, P.S. Harris, Marcel Dekker. New York **14**, 83 (1978)
8. G.G. Tibbetts, Vapor-grown carbon fibers, in *Carbon Fibers Filaments and Composites*, ed. by J.L. Figueiredo, C.A. Bernardo, R.T.K. Baker, K.J. Hüttinger (Kluwer, Dordrecht, 1990), pp. 73–94
9. P. Gadelle, The growth of vapour deposited carbon fibres, in *Carbon Fibers Filaments and Composites*, ed. by J.L. Figueiredo, C.A. Bernardo, R.T.K. Baker, K.J. Hüttinger (Kluwer, Dordrecht, 1990)
10. J.B. Donnet, R.C. Bansal, *Carbon Fibers* (Marcel Dekker Inc, New York, 1984)
11. J.B. Donnet, R.C. Bansal, *Carbon Blacks: Physics, Chemistry, and Elastomer Reinforcement* (Marcel Dekker Inc, New York, 1976)
12. H. Kroto, Science **242**, 1139 (1988)
13. S. Iijima, Nature **354**, 56 (1991)
14. A.K. Geim, K.S. Novoselov, Nat. Mater. **6**, 183 (2007)

15. T.A. Edison, U.S. Patent. 223, 898 (1879)
16. X.C. Hu, H.H. Yang, *Comprehensive Comp. Mater.* **1**, 327 (2000)
17. N. Bhardwaj, S.C. Kundu, Biotech. Adv. **28**, 325 (2010)
18. A. Marcinčin, Prog. Polym. Sci. **27**, 853 (2002)
19. B.C. Elisa, M.M.V. Salim, P.B. Cristiano, Carbon **41**, 1707 (2003)
20. P.J.M. Carrott, J.M.V. Nabais, M.M.L. Ribeiro Carrott, J.A. Pajares, Carbon **39**, 1543 (2001)
21. P.M. Sanjeev, C.H. Kim, S.Y. Kim, B.H. Kim, W.C. Kim, K.S. Yang, Synth. Metals **162**, 453 (2012)
22. W.M. Qiao, S.H. Yoon, Y. Korai, I. Mochida, S. Inoue, T. Sakurai, T. Shimohara, Carbon **42**, 1327 (2004)
23. E. Frank, F. Hermanutz, M.R. Buchmeiser, Macromol. Mater. Eng. **297**, 493 (2012)
24. J. Lin, K. Koda, T. Yamada, M. Enoki, Y. Uraki, J. Wood Chem. Technol. **34**, 111 (2014)
25. Y.Z. Wang, S.G. Wang, J.L. Liu, Key Eng. Mater. **575–576**, 151 (2014)
26. M. Zhang, A.A. Ogale, *Carbon.* Article in Press (2013)
27. A. Ju, H. Xu, M. Ge, *J. Therm. Anal. Calorim.* Article in press (2013)
28. M.R. Buchmeiser, J. Unold, K. Schneider, E.B. Anderson, F. Hermanutz, E. Frank, A. Müller, S. Zinn, J. Mater. Chem. **1**, 13154 (2013)
29. A. Ju, A. Liu, M. Luo, H. Xu, M. Ge, J. Polym. Res. **20**, 1 (2013)
30. Y.G. Ying, Y.P. Pan, R. Ren, J.M. Dang, C.L. Liu, Mater. Chem. Phys. **143**, 455 (2013)
31. Q. Hu, X. Wang, Z. Wang, Ceramics Int. **39**, 8487 (2013)
32. C. Dong, I.D. Davies, Mater. Design **54**, 955 (2014)
33. E. Fitzer, D.D. Edie, D.J. Johnson, Carbon fibers-present state and future expectation; Pitch and mesophase fibers; Structure and properties of carbon fibers, in *Carbon Fibers Filaments and Composites*, ed. by J.L. Figueiredo, C.A. Bernardo, R.T.K. Baker, K.J. Huttinger, 1st edn. (Springer, New York, USA, 1989)
34. E. Fitzer, in *Carbon Fibers Filaments and Composites*, ed. by J.L. Figueirede, C.A. Bernardo, R.T.K. Baker, K.J. Huttinger (Kluwer Academic, Dordrecht, 1990)
35. D. Jian, S. Wanci, Z. Baofa, L. Xuan, K. Feiyu, G. Jialin, L. Dongsheng, C. NanPing, Carbon **39**, 2325 (2001)
36. B.R. Growth, J. Appl. Phys. **31**, 283 (1960)
37. H. Herssler, Verstärkte Kunststoffe in der Luft- und Raumfahrttechnik. Kunststoffe und Elastomere in der Praxis (1986)
38. W. Bin, Z. Chun, X. Shijie, Z. Jing, X. Lianghua, J. Appl. Polym. Sci. **125**, 3545 (2012)
39. Y. Meijie, X. Yong, W. Chengguo, H. Xiuying, Z. Bo, Q. Kun, Y. Hua, J. Appl. Polym. Sci. **125**, 3159 (2012)
40. K. Kong, L.B. Deng, I.A. Kinloch, R.J. Young, S.J. Eichhorn, J. Mater. Sci. **47**, 5402 (2012)
41. S.Y. Gu, J. Ren, Q.L. Wu, Synth. Metals **155**, 157 (2005)
42. J.H. Kim, H.S. Ganapathy, S.S. Hong, Y.S. Gal, K.T. Lim, J. Supercritic, Fluids **47**, 103 (2008)
43. W. Meiyu, W. Qiaoying, L. Kaina, W. Yiqiong, L. Haiqing, Polym. Degrad. Stab. **97**, 1511 (2012)
44. Y. Ma, Z.H. Chen, Ceramic. Int. **38**, 4229 (2012)
45. F. Vautard, S. Ozcan, H. Meyer, Appl. Sci. Manufac. **43**, 1120 (2012)
46. P. Tsotra, K. Friedrich, Appl. Sci. Manufac. **34**, 75 (2003)
47. M.A. Montes-Morán, A. Martínez-Alonso, J.M.D. Tascón, M.C. Paiva, C.A. Bernardo, Carbon **39**, 1057 (2001)
48. L.M. Manocha, Y. Eiichi, T. Yasuhiro, T. Shiushichi, Carbon **26**, 333 (1988)
49. S. Wei, G. Aijuan, L. Guozheng, Y. Li, Appl. Surf. Sci. **257**, 4069 (2011)
50. B.M. Parker, R.M. Waghorne, Composites **13**, 280 (1982)
51. A.R. Bunsell, *Fibre Reinforcements for Composites Materials* (Elsevier Science Publishers B.V., Amsterdam, 1988)
52. D.D. Edie, Carbon **36**, 345 (1998)
53. N. Takami, A. Satoh, M. Hara, T. Ohsaki, J. Electrochem. Soc. **142**, 371 (1995)

54. N. Takami, A. Satoh, M. Hara, T. Ohsaki, J. Electrochem. Soc. **42**, 2654 (1995)
55. J.L. Figueiredo, C.A. Bernardo, R.T.K. Baker, K.J. Huttinger, *Carbon Fibers Filaments and Composites*, Kluwer Academic Publishers (1990)
56. I. Mochida, K. Shimizu, Y. Korai, S. Fujiyama, H. Toshima, T. Hono, Carbon **30**, 55 (1992)
57. S. Ohtani, Carbon **3**, 31 (1965)
58. S. Sudo, K. Shimizu, J. Appl. Polymer Sci. **44**, 127 (2003)
59. A.H. Wazir, L. Kakakhel, New Carbon Mater. **24**, 83 (2009)
60. E. Fitzer, K. Müller, W. Schäfer, *Chemistry and Physics of Carbon* (Marcel Dekker, New York, 1971)
61. J.L. Figueiredo, C.A. Bernardo, R.T.K. Baker, K.J. Huttinger, *Carbon Fibers Filaments and Composites*, Kluwer Academic Publishers (1990)
62. S. Otani, Y. Fukuoka, B. Igarashi, K. Sasaki, US Patent 3,461,082 (1969)
63. Y. Fukuoka, Jpn. Chem. Q **5**, 63 (1969)
64. Kayacarbon, Manufacturer's Brochure NKCL, 'Kayacarbon' Manufacturer's Brochure, Nippon Kayaku Co. Ltd., Japan
65. W.G. Glasser, S.S. Kelley, *Concise Encyclopedia of Polymer Science and Engineering* (Wiley, New York, 1990)
66. K. Sudo, K. Shimizu, N. Nakashima, A. Yokoyama, J. Appl. Polym. Sci. **48**, 1485 (1993)
67. S. Kubo, N. Ishikawa, Y. Uraki, Y. Sano, Mokuzai Gakkaishi **43**, 655 (1997)
68. Y. Uraki, S. Kubo, H. Kurakami, Y. Sano, Holzforschung-Int. J. Bio. Chem. Phys. Tech. Wood **51**, 188 (1997)
69. J.F. Kadla, S. Kubo, R.A. Venditti, R.D. Gilbert, A.L. Compere, W. Griffith, Carbon **40**, 2913 (2002)
70. S. Kubo, J.F. Kadla, J. Polym. Environ. **13**, 97 (2005)
71. D.A. Baker, N.C. Gallego, F.S. Baker, J. Appl. Polym. Sci. **124**, 227 (2012)
72. J.L. Braun, K.M. Holtman, J.F. Kadla, Carbon **43**, 385 (2005)
73. K. Sudo, K. Shimizu, J. Appl. Polym. Sci. **44**, 127 (1992)
74. J.M. Pickel, W.L. Griffith, A.L. Compere, Abs. Pap. Am. Chem. Soc. **231**, 133 (2006)
75. Y.H.P. Zhang, J. Int. Microbio. Biotech. **35**, 367 (2008)
76. W.M. Qiao, M. Huda, Y. Song, S.H. Yoon, Y. Korai, I. Mochida, O. Katou, H. Hayashi, K. Kawamoto, Energ. Fuels **18**, 2576 (2005)
77. N.M. Rodriguez, J. Mater. Res. **8**, 3233 (1993)
78. W. Xia, O.R.K. Schlüter, C. Liang, M.W.E. Berg, M. Curaya, M. Muhler, Catal. Today **102**, 34 (2005)
79. E. Hammel, X. Tang, M. Trampert, T. Schmitt, K. Mauthner, A. Eder, P. Pötschke, Carbon **42**, 1153 (2004)
80. G.T. Gary, Carbon **27**, 745 (1989)
81. R.M. Levit, Khim. Volokna **6**, 16 (1990)
82. A. Ahmadpour, D.D. Do, Carbon **35**, 1723 (1997)
83. S.J. Park, B.J. Kim, J. Colloid Interface Sci. **282**, 124 (2005)
84. S.J. Park, J.S. Shin, J.W. Shim, S.K. Ryu, J. Colloid Interface Sci. **275**, 342 (2004)
85. S.J. Park, S.Y. Jin, Carbon **42**, 2113 (2004)
86. S.J. Park, B.J. Kim, J. Colloid Interface Sci. **291**, 597 (2005)
87. L. Giraldo, M.F. González-Navarro, J.C. Moreno-Piraján, Carbon-Sci. Tech. **5**, 303 (2013)
88. H.A. Al-Aoh, M.J. Maah, R. Yahya, M.R.B. Abas, Asian J. Chem. **25**, 9573 (2013)
89. Y.F. Li, Y. Liu, H.Q. Liu, L. Li, Adv. Mater. Res. **807–809**, 1343 (2013)
90. X. Ma, N. Li, J. Jiang, Q. Xu, H. Li, L. Wang, J. Lu, J. Environ. Chem. Eng. **1**, 466 (2013)
91. S. Shrestha, G. Son, S.H. Lee, T.G. Lee, Chemosphere **92**, 1053 (2013)
92. C.L. Lin, Y.H. Cheng, Z.S. Liu, J.Y. Chen, J. Porus Mater. **20**, 883 (2013)
93. Castro-Muñiz, F. Suárez-García, A. Martínez-Alonso, J.M.D. Tascõn, T. Kyotani, Chemsuschem **6**, 1406 (2013)
94. C.H. Jung, I.H. Yoon, H.J. Woon, W.K. Choi, J.K. Moon, Asian J. Chem. **25**, 5602 (2013)
95. M.S. Berber-Mendoza, R. Leyva-Ramos, F.J. Cerino-Cordoba, J. Mendoza- Barron, H.J.A. Garcia, J.V. Flores-Cano, Water Air Soil Pollut. **224**, 1604 (2013)

96. W. Kempiński, D. Markowski, M. Kempiński, M. Śliwińska-Bartkowiak, Carbon **57**, 533 (2013)
97. H.A. Al-Aoh, M.J. Maah, R. Yahya, M.R.B. Abas, Asian J. Chem. **25**, 9582 (2013)
98. J. Yun, H.I. Kim, Y.S. Lee, J. Mater. Sci. **48**, 8320 (2013)
99. L. Huang, C. Shi, B. Zhang, S. Niu, B. Gao, Sep. Sci. Technol. **48**, 1356 (2013)
100. Y.C. Chiang, W.H. Lin, Appl. Mech. Mater. **284–287**, 72 (2013)
101. Y. Takahashi, H. Fujita, A. Sakoda, Adsorption **19**, 143 (2013)
102. C.A. Jeffs, M.W. Smith, C.A. Stone, C.G. Bezzu, K.J. Msayib, N.B. McKeown, S.P. Perera, Micro. Meso. Mater. **170**, 105 (2013)
103. J. Liu, Q. Zhou, J. Chen, L. Zhang, N. Chang, Chem. Eng. J. **215–216**, 859 (2013)
104. Z. Zhang, J.D. Atkinson, B. Jiang, M.J. Rood, Z. Yan, Appl. Cat. B: Environ. **148–149**, 573 (2014)
105. L. Sun, Y. Yao, L. Wang, Y. Mao, Z. Huang, D. Yao, W. Lu, W. Chen, Chem. Eng. J. **240**, 413 (2014)
106. F. Zhang, X.J. Ma, Appl. Mech. Mater. **469**, 117 (2014)
107. Y. Yao, V. Velpari, J. Economy, Fuel **116**, 560 (2014)
108. Z. Yue, C. Mangun, J. Economy, P. Kemme, D. Cropek, S. Maloney, Environ. Sci. Technol. **35**, 2844 (2001)
109. J.W. Shim, S.J. Park, S.K. Ryu, Carbon **39**, 1635 (2001)
110. S.J. Park, J.S. Shin, J. Colloid Interface Sci. **264**, 39 (2003)
111. S.J. Park, Y.M. Kim, Mater. Sci. Eng., A **391**, 121 (2005)
112. S.Y. Lee, S.J. Park, J. Colloid Interface Sci. **389**, 230 (2013)
113. J.S. Im, S.J. Park, Y.S. Lee, Mater. Res. Bull. **44**, 1871 (2009)
114. J.S. Im, S.J. Park, T.J. Kim, Y.S. Lee, Int. J. Hydrogen Energy **34**, 3382 (2009)
115. M.J. Jung, J.W. Kim, J.S. Im, S.J. Park, Y.S. Lee, J. Ing. Eng. Chem. **15**, 410 (2009)
116. S.J. Park, B.J. Kim, Y.S. Lee, M.J. Cho, Int. J. Hydrogen Energy **33**, 1706 (2008)
117. S.Y. Lee, S.J. Park, J. Colloid Interface Sci. **389**, 230 (2013)
118. M. Foston, G.A. Nunnery, X. Meng, Q. Sun, F.S. Baker, A. Ragauskas, Carbon **52**, 65 (2013)
119. M.S. Morales, A.A. Ogale, J. Appl. Polym. Sci. **130**, 2494 (2013)
120. M. Speiser, S. Henzler, U. Hageroth, A. Renfftlen, A. Müller, D. Schawaller, B. Sandig, M. R. Buchmeiser, Carbon **63**, 554 (2013)
121. A. Diaz, M. Guizar-Sicairos, A. Poeppel, A. Menzel, O. Bunk, *Carbon*. Article in press (2013)
122. S.C. Bennett, D.J. Johnson, Strength-structure relationship in PAN-based carbon fibers, in *5th Lond. Int. Carbon and Graphite Conf., Soc. Chem. Ind.* (London, 1978)
123. D.L. Chung, *Carbon Fiber Composites* (Butterworth-Heinemann, Boston, Massachusetts, USA, 1992)
124. J.G. Morley, *High-Performance Fiber Composites* (Academic Press, Orlando, Florida, USA, 1987)
125. W. Johnson, W. Watt, Nature **215**, 384 (1967)
126. X. Bourrat, E.J. Roche, J.G. Lavin, Carbon **28**, 435 (1990)
127. A. Oberlin, Carbon **22**, 521 (1984)
128. http://www.toray.com/
129. http://www.tohotenax.com/
130. http://www.m-kagaku.co.jp/
131. http://www.hexcel.com/
132. https://www.cytec.com/
133. http://www.fpcusa.com/

Chapter 2
Precursors and Manufacturing of Carbon Fibers

Abstract In this chapter, we will present the precursors and manufacturing of carbon fibers. Among the precursors used for the production of carbon fibers, polyacrylonitrile (PAN)-based and pitch-based precursors are the most important. A significant amount of work has been done on relating the fiber structure to the properties and translating that relationship into production for either reducing cost or increasing fiber properties. However, challenges, including cost reduction, improvement in tensile and compressive strength, and alternative precursor development, still remain. We will also introduce many linear and cyclic polymers for carbon fibers, which are expected to open the door for the low-cost carbon fibers.

2.1 Introduction

Carbon fibers are novel high-performance materials. They could be described as fibers containing at least 90% carbon obtained by the controlled pyrolysis of appropriate fibers. Edison in 1879 found that carbon fibers can be used as carbon filaments in electric lamps. Since the early work of Edison, the carbon fibers have been investigated and used intensively because they generally have excellent tensile properties, low densities, high thermal and chemical stabilities in the absence of oxidizing agents, good thermal and electrical conductivities, and excellent creep resistance [1–16].

In recent years, the carbon fiber industry has been growing steadily to meet the demands arising from different applications such as aerospace (aircraft and space systems), military, turbine blades, construction, lightweight cylinders and pressure vessels, medical, automobile, and sporting goods. Furthermore, in the case of the carbon fibers, the range of their applications would depend on the type of precursors used to produce the carbon fibers. Consequently, many types of precursors have been studied to produce the carbon fibers. The ideal features of precursors required to manufacture carbon fibers are easy conversion to carbon fiber, high carbon yield, and cost-effective processing. From this perspective, the following four types of

S.-J. Park, *Carbon Fibers*, Springer Series in Materials Science 210,
https://doi.org/10.1007/978-981-13-0538-2_2

precursors have been widely used, which have also proven to be the most popular ones [17–31]:

1. Acrylic precursors: They have been successfully used for carbon fiber preparation by most industrial manufacturers for quite some time now. These acrylic precursors contain >85% acrylonitrile (AN) monomer. In particular, polyacrylonitrile (PAN) is the most popular acrylic precursor, which is used widely to produce the carbon fibers.
2. Cellulosic precursors: They contain 44.4% carbon. However, in practice, the reaction is more complicated than mere dehydration and the carbon yield is only approximately 25–30%.
3. Pitch-based precursors: They have a yield of 85%, and the resultant carbon fibers from these precursors show a high modulus owing to the more graphitic nature. On the other hand, the pitch-based carbon fibers have poorer compression and transverse properties compared to the PAN-based carbon fibers.
4. Other forms of precursors: Vinylidene chloride and phenolic resins as precursors for the manufacture of carbon fibers have been investigated, but were not found to be commercially viable.

2.2 Acrylic Precursors

Various acrylic precursors have been utilized to produce carbon fibers for use in carbon fiber-based composite applications owing to certain desirable physical properties [32, 33]. The acrylic precursors for the carbon fiber industry originated from the companies that were established commercial-scale producers of textile-grade acrylic fibers. This was because the carbon fibers were produced through the pyrolysis of the acrylic fibers. Therefore, the manufacturers of carbon fibers could most readily adapt the existing technology for precursors to manufacture carbon fibers. The manufacturing processes of acrylic precursors by various manufacturers are listed in Table 2.1. In particular, the resultant carbon fibers from

Table 2.1 Manufacturing processes for acrylic precursors

Manufacturers	Trade name	Type of polymerization	Solvent	Typical % of polymer
Accordis	Courtelle	Continuous solution	NaCNS	10–15
Asahi	Cashmilon	Continuous aqueous dispersion	H_2O/ HNO_3	8–12
Hexcel	Exlan	Continuous aqueous dispersion	H_2O/ NaCNS	10–15
Mitsubishi	Finel	Continuous aqueous dispersion	H_2O/ DMAC	22–27
Toho	Beslon	Continuous solution	$ZnCl_2$	8–12
Toray	Toraylon	Batch solution	DMSO	20–25

acrylic precursors such as polyacrylonitrile (PAN)-based carbon fibers have been widely used as reinforcing materials in automobile, aerospace, recreational, and various other industries [34–37].

2.2.1 PAN Precursors

PAN-based polymer precursors for carbon fibers could be primarily classified into pure homopolymer and comonomers. Generally, the comonomers are widely used in PAN-based polymer precursors to manufacture carbon fibers [38–40].

The homopolymer PAN product is slightly difficult to process into carbon fibers because the initial oxidation stage of the process cannot be easily controlled owing to the sudden and rapid evolution of heat, coupled with a relatively high initiation temperature. Such heat could result in poor properties of carbon fibers owing to the chain scission from the thermal shock. It is known that the homopolymer PAN product has never been used as a precursor for manufacturing carbon fibers. To overcome the resultant poor properties in the carbon fibers due to the rapid evolution of heat, the exothermic reaction should be adequately controlled using suitable comonomers such as itaconic acid.

On the other hand, as mentioned above, the comonomers could exert a significant effect on the stabilization process, thereby enhancing the segmental mobility of the polymer chains resulting in better orientation and mechanical properties of the precursor and manufactured carbon fibers. In addition, the selection of suitable comonomers could reduce the initial temperature of cyclization in the manufacturing process of carbon fibers. Therefore, several researchers have discussed the effect of comonomer composition on the properties of the PAN precursor and resultant carbon fibers.

Vinyl esters such as vinyl acetate (VAc), methacrylate (MA), and methyl methacrylate (MMA) could be used as comonomers for acrylonitrile, though VAc might not be an appropriate carbon fiber precursor [41–44]. These comonomers act like a plasticizer and break up the structure to make the polymer more readily soluble in the spinning solvent, thereby improving the quality of spinning, modifying the fiber morphology and where appropriate, improving the rate of diffusion of the dye into the fibers. However, the comonomers should be minimally used for establishing good properties in the carbon fibers because they could affect the cyclization step in the manufacturing process of carbon fibers. Table 2.2 lists the possible comonomers. These monomers have similar reactivities, and the resulting polymer compositions will have more or less the same composition as the monomers in the feed.

Carboxylic acids could be used as effective comonomers because their presence affects the ease of oxidation, exothermicity, and carbon yield of the precursor. In addition, itaconic and methacrylic acids have been confirmed to be most effective comonomers for reducing the exothermicity. Previous studies involving acrylic acid, acrylamide, sodium acrylate, and different acidic comonomers confirmed that

Table 2.2 Comonomers for PAN precursors

Class	Comonomer	Structure
Acids	Acrylic acid	$CH_2{=}CHCOOH$
	Itaconic acid	$CH_2{=}C(COOH)CH_2COOH$
	Methacrylic acid	$CH_2{=}C(CH_3)COOH$
Vinyl esters	Methyl acrylate	$CH_2{=}CHCOOCH_3$
	Ethyl acrylate	$CH_2{=}CHCOOC_2H_5$
	Butyl acrylate	$CH_2{=}CHCOO(CH_2)_3CH_3$
	Methyl methacrylate	$CH_2{=}C(CH_3)COOCH_3$
	Ethyl methacrylate	$CH_2{=}C(CH_3)COOC_2H_5$
	Propyl methacrylate	$CH_2{=}C(CH_3)COO(CH_2)_2CH_3$
	Butyl methacrylate	$CH_2{=}C(CH_3)COO(CH_2)_3CH_3$
	β-Hydroxyethyl methacrylate	$CH_2{=}C(CH_3)$ $COOCH_2CH_2OH$
	Dimethylaminoethyl methacrylate	$CH_2{=}C(CH_3)COOCH_2CH_2N$ $(CH_3)_2$
	2-Ethylhexyl acrylate	$CH_2{=}CHCOOCH_2(C_2H_5)$ $(CH_2)CH_3$
	Isopropyl acetate	$CH_3COOC(CH_3){=}CH_2$
	Vinyl acetate	$CH_3COOCH{=}CH_2$
	Vinyl propionate	$C_2H_5COOCH{=}CH_2$
Vinyl amides	Acrylamide	$CH_2{=}CHCONH_2$
	Diacetone acrylamide	$CH_2{=}CHONHC$ $(CH_3)_2CH_2COCH_3$
	N-methylolacrylamide	$CH_2{=}CHONHCH_2OH$
Vinyl halide	Allyl chloride	$CH_2{=}CHCH_2Cl$
	Vinyl bromide	$CH_2{=}CHBr$
	Vinyl chloride	$CH_2{=}CHCl$
	Vinylidene chloride (1,1-dichloroethylene)	$CH_2{=}CHCl_2$
Ammonium salts of vinyl compounds	Quaternary ammonium salt of aminoethyl-2-methylpropenoate	$CH_2{=}CH(CH_3)COOC_2H_4NH_2$
Sodium salts of sulfonic acids	Sodium vinyl sulfonate	$CH_2{=}CHSOONa$
	Sodium p-styrene sulfonate (SSS)	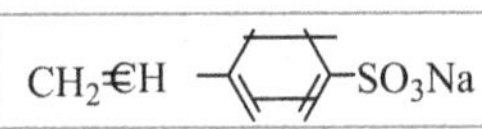
	Sodium methallyl sulfonate (SMS)	$CH_2{=}C(CH_3)CH_2SO_3Na$
	Sodium-2-acrylamido-2-methyl propane sulfonate (SAMPS)	$CH_2{=}CHCONH(CH_2)_3SO_3Na$
Others	Methacrylonitrile	$CH_2{=}C(CH_3)CN$
	2(1-Hydroxyalkyl) acrylonitrile	$CH_2{=}C{-}X$ (with $R{-}CHOH$ on C) (where R = $-CH_3$ or $-C_2H_5$ and X = -CN or –COOH)

(continued)

Table 2.2 (continued)

Class	Comonomer	Structure
	Allyl alcohol	$CH_2{=}CHCH_2OH$
	Methallyl alcohol	$CH_2{=}C(CH_3)CH_2OH$
	1-Vinyl-2-pyrrolidone	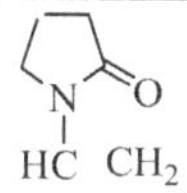
	4-Vinylpyridine	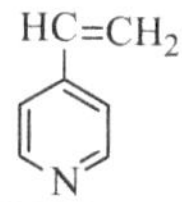
	2-Methylene glutaronitrile	$NCCH_2CH_2C{=}CH_2CN$

the effectiveness of comonomers was in the following order: itaconic acid > methacrylic acid > acrylic acid > acrylamide [40].

The superiority of itaconic acid results from the presence of two carboxylic acid groups, which increases the possibility of interacting with a nitrile group. In other words, if one carboxylic acid group of itaconic acid moved away from an adjacent nitrile group owing to the dipole–dipole repulsion, the other carboxylic acid group could move into the vicinity of a nitrile group, thereby facilitating participation in the cyclization process [45, 46].

2.2.2 *Polymerization Methods for Production of PAN-Based Precursors*

Acrylic precursors such as acrylonitrile and comonomers are initiated by free radical reaction and can be polymerized by one of the several methods listed below:

1. Solution polymerization: This method is used to prepare acrylonitrile and comonomers by dissolving a monomer and catalyst in a nonreactive solvent. During this reaction, the solvent absorbs the heat generated by the chemical reaction and controls the reaction rate. The solvent used in the polymerization process usually remains as a solvent for the acrylic precursors (reactant). This process is only suitable for the creation of wet polymer types because the excess solvent is difficult to remove. During this reaction, the solvent absorbs the heat generated by the chemical reaction and controls the reaction rate. The solvent used in the polymerization process usually remains as a solvent for the acrylic precursors (reactant). This process is only suitable for the creation of wet polymer types because the excess solvent is difficult to remove. Although the removal of excess solvent is possible using distillation, this method is usually not commercially viable. This polymerization method for the preparation of PAN-based precursors offers a few advantages as along with one major disadvantage [47].

2. Bulk polymerization: This method is the simplest and direct way of synthesizing polymers. This polymerization is carried out by adding a soluble initiator to a pure, liquid monomer. The initiator dissolves in the monomer, and the reaction is initiated by either heating or exposure to radiation. The mixture becomes viscous as the reaction proceeds. At that instant, the reaction is exothermic and a reactant with a broad molecular weight distribution is produced. Therefore, this polymerization method for the preparation of PAN-based precursors is used in the small-scale manufacturing process, wherein it is easy to remove the reaction heat [48].
3. Emulsion polymerization: This method is a type of radical polymerization, which is usually carried out with an emulsion containing water, monomer, and surfactant. The PAN-based precursors are often made commercially using emulsion polymerization [49].
4. Aqueous dispersion polymerization: This method is useful to prepare micro- or submicron-scale monodisperse polymer particles in a single step. In this polymerization method, all reaction materials are dissolved in the reaction medium in the initial stage of the polymerization. Then, insoluble spherical polymer particles, stabilized using steric molecules, are formed and dispersed in the reaction medium. PAN-based precursors could be prepared using this polymerization method with an ionic monomer to attain a narrow particle size distribution with a mean diameter as low as approximately 3 $\pm$ 1.5 microns [50].

2.2.3 Manufacture of Carbon Fibers from PAN-Based Precursors

As mentioned earlier, the PAN-based polymers are the optimum precursors for the carbon fibers owing to a combination of tensile and compressive properties as well as the carbon yield. The PAN-based fibers were first developed by Dupont in the 1940s for use in the textile fiber. The thermal stability of PAN-based fibers was an important factor in expanding the application of fibers. Later, this property led to further research on the heat treatment of PAN fibers. In the early 1960s, PAN fibers were first carbonized and graphitized by Shindo at the Government Industrial Research Institute, Osaka, Japan, and these fin the early 1960s, PAN fibers were first carbonized and graph 112 GPa, respectively [51]. The process involved employing tension in both stabilization and carbonization steps. According to Toray, Shindo's patent was licensed to Toray in 1970 by the Japanese Ministry of International Trade and Industry (MITI) to produce PAN-based Torayca carbon fibers. During the 1960s, Watt and Johnson at the Royal Aircraft Establishment, England, and Bacon and Hoses at Union Carbide, the USA, also developed a method for producing carbon fibers from PAN [52–54].

The manufacturing steps for producing the carbon fibers from PAN could be categorized as follows as shown in Fig. 2.1: polymerization of PAN-based precursors, spinning of fibers, thermal stabilization, carbonization, and graphitization. The manufacturing steps for producing the carbon fibers from PAN-based precursors could be categorized as follows as shown in Fig. 2.1: polymerization of PAN-based precursors, spinning of fibers, thermal stabilization (oxidation), carbonization, and graphitization Fig. 2.1. In addition, these PAN-based copolymers containing 2–15% acrylic acid, methacrylic acid, methacrylate, and/or itaconic acid are generally used as precursors of carbon fibers produced through carbonization because the use of comonomers affects the molecular alignment and stabilization conditions. The typical carbon yield from PAN-based precursors is approximately 50–60% [55].

The typical manufacturing steps involved in the production of carbon fibers from PAN-based precursors are listed below:

1. Polymerization of PAN-based precursors and spinning of fibers

The PAN polymer precursor has been widely used as the basic backbone of chemical structure for spinning precursor fibers. Figure 2.2 shows the chemical structure of PAN. In addition, the commercial PAN-based polymer precursors used for spinning the precursor fibers usually contain approximately 2–10% of a comonomer such as methyl acrylate (MA), methyl methacrylate (MMA), or itaconic acids (ITAs). Most carbon companies manufacture their own precursors by in-house technologies. Consequently, the composition of the PAN polymer precursor is not well known because it can control the properties of the final products.

Generally, the PAN polymer precursors contain polar nitrile groups and have a high melting point, resulting from strong intermolecular interactions. Therefore, the

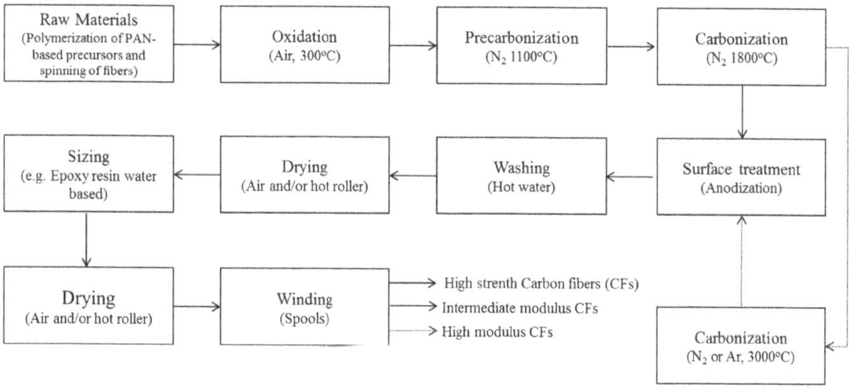

Fig. 2.1 Schematic for manufacturing process of carbon fibers from PAN-based precursors

Fig. 2.2 Chemical structure of PAN

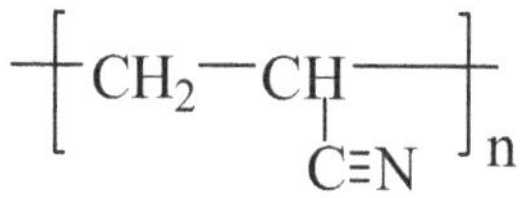

PAN polymer precursors tend to degrade before the temperature reaches their melting point. The spinning of PAN fibers in the carbon fiber industry is performed using traditional manufacturing techniques of acrylic textile fibers. Wet spinning is used in most of the commercial manufacturing processes of carbon fibers produced from PAN-based polymer precursors. However, it is gradually being replaced by dry-jet wet (air gap) spinning [56]. The melt spinning of PAN-based polymer precursors has been previously practised; however, it has yet to become an acceptable manufacturing process of carbon fibers, commercially [57–59]. Figure 2.3 shows the typical layout of a plant for processing PAN-based fibers.

2. Thermal stabilization (oxidation)

This process is critical to obtaining high-quality carbon fibers and could take up to several hours, depending on the temperature, precursor diameter, and precursor

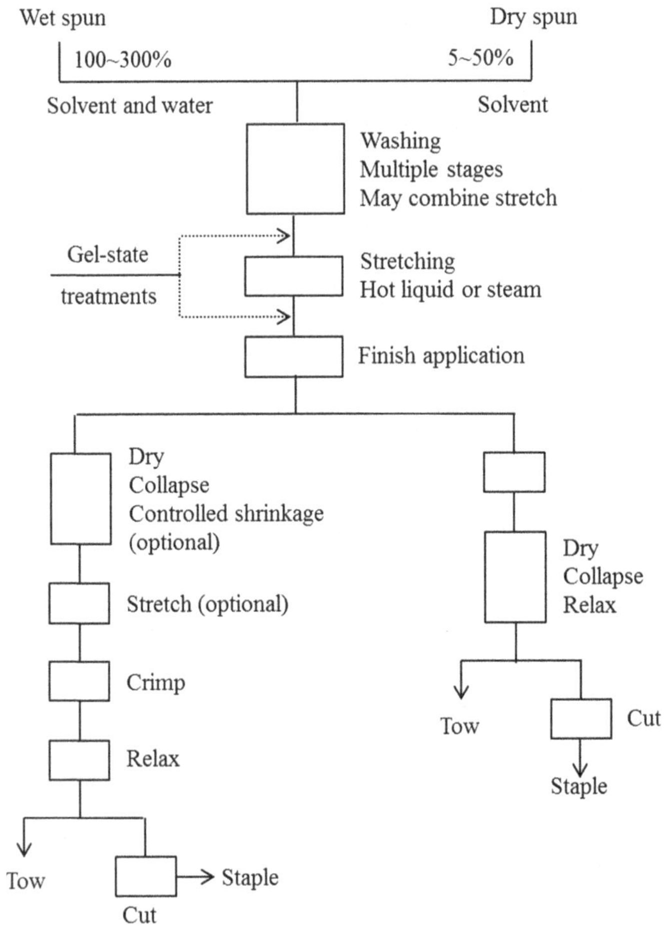

Fig. 2.3 Typical layout of a plant for processing PAN-based fibers

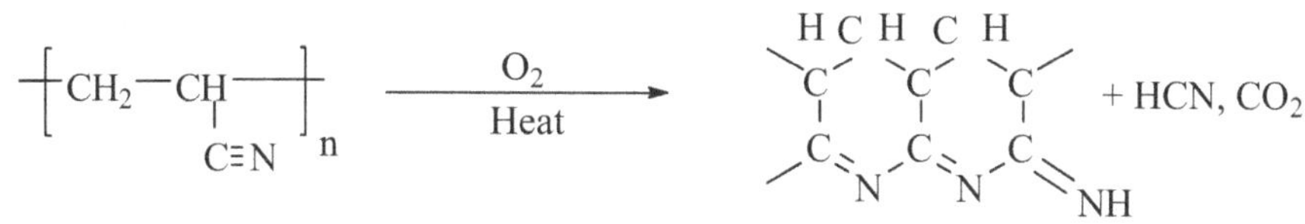

Fig. 2.4 Mechanism of cyclization of PAN polymer precursor

fiber characteristics [60–70]. Proper conditions such as heating rate, time, and temperature of heating should be established for the optimum stabilization of each precursor. The PAN-based polymer precursor is stabilized by controlled low temperature heating over the range 200–300 °C in air to convert the precursor to a form, which could be further heat-treated without either the melting or fusion of the fibers [71].

In this process, the linear molecules of PAN-based polymer precursor are first converted into cyclic structures. However, the cyclization is a complicated process, the mechanism of which is still unclear. In general, the most widely known reaction mechanism is shown in Fig. 2.4.

3. Carbonization and graphitization

The carbonization and graphitization of thermally stabilized fibers are carried out in an inert atmosphere containing gases such as nitrogen (N_2) or Ar [72]. Generally, N_2 is the preferred gas but Ar is used despite being eight times more expensive. This is because Ar provides improved strength to the carbon fiber owing to the higher density and viscosity of argon. The temperature of carbonization is usually determined by the type of application of the resulting carbon fibers. For high-strength applications, the carbonization temperature over the range 1500–1600 °C is preferred because at temperatures above 1600 °C, a decrease in the tensile strength occurs. On the other hand, an additional heat treatment above 1600–1800 °C and up to 3000 °C, i.e., graphitization process, is required to obtain a high modulus in the carbon fibers [73, 74]. Nitrogen cannot be used at temperatures above approximately 2000 °C owing to its reaction with carbon to form cyanogen. The heating rate and retention time during carbonization are different depending on the type of the precursor and stabilization conditions. Figure 2.5 shows the typical tensile strength (GPa) of the PAN-based carbon fibers depending on maximum carbonization temperature.

As mentioned above, the carbonization and graphitization of the thermally stabilized fibers is a two-step process, i.e., low-temperature carbonization and high-temperature graphitization, depending on the requisite properties of the carbon fibers. Bromley et al. [75] confirmed that the gases evolved during the carbonization of PAN-based carbon fibers over the low-temperature range 200–1000 °C, and the observed gases are listed in Table 2.3.

Figure 2.6 shows the schematic of the graphite structure. Dehydrogenation joined the ladder molecules to form graphite-like ribbons; however, denitrogenation

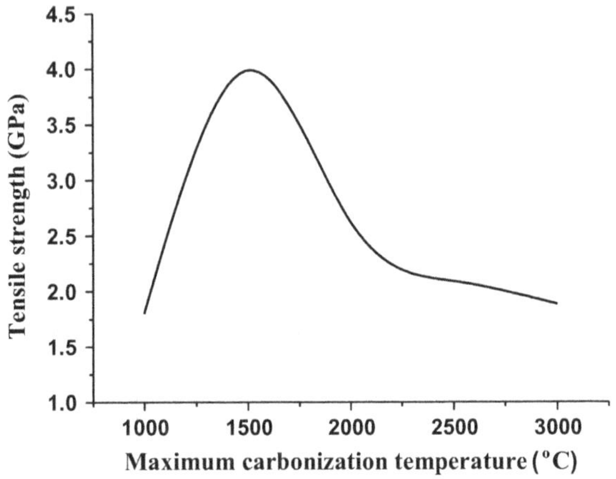

Fig. 2.5 Typical tensile strength (GPa) of PAN-based carbon fibers depending on maximum carbonization temperature

Table 2.3 Gases evolved during carbonization of PAN-based carbon fibers

Temperature (°C)	Observation	Interpretation
220	HCN evolved and O_2 chemically bonded	Ladder polymer formation and oxidation of polymer.
260	Little changed. No modulus increased	No chain scission.
300	Large CO_2 and H_2O evolution; also CO, HCN and some nitriles. No modulus increased	CO_2 from –COOH groups in oxidized polymer No cross-linking
400	CO_2, H_2O, CO, HCN, and NH_3 evolved. Small evolution of C3 hydrocarbons and nitriles Modulus increased	Cross-linking by intramolecular H_2O elimination
500	Increased H_2 evolution. Some NH_3 and HCN evolved. Modulus increased	Cross-linking by dehydrogenation
600	Reduced H_2 evolution. HCN and trace N_2	Cross-linking by dehydrogenation
700	N_2, HCN, and H_2 evolution. Modulus increased	Cross-linking by dehydrogenation and evolution of N_2
800	Large increase in N_2, H_2, and HCN still evolved. Modulus increased	Cross-linking by evolution of N_2
900	Maximum evolution of N_2, some H_2, and traces of HCN. Modulus increased	Cross-linking by N_2 elimination
1000	N_2 evolution decreases to approximately the same level as that at 800 °C. Trace H_2 evolved. Modulus increased	Cross-linking by N_2 elimination

Fig. 2.6 Schematic of graphite structure

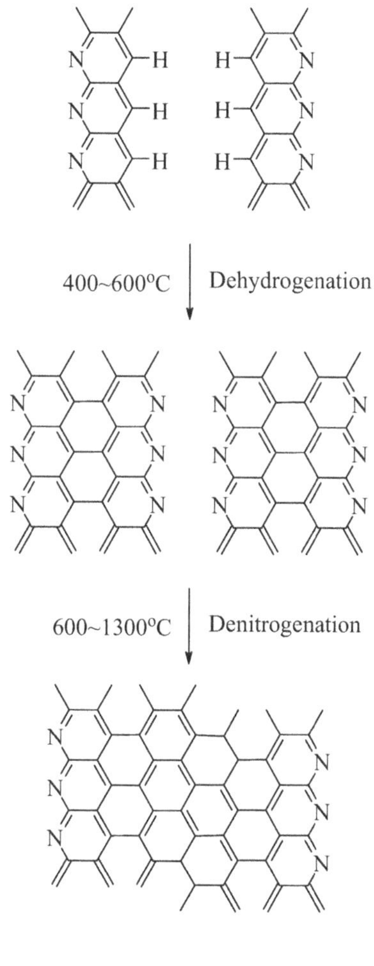

made the ribbons to form sheet-like structures. On the other hand, the high carbonization temperature caused the ordered structure to grow in both thickness and area, increased the crystalline orientation in the fiber direction, and reduced the interlayer spacing and void content. In addition, the graphite structures could further grow at higher temperatures resulting from the elimination of N_2.

4. Surface treatment and washing

Generally, the surface treatment of carbon fibers was performed to improve the mechanical properties of the composite through alteration of the fiber surface. In many companies, the treatment method of the surface of carbon fibers is still kept confidential. The most often used surface-treatment methods for carbon fibers could be categorized as liquid and gaseous oxidation treatments. The liquid oxidation treatment is well known and could double the composite shear strengths with slight reductions (4–6%) in fiber tensile strengths [76].

Among the oxidation-treatment methods of the liquid type, the anodic oxidation-treatment method has been widely used in the surface treatment of commercial carbon fibers because it is inexpensive, fast, and efficient. Figure 2.7 shows the schematic of the surface treatment and washing baths. In this method, Faraday's Law applies and 96,500 C liberates 1 g-equiv of O_2. The duration of the surface treatment is related to the line speed. In addition, the current density as a standard variable is used to control the treatment level per unit length of carbon fiber during the surface treatment, usually expressed as C/m.

The electrically conductive carbon fibers form the anode during the electrolysis of an acid or a salt solution such as nitric acid (HNO_3), sulfuric acid (H_2SO_4), ammonium sulfate ($(NH_4)_2SO_4$), and ammonium bicarbonate (NH_4HCO_3) [77]. One of the electrolytes ammonium sulfate is usually used in the commercial surface-treatment processes of carbon fibers. This causes the carbonyl containing groups such as COOH to form on the smooth fiber surface. The carbonyl groups improve the cohesion between the fiber and resin used in the final composite.

After this surface treatment, the excess electrolyte is removed using warm water wash treatment. The carbon fibers are then passed onto the next process through one or more water baths constantly flowing with water.

Currently, the demand for the surface treatment of carbon fibers has significantly increased with increasing necessity for high-performance carbon fiber composites.

5. Drying, sizing, and winding.

Carbon fibers require some protection or lubrication for the ease of handling because of their brittleness. The carbon fibers are predried for the sizing treatment,

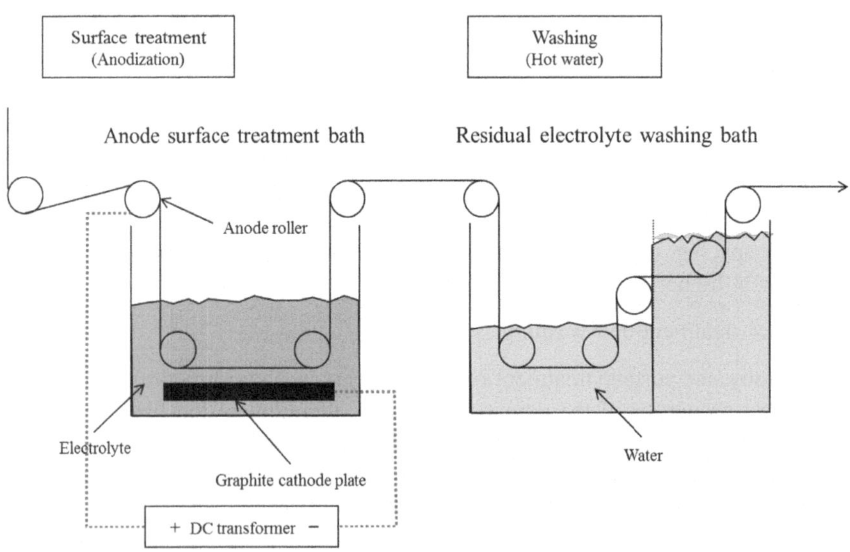

Fig. 2.7 Schematic of surface treatment and washing baths

and the sizing materials are selected such that they protect the physical characteristics of carbon fibers. These sizing materials need to provide consistent handling and not build up residue on the processing equipment. The sizing materials also need to be compatible with the matrix resin. This includes solubility in and/or reactivity with the formulated resin. This allows the resin to penetrate the fiber bundle and interact with the fiber surface. Generally, the epoxy resins or epoxy formulations are used as sizing materials. The sizing materials should not change either the chemical or physical characteristics of the carbon fibers during storage. Some sizing materials are water soluble, and washable after either weaving or braiding [78].

The fiber sizing, process to apply sizing, and sizing content are considered to be critical factors in the carbon fiber specification. The type of size material and particle size of the aqueous dispersion must be controlled to establish good properties in the carbon fibers after sizing. From these viewpoints, the types of emulsifier and resin, and their respective concentrations are key to improving the characteristics of the carbon fibers. The control of wetting in the sizing bath is needed to control the level of size on the carbon fibers. All the steps pertaining to the application of the sizing to the carbon fiber and drying must also be consistent (Fig. 2.8).

Many sizing materials such as epoxy resins are not soluble in water and must be applied as a dispersion of emulsion in water. This could result in the sizing being uniformly distributed on the surface of the fibers. Alternately, the sizing materials could exist as either droplets on the fiber surface or by sticking together a number of individual fibers. The particle size of the emulsion in the sizing bath is controlled to provide a dependable product.

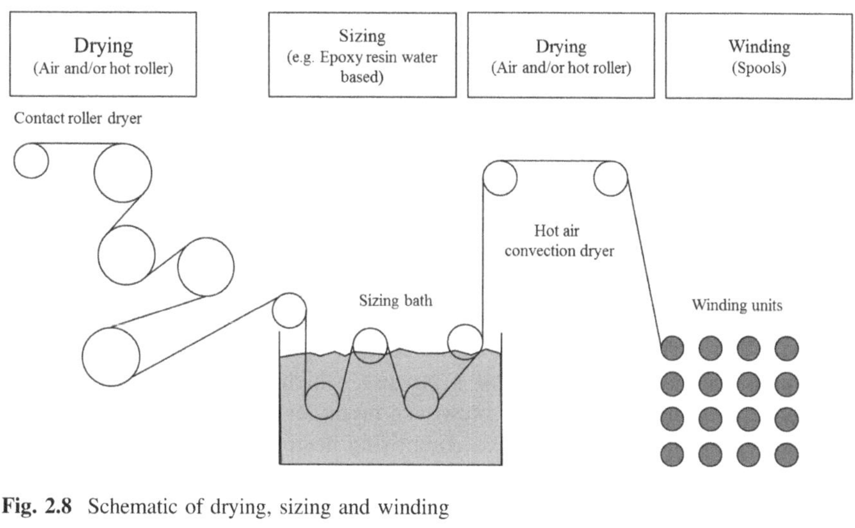

Fig. 2.8 Schematic of drying, sizing and winding

Meanwhile, the dried carbon fibers after the sizing treatment are collected using winders. The winding machines generally operate automatically. In addition, the winders could usually produce finished spools of up to 12 kg in weight.

2.2.4 Types of Polyacrylonitrile-Based Carbon Fibers

There are three basic categories of PAN-based carbon fibers:

1. Large tow: Inexpensive to manufacture and can be conveniently chopped to a staple form
2. General-purpose grades: Less stringent product qualification
3. Aerospace grades: Premium-grade products.

Each category of carbon fibers is available in several production types, which are based on their tensile strength and modulus:

- general purpose
- high strength
- intermediate modulus
- high modulus.

2.3 Cellulosic Precursors

2.3.1 Cellulosic Precursors

Cellulose as a precursor for making carbon fibers was first used by Thomas Edison in the 1880s for his revolutionary electric lamp filament. Almost 80 years later, in 1959, the National Carbon Company (a division of Union Carbide) produced a carbon cloth from a rayon precursor, and two years later, in 1961, a carbon yarn was introduced [79]. In 1965, Thornel series—carbon fibers with a tensile strength of 1.25 GPa and a Young's modulus of 170 GPa—was introduced [80]. The properties of these carbon fibers were improved by postcarbonization treatment involving stretching at 2500 °C. However, the production of these carbon fibers was delayed by more than 10 years because of the high cost of the heat stretching process, lower yield, and properties of the cellulose precursor.

Ford and Mitchell [81] proposed a process, which could be used for rayon monofilaments, cellulosic yarns, or prewoven rayon textile material. Their process involved a controlled heating process comprising heating at 10 °C per hour up to 100 °C, 50 °C per hour up to 400 °C and 100 °C per hour up to approximately 900 °C, followed by heating up to 3000 °C until the graphitization had occurred. A protective atmosphere of nitrogen or other inert gases was used when heating over the range 900–3000 °C. The resultant carbon fibers from this process showed improved tensile strength compared to the earlier carbon fibers. Furthermore, the stronger graphitic structure in these carbon fibers was confirmed using X-ray patterns [82, 83].

Tang and Bacon [84] proposed the four simplified stages mentioned below for the conversion of cellulose to carbon:

1. Stage I: Physical desorption of approximately 12% absorbed water (25–150 °C) with a small degree of change in the lateral order.
2. Stage II: Dehydration from the –H and –OH fragments present in the cellulose unit (150–240 °C). IR shows that –C=O and –C–C– are involved, and hence, dehydration is essentially intramolecular.
3. Stage III: Thermal cleavage of the glycosidic linkage and scission of other C=O and some C–C bonds via a free radical reaction (240–400 °C) which leads to the formation of large amounts of tar, H_2O, CO, and CO_2.
4. Stage IV: Aromatization (400 °C and above), wherein each cellulose unit breaks down into a residue containing four C atoms, which then polymerize through condensation reactions involving the removal of –H above 400 °C into a C-polymer with a graphite-like structure.

Strong [85] described the small-scale heat treatment of rayon precursors for stress graphitization. Because stress graphitization was to be accomplished at approximately 2800 °C, any nonuniformity in the yarn could lead to either a nonuniform stretch or breakage during the stretching process. Therefore, it was essential that the fiber be well supported and transported by rollers. The rayon textile finish was removed by extraction with boiling water and as far as could be ascertained, no flame resistant finish was applied.

The pyrolysis process was undertaken with either: (i) a continuous treatment for approximately 7 min in oxygen at 260–280 °C; or (ii) a batch treatment for approximately 20 h in air at 225 °C with 40% weight loss, followed by 7 min in O_2 to afford a total weight loss of 50–55%. To initiate the development of the final carbon structure and sufficiently improve the strength to enable stress-carbonization, treatment in N_2 for 1.5 min at 350 °C was carried out. The material was carbonized over the range 900–2000 °C and graphitized over the range 2800–2900 °C. X-ray diffraction studies showed that the starting material Cellulose II was converted to the Cellulose IV structure before degradation commenced. It was observed that during pyrolysis in air, shrinkage occurred at approximately 25–45% weight loss, associated with a tendency to kink if the yarn was not under tension. So, the pyrolysis stage was deliberately limited to postpone the second weight loss stage to the carbonization step, during which sufficient tension could be applied to prevent kinking.

2.3.2 *Rayon Precursor for Production of Cellulose-Based Carbon Fibers*

Cellulose is a promising precursor for carbon fiber production. In addition, through pyrolysis it forms strong carbon fibers and the cellulosic precursors have high thermal conductivity, high purity, mechanical flexibility, and low precursor cost [86]. The regenerated cellulose fiber precursors used to make carbon fibers are viscose, cuprammonium rayon or rayon textile-grade rayon.

1. Viscose rayon [87]

The process of manufacturing viscose rayon consists of the following steps mentioned, in the order in which they are carried out: (1) steeping, (2) pressing, (3) shredding, (4) aging, (5) xanthation, (6) dissolving, (7) ripening, (8) filtering, (9) degassing, (10) spinning, (11) drawing, (12) washing, and (13) cutting. The various steps involved in the process of manufacturing viscose are shown in Fig. 2.9.

(1) Steeping

Cellulose pulp is immersed in 17–20% aqueous sodium hydroxide (NaOH) over the temperature range 18–25 °C to swell the cellulose fibers and convert the cellulose to alkali cellulose.

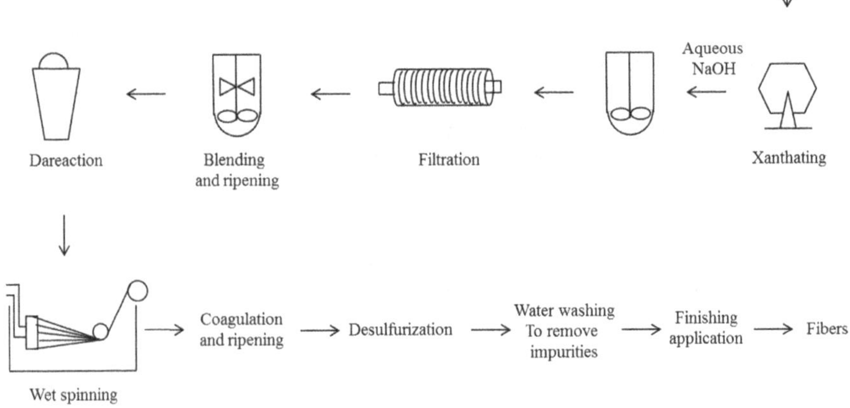

Fig. 2.9 Process of manufacture of viscose rayon fibers

$$\underset{\text{cellulose}}{(C_6H_{10}O_5)_n} + nNaOH \longrightarrow \underset{\text{sodium cellulose}}{C_6H_9O_5Na} + nH_2O$$

(2) Pressing

The swollen alkali cellulose mass is pressed to a wet weight equivalent of 2.5–3.0 times the original pulp weight to obtain an accurate ratio of alkali to cellulose.

(3) Shredding

The pressed alkali cellulose is shredded mechanically to yield finely divided, fluffy particles called "crumbs." This step provides increased surface area of the alkali cellulose, thereby increasing its ability to react in the steps that follow.

(4) Aging

The alkali cellulose is aged under controlled conditions of time and temperature (between 18 and 30 °C) in order to depolymerize the cellulose to the desired degree of polymerization. In this step, the average molecular weight of the original pulp is reduced by a factor of two to three. The reduction of the cellulose is done to get a viscose solution of right viscosity and cellulose concentration.

(5) Xanthation

In this step, the aged alkali cellulose crumbs are placed in vats and are allowed to react with carbon disulfide under controlled temperature (20–30 °C) to form cellulose xanthate.

$$(C_6H_9O_4ONa)_n + nCS_2 \rightarrow (C_6H_9O_4O{-}SC{-}SNa)_n$$

Side reactions, which occur along with the conversion of alkali cellulose to cellulose xanthate, are responsible for the orange color of the xanthate crumb as well as the resulting viscose solution. The orange cellulose xanthate crumb is dissolved in dilute sodium hydroxide at 15–20 °C under high-shear mixing conditions to obtain a viscous orange-colored solution called "viscose," which is the basis for the manufacturing process. The viscose solution is then filtered (to remove the insoluble fiber material) and deaerated.

(6) Dissolving

The yellow crumb is dissolved in an aqueous caustic solution. The large xanthate substituents on cellulose force the chains apart, reducing the interchain hydrogen bonds and allowing water molecules to solvate and separate the chains, thereby leading to a solution of the otherwise insoluble cellulose. Because of the blocks of the unxanthated cellulose in the crystalline regions, the yellow crumb is not completely soluble at this stage. Because the cellulose xanthate solution (or more accurately, suspension) has high viscosity, it has been termed "viscose."

(7) Ripening

The viscose is allowed to stand for a period of time to "ripen." Two important processes occur during ripening: redistribution and loss of xanthate groups. The reversible xanthation reaction allows some of the xanthate groups to revert to the cellulosic hydroxyls and free CS_2. This free CS_2 can then either escape or react with other hydroxyls on other portions of the cellulose chain. In this way, the ordered, or crystalline, regions are gradually broken down and a more complete solution is achieved. The CS_2, which is lost, reduces the solubility of the cellulose and facilitates the regeneration of the cellulose after it is formed into a filament.

$$(C_6H_9O_4O{-}SC{-}SNa)_n + nH_2O \rightarrow (C_6H_{10}O_5)_n + nCS_2 + nNaOH$$

(8) Filtering

The viscose is filtered to remove the undissolved materials, which might either disrupt the spinning process or cause defects in the rayon filament.

(9) Degassing

Bubbles of air entrapped in the viscose must be removed prior to extrusion or else they cause voids, or weak spots, in the fine rayon filaments.

(10) Spinning (Wet Spinning)

Production of Viscose Rayon Filament: The viscose solution is metered through a spinneret into a spin bath containing sulfuric acid (to acidify the sodium cellulose xanthate), sodium sulfate (to impart a high salt content to the bath, which is useful in the rapid coagulation of viscose), and zinc sulfate (for exchange with sodium xanthate to form zinc xanthate, to cross-link the cellulose molecules). Once the cellulose xanthate is neutralized and acidified, a rapid coagulation of the rayon filaments occurs, followed by simultaneous stretching and decomposition of cellulose xanthate to regenerated cellulose. The stretching and decomposition processes are vital for obtaining the desired tenacity and other properties of rayon. The slow regeneration of cellulose and stretching of rayon leads to greater areas of crystallinity within the fiber, as is done with high-tenacity rayon.

The dilute sulfuric acid decomposes the xanthate and regenerates cellulose by wet spinning. The outer portion of the xanthate is decomposed in the acid bath, forming a cellulose skin on the fiber. Sodium and zinc sulfates control the rate of decomposition (of cellulose xanthate to cellulose) and fiber formation.

$$(C_6H_9O_4O{-}SC{-}SNa)_n + (n/2)H_2SO_4 \rightarrow (C_6H_{10}O_5)_n + nCS_2 + (n/2)Na_2SO_4$$

Elongation at break is seen to decrease with increasing degree of crystallinity and orientation of rayon.

(11) Drawing

The rayon filaments are stretched while the cellulose chains are still relatively mobile. This causes the chains to stretch out and orient along the fiber axis. As the chains become more parallel, the interchain hydrogen bonds form, providing the filaments with the properties necessary for use as textile fibers.

(12) Washing

The freshly regenerated rayon contains many salts and other water-soluble impurities, which need to be removed. Several different washing techniques may be used.

(13) Cutting

If the rayon is to be used as staple (i.e., discreet lengths of fiber), the group of filaments (termed "tow") is passed through a rotary cutter to yield a fiber, which can be processed analogously to cotton.

2. Currammonium rayon [88]

In this process, the fibers are produced in a solution of cellulosic material in cuprammonium hydroxide at a low temperature in nitrogen atmosphere, followed by extruding through a spinneret into a sulfuric acid solution, necessary to decompose the cuprammonium complex to cellulose. This is a relatively more expensive process than that used for viscose rayon. However, the cross sections of the resultant fibers are almost round (Fig. 2.10).

3. Saponified cellulose acetate rayon [88, 89] Rayon can be produced from cellulose acetate yarns through saponification. Purified cotton is steeped in glacial acetic acid to make it more reactive. It is then acetylated with an excess of glacial acetic acid and acetic anhydride, followed by sulfuric acid to promote the reaction. The cellulose triacetate formed by acetylation is hydrolyzed to convert the triacetate to diacetate. The resultant mixture is poured into water, which precipitates the cellulose acetate. For spinning, it is dissolved in acetone, filtered, deaerated, and extruded into hot air, which evaporates the solvent. A high degree of orientation can be given to the fiber by drawing because cellulose acetate is more plastic. Its fiber cross section is nearly round, but lobed (Fig. 2.11).

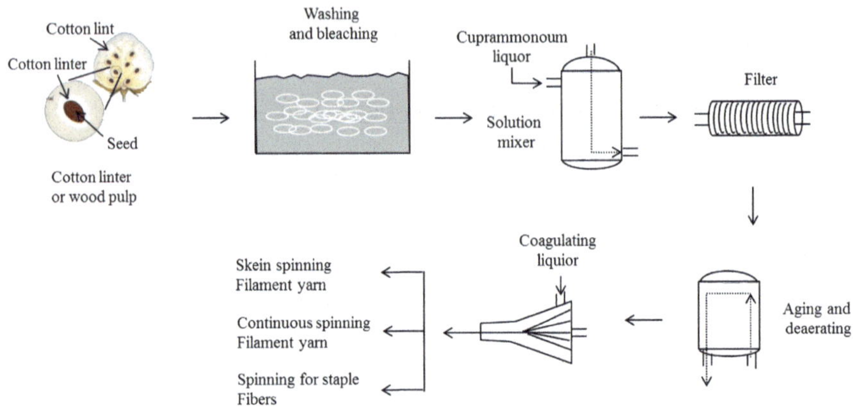

Fig. 2.10 Process of manufacture of currammonium rayon fibers

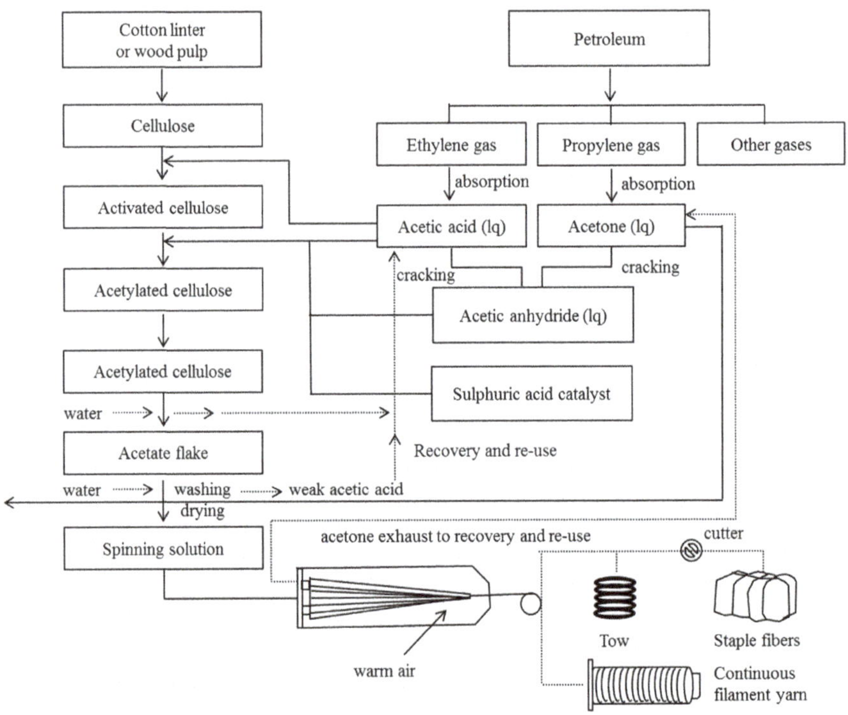

Fig. 2.11 Process of manufacture of saponified cellulose acetate rayon

2.3.3 *Manufacture of Carbon Fibers from Cellulosic Precursors*

Rayon fibers, as mentioned above, can be converted into carbon fibers with chemical, physical, mechanical, and microstructural changes, through stabilization and carbonization processes. With regard to chemical and mechanical properties, the stabilization of the precursor fibers is important to produce stable carbon fibers through the subsequent carbonization process. In particular, the thermal shrinkage of the cellulose fibers occurs owing to the weight reduction of fibers during the stabilization process [90]. Therefore, in this section, only the stabilization, carbonization, and graphitization of cellulosic precursors for manufacturing of carbon fibers will be described. Other processes such as surface treatment, sizing, and winding are identical to the processing of the PAN-based carbon fibers.

The typical stabilization, carbonization, and graphitization processes for the production of carbon fibers from cellulosic precursors are mentioned below:

1. Stabilization

Cellulose is a glucose-based, linear polymer connected by β-(1-4) glycosidic linkages (Fig. 2.12). The hydroxyl groups can easily form intra- and intermolecular hydrogen bonds in the cellulose structure, and the hydrogen bonds can lead to various ordered crystalline arrangements. From the molecular stoichiometry of $(C_6H_{10}O_5)n$, it was found that the theoretical carbon yield of the carbonization process for the cellulosic structure is 44.4%. However, the actual carbon yield is only between 10 and 30%, resulting from the depolymerization of the macromolecular chains and the elimination of carbon by oxygen in the forms of carbon monoxide (CO), carbon dioxide (CO_2), aldehydes, organic acids, and tars. Therefore, the selection of suitable stabilizer materials is essential for improving the carbon yield and properties of the products in the carbonization process, which can lead to higher efficiency [91–94].

The degradation of general cellulose fibers under an inert atmosphere initiates at 200 °C and ends at around 380 °C. In addition, the thermal stability of the cellulose sources depends on the process conditions for cellulose fiber production. Although the physicochemical processes taking place during the transformation of cellulose into carbon are complex, it is certain that the depolymerization of the macromolecular chains produces a variety of oxygenated compounds. This leads to the

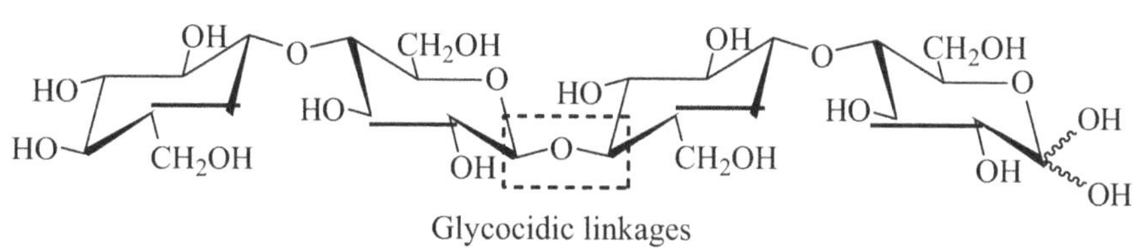

Fig. 2.12 Chemical structure of cellulose polymer chain

major mass loss of the solid residue through the production of volatile substances. Various methods can help in reducing the burning loss. One method involves the modification of cellulose precursors to improve the yield and properties of the carbon fibers. The other method involves the pyrolysis of the cellulose with slow heating rates of a few °C/h or the treatment of cellulose fibers with suitable impregnators [95, 96].

The pyrolysis of cellulose is mainly controlled by two predominant reactions, dehydration, and depolymerization (cleavage) (Fig. 2.12). The main reaction at low temperatures (300 °C) is the dehydration reaction for the stabilization of the cellulose structure. During the dehydration, the elimination of the hydroxyl groups leads to conjugated double bonds, and subsequently, the dehydrated cellulose ring becomes less accessible for cleavage compared to the original cellulose structure in an aromatic form in the carbonization step. The polymeric structure is basically retained through dehydration, and the weight loss at this temperature range is usually owing to the evaporation of water. Depolymerization in the early stages of pyrolysis with the incomplete dehydration of the cellulose structure causes major mass loss at higher temperatures. Therefore, a slow heating rate of a few °C/h can be used to increase the carbon yield of the cellulose fibers. In addition, the slow pyrolysis influences the properties of the final fibers such as improved density, porosity, and microstructure, compared to fast pyrolysis [97].

2. Carbonization and graphitization

As shown in Fig. 2.13, the first H_2O dehydration occurs between 25 and 150 °C (Stage I), and the physical desorption of water and dehydration of the cellulosic unit occurs between 150 and 240 °C (Stage II). This leads to the formation of the double-bonded intermediates [98]. Carbonization of cellulose refers to the conversion process from this depolymerized structure into graphite-like layers through repolymerization. The process begins at approximately 300 °C and continues up to 900 °C, as shown in the Stages III and IV of Fig. 2.13. The basic microstructure of the carbon is formed during the Stage III [82]. As shown in Fig. 2.13, the thermal cleavage of the glycosidic linkage and the scission of ether bonds occur over the range 240–400 °C. Moreover, depolymerization to monosaccharide derivatives occurs during this stage of carbonization. Then, these intermediates form aromatic structures, releasing gases containing noncarbon atoms (O, H) [99]. The carbonaceous residue is converted into a more ordered carbon structure through heat treatment between 400 and 900 °C under an inert atmosphere. The full mechanism of aromatization related to the graphitic products is still unknown, owing to the complex characterization of the cellulose decomposition and existence of many competing reactions during the different stages of carbonization.

The heat treatment up to 900 °C causes the formation of semiordered carbonaceous structures under an inert atmosphere. After this stage, the carbonized fibers can be treated with heat at higher temperatures to initiate graphitization. In general, graphitization is carried out under stress at 900–3000 °C to obtain high-modulus fibers through the development of an enhanced order of the graphene

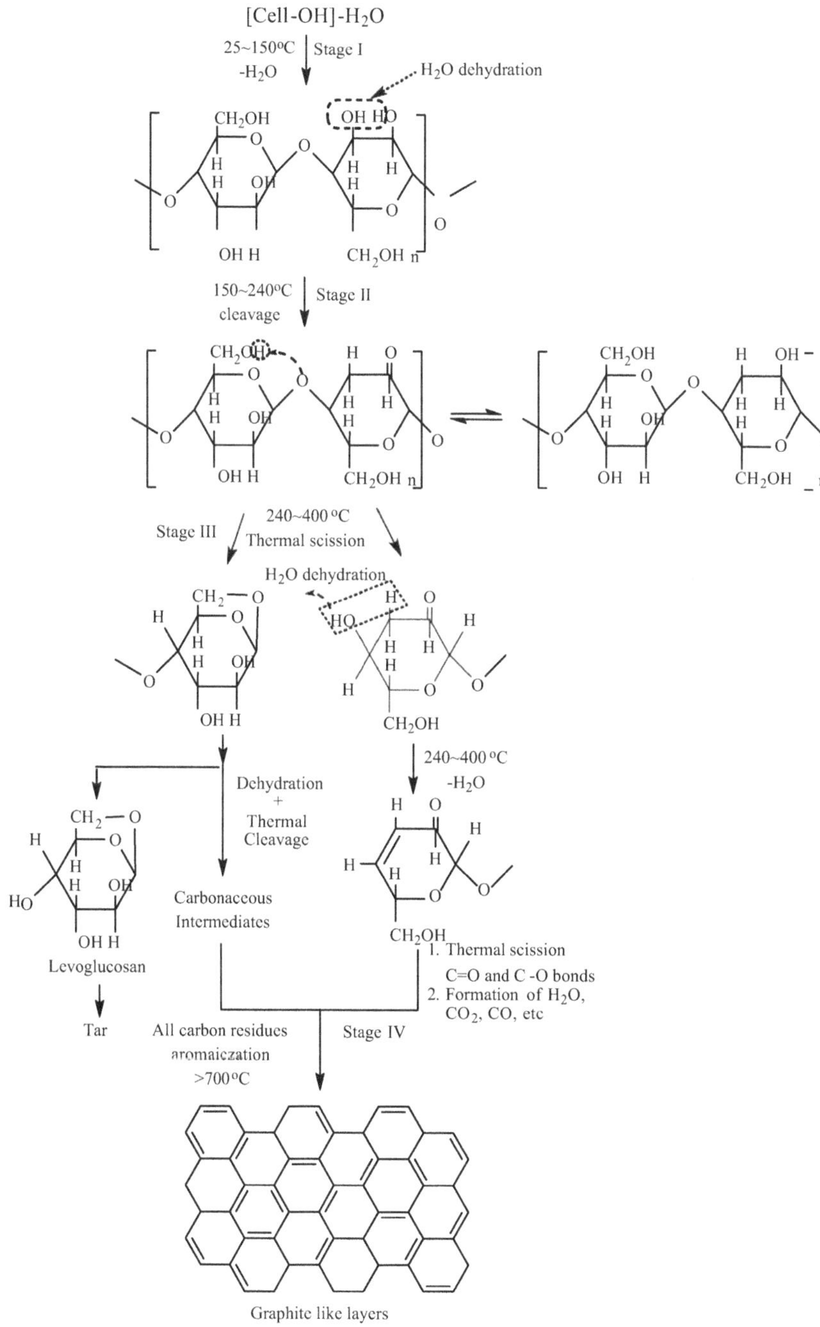

Fig. 2.13 Reactions involved in conversion of cellulose into carbon fibers

stacks, both laterally between the layers (crystallographic register) and in terms of the preferred orientation along the fiber axis. The Young modulus is increased by the increasing treatment temperature, if the graphitization is conducted under tension. After graphitization, the carbon content of the fibers usually increases to above 99%, and the fiber density increases resulting from the growth of crystallites [100].

2.4 Pitch Precursors

Pitches are complex blends of polyaromatic molecules and heterocyclic compounds, which can be used as precursors of carbon fibers or carbon fillers in carbon composites. These pitches can contain more than 80% carbon, and the composition of a pitch varies with the source tar and processing conditions [101]. In addition, these pitches can be obtained from one of several sources mentioned below:

1. Petroleum refining, normally called bitumen, or asphalt
2. Destructive distillation of coal
3. Natural asphalt, e.g., from Trinidad
4. Pyrolysis of PVC.

As mentioned above, the natural pitches are produced by the refining of petroleum and destructive distillation of coal, while the synthetic pitches are produced by the pyrolysis of synthetic polymers. Generally, among the prepared pitches using several sources, the petroleum pitch and coal pitch are widely used in the production of carbon fibers [102].

In terms of the components of pitch, Riggs et al. considered that the pitch is composed of four main classes of chemical compounds [103].

1. Saturates: Low-molecular-weight aliphatic compounds
2. Naphthene aromatics: Low-molecular-weight aromatics and saturated ring structures
3. Polar aromatics: Higher molecular weight and more heterocyclic in nature
4. Asphaltenes: Highest molecular weight fraction in pitch with the highest aromaticity and thermally most stable.

The compositions of various oils and pitches are listed in Table 2.4. Several researchers have confirmed that the asphaltene-rich materials are the most suitable for conversion into carbon fibers.

2.4.1 Petroleum Pitch Precursors

Petroleum pitch can be obtained from various sources such as heavy residue obtained from a catalytic cracking process, steam cracker tar—a by-product of the steam cracking of naphtha or gas oils to produce ethylene or any residues from

Table 2.4 Compositions of various oils and pitches

Compound	Asphaltene (%)	Polar aromatic (%)	Naphthene aromatic (%)	Saturate (%)	Softening point (°C)
Carbon black oil	2.5	10.6	69.0	17.9	
EXXON (DAU) bottoms (refinery sludge)	14.5	41.1	18.1	26.3	29
Ashland 240 petroleum pitch	64.4	8.6	25.4	1.6	119
Ashland 260 petroleum pitch	82.7	5.9	11.4	0.0	177

crude oil distillation or refining [104]. Many methods can be used for the production of pitch and are based on an initial refining process, which can include either one method or combination of several treatment methods listed below:

1. Prolonged heat treatment to advance the molecular weight of the components
2. Air bowing at approximately 250 °C
3. Steam stripping and application of vacuum to remove low boiling components
4. Distillation.

In common with the coal-tar pitch, the chemical and physical characteristics of petroleum pitch are dependent on the process and conditions employed, especially the process temperature and heat-treatment time. Generally, longer times and higher temperatures produce pitches with increased aromaticity and higher anisotropic contents. The petroleum pitches are usually less aromatic compared to the coal-tar pitch [105].

2.4.2 *Coal-Tar Pitch Precursors*

Coal tar is a by-product of the coking of bituminous coals to produce cokes [106]. The metallurgical cokes are produced at high temperatures (between 900 and 1100 °C), but it produces smokeless fuels at lower temperatures (approximately 600 °C). The low-temperature process affords a smaller amount of tar compared to the high-temperature process. Coal-tar pitch is obtained from the coal tar using distillation and heat-treatment processes. The pitch is the residue, which follows the removal of the heavy (creosote or anthracene) oil fractions. The pitches are complex mixtures containing many different individual organic compounds, and the precise compositions and properties vary according to the source of the tar and method of removal of low-molecular-weight fractions. Smith et al. [107, 108] reported that roughly two-thirds of the compounds isolated hitherto from the coal-tar pitch are aromatic, and the rest are heterocyclic. Most of the compounds are substituted with the methyl group. The majority of the coal-tar pitch components contains between

Table 2.5 Aromatic hydrocarbon components present in coal-tar pitch

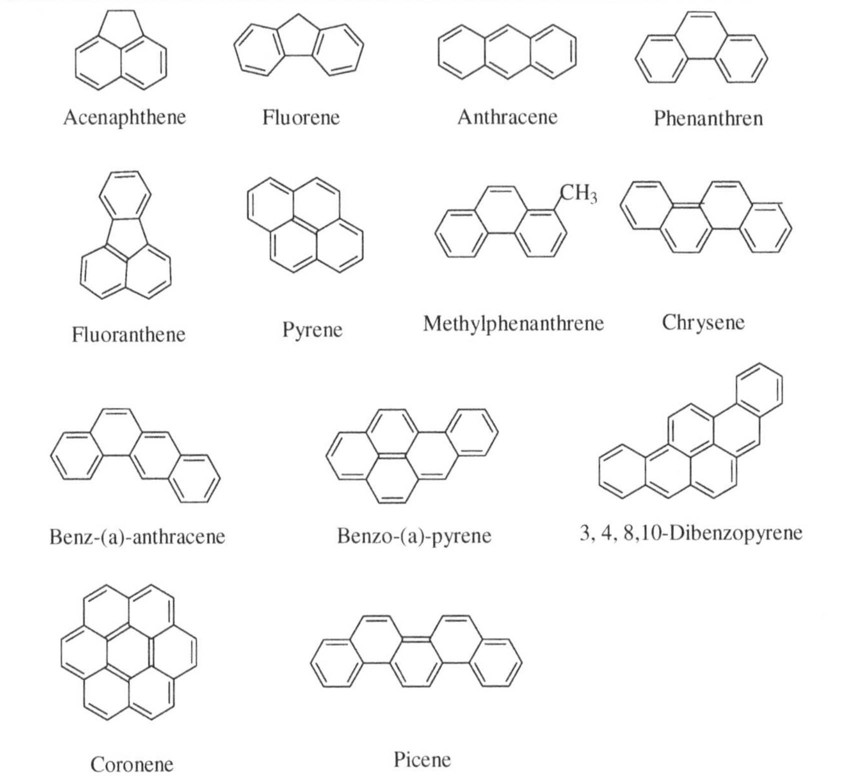

three and six rings, having boiling points over the range 340–550 °C. Table 2.5 lists the major aromatic hydrocarbon compounds, which have been quantitatively found in a typical coal-tar pitch [106–109].

2.4.3 Preparation Methods of Pitch-Based Precursors

1. Preparation methods of petroleum pitch

Many petroleum products are referred to as "pitch" by the petroleum industry. This could cause considerable confusion outside the refining community. In most cases, the different types of petroleum pitch exist as black solids at room temperature. The individual characteristics of the petroleum pitches vary with the functions of the feedstock and specific processes used in their manufacture. Feedstock can range from being predominantly aliphatic to predominantly aromatic-type chemical structures. A reaction step is used to generate and/or concentrate the large molecules typically observed in petroleum pitch. The most common processes used to generate petroleum pitches are either singular processes or a combination of: (a) solvent deasphalting, (b) oxidation, and (c) thermal processes [110].

Solvent deasphalting is used to separate the fractions of various heavy oils. It involves mixing the feedstock with a paraffinic solvent such as propane, butane, or pentane. The mixing of the feedstock with these light paraffinic solvents causes the precipitation of molecules with higher molecular weights and aromaticities. The chemical and physical properties of this type of petroleum pitch are more closely associated with the asphalt cements used for paving roads. Typical properties include specific gravity of approximately 1.0 g/cc at 60 °F, with the chemical composition containing significant amounts of nonaromatic hydrocarbons and high levels of iron, nickel, and vanadium [111].

Various grades of pitch can be produced by the oxidation of heavy petroleum hydrocarbons. Although oxygen is used in the process, the products typically do not contain significant amounts of oxygen. During this reaction, the presence of oxygen is successful in generating free radicals, which induce the polymerization reactions. The chemical properties of these products will depend upon the starting material and degree of reaction, but the pitches produced typically have low coking values and high viscosities [112].

Thermal processing is used to produce the petroleum pitch, as noted in several patents. Thermal processing has traditionally been used to produce the high specific gravity and aromatic petroleum pitches referred to in the introduction of this chapter. The thermal processes typically employ heat-treatment temperatures over the range 300–480 °C. A typical schematic for producing the petroleum pitch from crude oil using thermal processing is shown in Fig. 2.14.

2. Preparation methods of coal-tar pitch

Coal tar is a by-product of the coking of coal, used to produce metallurgical coke [113]. Coal is heated to approximately 1100 °C in a coke oven to produce coke (the primary product) and by-products such as coke oven gas, coal-tar light oil, and coal tar. The typical yields are 70% solid products and 30% liquid products. The yield of coal tar, the feedstock for producing coal-tar pitch, from a ton of coal is 30–45 L (8–12 gallons). Coal-tar pitch has several uses but the majority of the pitch produced is used as a binder for petroleum coke to produce anodes and graphite electrodes [114]. Figure 2.15 shows the schematic for coal-tar pitch production. As indicated in the figure, the coal-tar pitch is produced by the distillation of coal tar.

Pitch as a precursor has the advantage of lower material cost, higher char yield, and higher degree of orientation compared to those of PAN [115]. The graphitic structure also affords pitch-based carbon fibers higher elastic modulus and higher thermal and electrical conductivity along the fiber direction [116]. However, the processing cost (mainly pitch purification, mesophase formation, and fiber spinning) to achieve high-performance carbon fibers is higher. The pitch from petroleum and coal tar is isotropic. By evaporating the low-molecular-weight fractions,

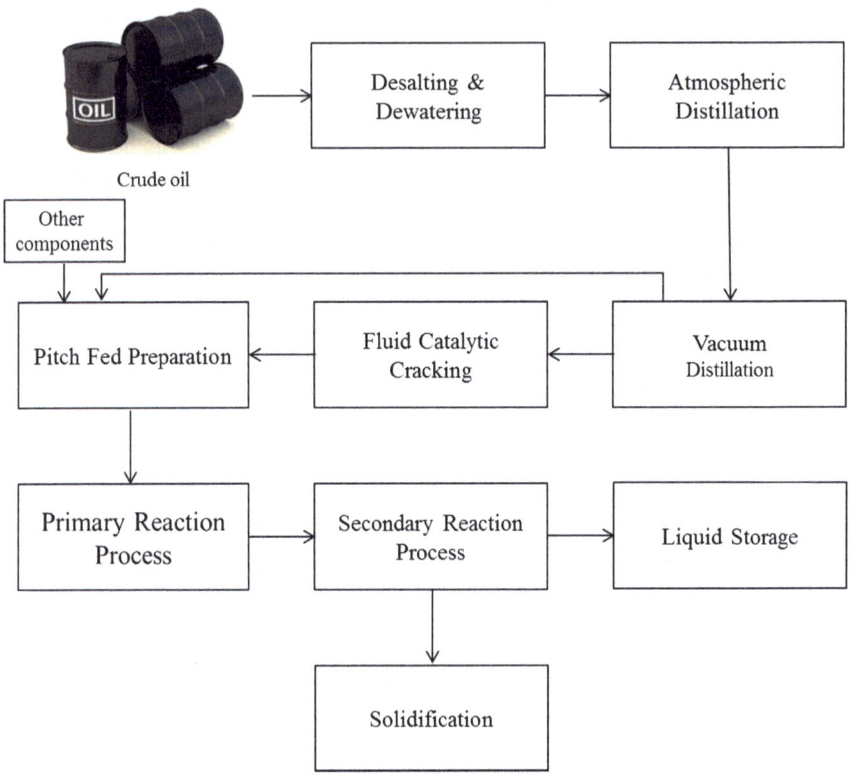

Fig. 2.14 Schematic for the manufacturing process of petroleum pitch from crude oil

the isotropic pitch can be melt-spun into low-cost general-purpose (low strength and low modulus) carbon fibers. To produce high-performance fibers, an expensive hot stretching process (explained in the following section) needs to be applied. A more common method to produce high-performance carbon fibers from the pitch is to use an anisotropic pitch such as mesophase pitch [117].

1. Isotropic pitch

Isotropic pitches are used to make a general-purpose (GP) grade of pitch carbon fibers, which are not graphitic and have poorer properties compared to the high-performance grade (HP), which requires a special treatment process to convert the pitch to a mesophase grade, i.e., an optically anisotropic and graphitic material. The isotropic pitch has to be treated to generate a product suitable for melt spinning, with low volatility and filtered to remove the solid particles. A good starting material would be Ashland 240, which has low quinoline content. The volatile components can be effectively removed with a wiped film evaporator, wherein a thin film of the molten pitch in the evaporator is continuously wiped over the heating surface, exposing a fresh surface, thus permitting the efficient removal of

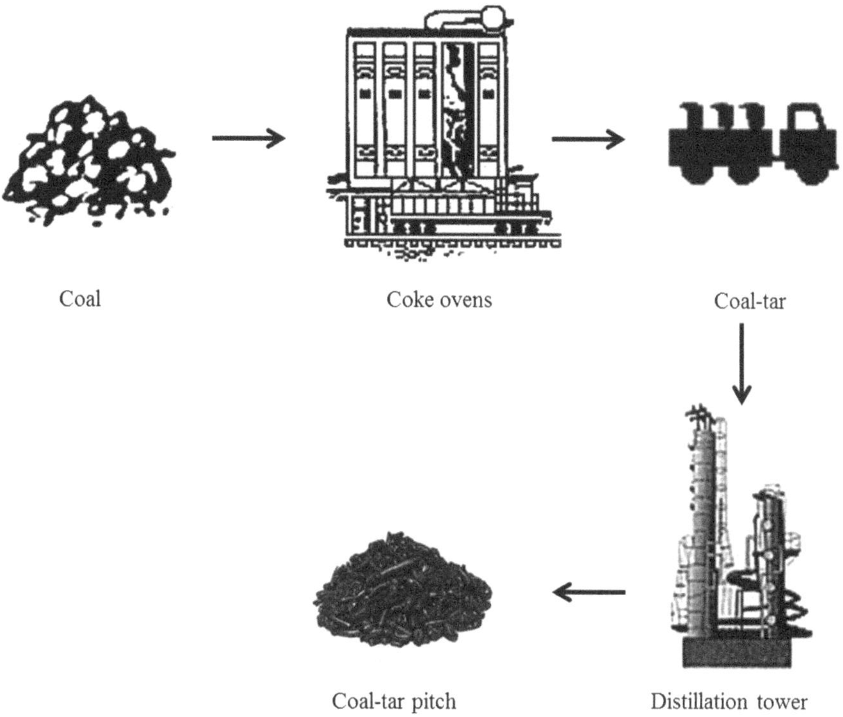

Fig. 2.15 Schematic for the manufacturing process of coal-tar pitch

some of the volatile components without overheating. Also, while the pitch is molten, any solid impurities are removed by filtration. These refining processes raise the softening point and avoid the formation of mesophase. Hence, increasingly, pitches such as Ashland's Aerocarb 60 and 70 can be used. The preparations of isotropic pitches have been undertaken to make general-purpose carbon fibers [118, 119].

2. Mesophase pitch

The mesophase pitches used for high-modulus carbon fiber production can be formed either by the thermal polymerization of petroleum- or coal-tar-based pitches, or the catalytic polymerization of pure compounds such as naphthalene. The mesophase transformation results in an intermediate phase, formed between 400 and 550 °C, by the thermal treatment of aromatic hydrocarbons. During the mesophase formation, domains of highly parallel, plate-like molecules form and coalesce until 100% anisotropic material may be obtained in due course. It has been well established that, when the mesophase pitch is carbonized, the morphology of the pitch is the primary factor that determines the microstructure of the resulting graphitic material [120, 121].

The typical methods used for the production of the mesophase pitch are listed below:

(1) pyrolysis of isotropic pitch
(2) solvent extraction
(3) hydrogenation
(4) catalytic modification.

2.4.4 *Manufacture of Carbon Fibers from Pitch-Based Precursors*

The production of carbon fibers from pitch-based precursors proceeds in several stages: production of precursor fibers; stabilization of the precursor fibers; carbonization of the stabilized precursor fibers and graphitization of the carbonized precursor fibers. The stabilization process is the most important stage during which the thermal oxidation takes place in the precursor carbon fibers. Only properly stabilized precursor fibers can assure satisfactory performance of the final carbon fiber product. Other processes such as surface treatment, sizing, and winding are identical to the processing of PAN-based carbon fibers.

The typical manufacturing steps for the production of carbon fibers from pitch-based precursors are listed below:

1. Production of precursor fibers by melt spinning [122]

Melt spinning involves three steps: melting the precursor, extrusion through a spinneret capillary, and drawing the fibers as they cool. Many investigators have concluded that this process is the primary source of structure in the mesophase pitch-based carbon fibers, and that thermal treatment only reinforces this structure. The effects of melt-spinning conditions on the structure and properties of carbon fibers will be discussed in this section.

Several process variables are important in determining the fibers' potential for developing an ordered graphitic structure. The first is melt temperature. Each mesophase has a range of temperature over which the melt spinning is possible.

Spinning at temperatures below this range results in high viscosities and brittle fracture during drawdown, while at temperatures above this range, thermal degradation of the pitch and dripping owing to low viscosity occur. The viscosities of all the mesophases are highly dependent upon temperature, as shown in Fig. 2.16; therefore, the temperature range for successful melt spinning is quite narrow. Using the AR mesophase, Mochida found that decreasing the melt-spinning temperature by even 15° resulted in a fourfold increase in viscosity and, in turn, significant decreases in the tensile strength and Young's modulus. It was, therefore, concluded that a low melt viscosity was necessary for the production of high-quality carbon fibers.

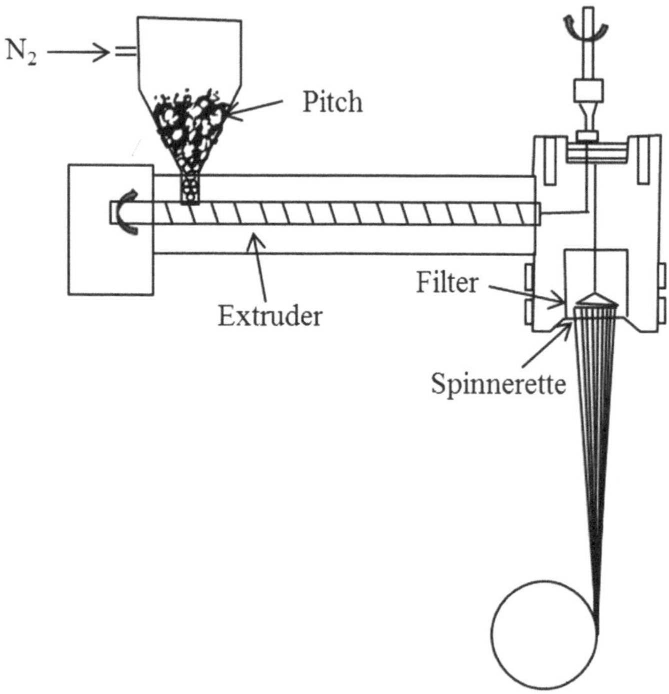

Fig. 2.16 Schematic of melt-spinning process for production of precursor fibers

2. Stabilization of precursor fibers

Analogous to the PAN precursor fibers, the pitch fibers are either infusibilized or oxidized in air at elevated temperatures before being exposed to the final high-temperature carbonization treatment. The oxidization temperature should be below the fiber softening point to keep the orientated structure. Depending on the composition, the mesophase pitch precursor is stabilized in air at 250–350 °C for a duration ranging from 30 min to several hours. There has been no consensus hitherto on the purpose of the fiber stretching in this step. The oxidized pitch molecules contain ketone, carbonyl, and carboxyl groups, which lead to the stronger hydrogen bonding between the adjacent molecules. The introduction of oxygen containing groups and formation of hydrogen bonding between the molecules facilitate the three-dimensional cross-linking, but hinder the growth of crystallites. Iodine has been used to reduce the stabilization time and increase the carbon yield of carbon fibers from the natural pitch. In a patent by Sasaki and Sawaki [123], the pitch fiber was soaked in a methanol solution of iodine till at least 0.05 wt% of iodine was imbibed. The fiber was then heated under an oxidizing atmosphere for infusibilization. The infusibilization time was affected by the amount of imbibed iodine but generally could be completed within approximately 10 min [123, 124]

3. Carbonization and graphitization of stabilized precursor fibers.

Stabilized fibers are then carbonized and graphitized. The greatest weight loss takes place in the early stages of carbonization. In order to avoid the defects created by the excessive release of volatiles, the fibers are precarbonized for a brief period of 0.5–5 min at 700–900 °C. The carbon fibers can be produced by carbonizing the stabilized fibers to 1500–1800 °C. Bright and Singer [125] reported that because of the degradation of the structure, the modulus decreased at temperatures up to approximately 1000 °C, but increased significantly upon further increase in temperature. The carbon fibers can be graphitized at temperatures close to 3000 °C for enhanced Young's modulus. Barr et al. [126] have shown that increasing the heat-treatment temperature could increase the preferred alignment of the crystalline lamellae.

Although graphite layers are aligned along the fiber axis, the transverse structures of the carbon fibers can be different. The velocity gradients orient the layers radially, circumferentially, or randomly. It has been reported that a radial crack can form in the mesophase carbon fibers with layer planes distributed radially. The alignments in the precursor fiber are retained in the resultant carbon fiber. Therefore, carbon fiber strength could be improved by adjusting the microstructure in the precursor fiber. Research has shown that the flaw sensitivity of mesophase carbon fibers is reduced by varying the microstructure of the pitch precursor fibers. The microstructure can be modified by changing the flow profiles during melt spinning. A radial cross section is usually formed through the laminar flow of the pitch melt. Petoca Oil Company has employed agitation in the spinneret to impart a randomized distribution for the folded graphene layer planes in the transverse direction. The agitation created a turbulent flow, and the resultant carbon fibers showed increased tensile strength. The turbulent flow can also be obtained by different die designs because the flow behavior heavily depends on the shape of the spinneret. Spinnerets containing sections with different diameters have shown to be able to change the fiber microstructure. Because melt flow is dependent on melt viscosity, the change in the microstructure can also be obtained simply by changing the spinning temperature. Otani and Oya [127] have shown that when the spinning temperature was raised to above 349 °C, the radial structure changed to either a random-type or radial-type, surrounded by an onion- skin-type structure.

2.5 Other Forms of Precursors

Many other polymers have also been investigated for their potential as carbon fiber precursors [128–137]. In addition to the cellulosic fibers discussed earlier, other natural fibers have been investigated and considered as the precursors for carbon fibers such as silk, chitosan, and eucalyptus [128–136]. They can lower the production cost; however, most of them are used for general-purpose carbon fibers, which do not afford strong mechanical properties.

In addition, many linear and cyclic polymers have been investigated as a precursor for the production of carbon fibers, including phenolic polymers, polyacenephthalene, polyamide, polyphenylene, poly-p-phenylene benzobisthiazole (PBBT), polybenzoxazole, polybenzimidazole, polyvinyl alcohol, polyvinylidene chloride, and polystyrene. Linear precursors require heat stretching to obtain high-performance carbon fibers, and their carbon yields are usually extremely low [137]. The polymers with high aromatic content can generally offer a high carbon yield and in some cases, easy stabilization. However, these polymers either have high costs or do not produce high-performance carbon fibers. Further research needs to be conducted to reduce the processing cost while improving the mechanical properties of the resultant carbon fibers.

2.6 Summary

In this chapter, we have given a brief overview of the precursors and manufacturing of carbon fibers. Among the precursors used for the production of carbon fibers, polyacrylonitrile (PAN)-based and pitch-based precursors are the most important. A significant amount of work has been done on relating the fiber structure to the properties and translating that relationship into production for either reducing cost or increasing fiber properties. However, challenges, including cost reduction, improvement in tensile and compressive strength, and alternative precursor development, still remain.

Many polymers such as linear and cyclic polymers have been evaluated as low-cost carbon fiber precursor materials. It will open the door for low-cost carbon fibers. However, more research needs to be conducted to optimize the processing conditions to enhance the mechanical properties and carbon yield of the resultant carbon fibers.

References

1. O.J. Paranna, in *Engineering Chemistry* (Tata Mcgraw Education, 2009) p. 248
2. W.T. Sedgwick, H.W. Tyler, in *A Short History of Science* (The Macmillan Company, 1969) p. 190
3. T. Roberts, in *The Carbon Fiber Industry: Global Strategic Market Evaluation 2006–2010* (Materials Technology Publications, Watford, UK, 2006) pp. 10, 93–177
4. J.B. Donnet, R.C. Bansal, in *Carbon Fibers*, 2nd edn. (Marcel Dekker, New York, NY, USA, 1990) pp. 1–145
5. X. Huang, Materials **2**, 2369 (2009)
6. M C. Red, Compos. Manuf., **7**, 24 (2006)
7. M. Wang, T. Wang, G. Zhao, Int. J. Appl. Environ. Sci. **8**, 1547 (2013)
8. S. Arbab, A. Zeinolebadi, Polym. Degrad. Stab. **98**, 2537 (2013)
9. M. Szala, in Fullerenes, Nanotubes, Carbon Nanostruct. **21**, 879 (2013)
10. M. Thunga, K. Chen, D. Grewell, M.R. Kessler, Carbon (2013), Article in press

11. D. Sun, R. Ban, P.H. Zhang, G.H. Wu, J.R. Zhang, J.J. Zhu, Carbon **64**, 424 (2013)
12. J.K. Lin, S. Kubo, T. Yamada, M. Enoki, Y. Uraki, J. Wood Chem. Technol. **34**, 111 (2014)
13. Y.Z. Wang, S.G. Wang, J.L. Liu, Key Eng. Mater. **575–576**, 151 (2014)
14. Y. Bai, Z.H. Huang, F. Kang, Carbon **66**, 705 (2014)
15. M. Vukčević, B. Pejić, A. Kalijadis, I. Pajić-Lijaković, M. Kostić, Z. Laušević, M. Laušević, Chem. Eng. J. **235**, 284 (2014)
16. Y. Yao, J. Chen, L. Liu, Y. Dong, A. Liu, J. Mater. Sci. **49**, 191 (2014)
17. E. Hammel, X. Tnag, M. Trampert, T. Schmitt, K. Mauthner, A. Eder, Carbon **42**, 1153 (2004)
18. D.L. Chung, in *Carbon Fiber Composites* (Butterworth-Heinemann, Boston, MA, USA 1994), pp. 3–65
19. M.K. Seo, S.J. Park, Mat. Sci. Eng. B **164**, 106 (2009)
20. S.S. Chair, O.P. Bahl, R.B. Mathur, Fiber Sci. Technol. **15**, 153 (1981)
21. E. Zussman, X. Chen, W. Ding, L. Calabri, D.A. Dikin, J.P. Quintana, R.S. Ruoff, Carbon **43**, 2175 (2005)
22. H.C. Shao, H.Y. Xia, G.W. Liu, G.J. Qiao, Z.C. Xiao, J.M. Su, X.H. Zhang, Y.J. Li, J. Mater. Eng. Perform (2013), Article in press
23. M.R. Buchmeiser, J. Unold, K. Schneider, E.B. Anderson, F. Hermanutz, E. Frank, A. Müller, S. Zinn, J. Mater. Chem. A **1**, 13154 (2013)
24. V. Kuzmenko, O. Naboka, P. Gatenholm, P. Enoksson, Carbon (2013), Article in press
25. G. Saparaga, T. Mikolajczyk, A. Frczek-Szczypta, Fibers Text. East Europe **21**, 33 (2013)
26. H. Bi, Z. Yin, X. Cao, X. Xie, C. Tan, X. Huang, B. Chen, F. Chen, Q. Yang, X. Bu, X. Lu, L. Sun, H. Zhang, Adv. Mater. **25**, 5916 (2013)
27. Y. Aykut, B. Pourdeyhimi, S.A. Khan, J. Mater. Sci. **48**, 7783 (2013)
28. S. Hu, Y.L. Hsieh, J. Mater. Chem. A **1**, 11279 (2013)
29. S.H. Park, S.G. Lee, S.H. Kim, J. Mater. Sci. **48**, 6952 (2013)
30. H.J. Ko, C.U. Park, H.H. Cho, Y.S. Lim, M.J. Yoo, M.S. Kim, Kor. J. Mater. Res. **48**, 6952 (2013)
31. A. Burkanudeen, G. Santhana Krishnan, N. Murali, J. Therm. Anal. Calorim. **112**, 1261 (2013)
32. R. Eslami Farsani, A. Shokuhfar, A. Sedghi, Int. J. Aerosp. Mech. Eng. **1**(4), 184 (2007)
33. P. Rajalingam, G. Radhakrishnan, JMS-Rev. Macromol. Chem. Phys. **C31**, 301 (1991)
34. E. Fitzer, Carbon **27**, 621 (1989)
35. R.B. Mathur, O.P. Bahl, J. Mittal, Compos. Sci. Technol. **51**, 223 (1994)
36. J. Riggs, in *Encyclopedia of Polymer Science and Engineering*, 2nd edn, vol 2, ed by H.F. Mark, J.I. Kroschwitz (Wiley, New York, 1985), p. 17
37. E. Fitxer, M. Heine, *In Fiber Reinforcements for Composite Myers: Composite Material Series*, vol 2, ed by A. R. Bunsell (Elsevier Publishers, Amsterdam, 1988), pp. 73–148
38. J.B. Donnet, O.P. Bahl, *Encyclopedia of physical science and technology*, vol. 2 (Academic Press, New York, 1987), p. 515
39. S. Dalton, F. Heatley, P.M. Budd, Polymer **40**, 5531 (1999)
40. N. Grassie, R. McGuchan, Eur. Polym. J. **8**, 257 (1972)
41. G.T. Sivy, M.M. Coleman, Carbon **19**, 127 (1981)
42. B.N. Frushover, in, *Acrylic Fiber Technology and Applications*, ed by J.C. Masson (Marcel Dekker, New York 1995), pp. 197–258 (Chapter 7)
43. V.A. Bhanu, P. Rangarajan, K. Wiles, M. Bortner, M. Sankarpandian, D. Godshall, T.E. Glass, A.K. Banthia, J. Yang, G. Wilkes, D. Baird, J.E. McGrath, Polymer **43**, 4841 (2002)
44. O.P. Bahl, in, *Carbon Fibers*, 3rd ed, ed by J.B. Donnet, T.K. Wang, S. Rebouillat, J.C.M. Peng (Marcel Dekker, New York, 1998), pp. 1–84 (Chapter 1)
45. A.K. Gupta, D.K. Paliwal, P. Bajaj, *JMS-Rev. Part C: Polym, Rev.*, **31**, 1 (1991)
46. R. Devasia, C.P.R. Nair, K.N. Ninan, Eur. Polym. J. **39**, 537 (2003)
47. C. Hou, R. Qu, Y. Liang, C. Wang, J. Appl. Polym. Sci. **96**, 1529 (2005)
48. G. Burillo, A. Chapiro, Z. Mankowski, J. Polym. Sci., Polym. Chem. Edi., **18**, 327 (1980)
49. L. Boguslavsky, S. Baruch, S. Margel, J. Colloid Interface Sci. **289**, 71 (2005)

50. P. Bajaj, D.K. Paliwal, A.K. Gupta, J. Appl. Polym. Sci. **49**, 823 (1993)
51. A. Shindo, Japan Patent Publication No. (1962)4405 (1962), Patent Application No. (1959) 28287 (1959)
52. R. Bacon, T. N. Hoses, *High Performance Polymers, Their origin and Development*, ed by R.B. Sanymour, G. S. Kirshambaum (Elsevier, New York, 1986), p. 342
53. K. Othmer, *Encyclopedia of Chemical Technology: Vol 26 Carbon Fibers*, 5th edn. (Wiley Blackwell, 2004), p. 730
54. R. Bacon, M.M. Tang, Carbon **2**, 211 (1964)
55. J.C. Chen, I.R. Harrison, Carbon **40**, 25 (2002)
56. V.B. Gupta, V.K. Kothari (eds.), *Manufactured Fibre Technology* (Chapman & Hall, London, 1997)
57. R.J. Jokarsky, L.E. Ball, M.M. Wu, C.E. Uebele, US Pat 6,114,034 (2000)
58. M.C Paiva, P Kotasthane, D.D Edie, A.A Ogale, Carbon, **41**, 1339 (2003)
59. A. Takaku, J. Shimizu, J. Appl. Polym. Sci. **29**, 1319 (1984)
60. O.P. Bahl, L.M. Manocha, Carbon **12**, 417 (1974)
61. P.C. Kang, G.Q. Chen, B. Zhang, G.H. Wu, S. Mula, C.C. Koch, Surf. Coat. Tech. **206**, 305 (2011)
62. D.H. Kim, B.H. Kim, K.S. Yang, Y.H. Bang, S.R. Kim, H.K. Im, J. Kor. Chem. Soc. **55**, 819 (2011)
63. W. Shen, S. Zhang, Y. He, J. Li, W. Fan, J. Mater. Chem. **21**, 14036 (2011)
64. V. Sridhar, J.H. Jeon, I.K. Oh, Carbon **49**, 222 (2011)
65. S.J. Park, Y. Jung, S. Kim, J. Fluorine Chem. **144**, 124 (2012)
66. Z. Li, J. Wang, Y. Tong, L. Xu, J. Mater. Sci. Technol. **28**, 1123 (2012)
67. Z. Zhang, J.D. Atkinson, B. Jiang, M.J. Rood, Z. Yan, Appl. Cat. B: Environ. **148–149**, 573 (2014)
68. A. Ju, Z. Liu, M. Luo, H. Xu, M. Ge, J. Polym. Res. **20**, 1 (2013)
69. Y.X. Yu, B.L. He, L. Li, Adv. Mater. Res. **791**, 506 (2013)
70. X.C. Wang, L. Zhou, J.W. Wang, Adv. Mater. Res. **815**, 827 (2013)
71. L.H. Peebles, *Carbon Fiber—Formation, Structure, and Properties* (CRC Press, Boca Raton, LA, 1995)
72. M.S.A. Rahaman, A.F. Ismail, A. Mustafa, Polym. Degrad. Stab. **92**, 1421 (2007)
73. J. Mittal, R.B. Mathur, O.P. Bahl, M. Inagaki, Carbon **36**, 893 (1998)
74. N. Yusof, A.F. Ismail, J. Anal. Appl. Pyrol. **93**, 1 (2012)
75. J. Bromley, in *Proceedings of the International Conference on Carbon Fibres, their Composites and Applications*, Plastic Institute, London, 1971, p. 3
76. L.T. Drzal, M.J. Rich, P.F. Lloyd, J. Adhes. **16**, 1 (1983)
77. A. Bismarck, M.E. Kumru, J. Springer, J. Simitzis, Appl. Surf. Sci. **143**, 45 (1999)
78. T.H. Cheng, J. Zhang, S. Yumitori, F.R. Jones, C.W. Anderson, Composites **25**, 661 (1994)
79. R. Bacon, US Pat 2,957,756 (1959)
80. W.A. Schalamon, R. Bacon, US Pat 3,716,331 (1973)
81. C.E. Ford, C.V. Mitchell, UNION CARBIDE CORP fibrous graphite, US Patent 3,107,152 (1963)
82. R. Bacon, M.M. Tang, Carbon **2**, 221 (1964)
83. O. Paris, C. Zollfrank, G.A. Zickler, Carbon **43**, 53 (2005)
84. M.M. Tang, R. Bacon, Carbon **2**, 211 (1964)
85. S.L. Strong, J. Mater. Sci. **9**, 993 (1974)
86. Q. Wu, D. Pan, Text. Res. J. **72**, 405 (2002)
87. B. Richard, L.A. Millington, C. Robert, US Pat 3,322,489 (1967)
88. J.G. Cook, in *Handbook of Textile Fibres: Volume 2: Man-Made Fibres* (Woodhead Publishing, 1984) p. 82, 100
89. H.L. Needles, in *Textile Fibers, Dyes, Finishes, and Processes* (Noyes Publications, Park Ridge, New Jersey, 1986) p. 46
90. R. Bacon, Carbon fibers from rayon precursors, in *Chemistry and Physics of Carbon*, vol. 9, ed. by P.L. Walker Jr., P.A. Thrower (Marcel Dekker, New York, 1973), pp. 1–102

91. X. Huang, Materials **2**, 2369 (2009)
92. F. Zeng, D. Pan, N. Pan, J. Inorg. Org. Polym. Mater. **15**, 261 (2005)
93. J.B. Tomlinson, C.R. Theocharis, Carbon **30**, 907 (1992)
94. D.Y. Kim, Y. Nishiyama, M. Wada, S. Kuga, Cellulose **8**, 29 (2001)
95. F.J. Kilzer, A. Broido, Pyrodynamics **2**, 151 (1965)
96. C.B. Cross, D.R. Ecker, O.L. Stein, US Pat 3,116,975 (1964)
97. P.H. Brunner, P.V. Roberts, Carbon **18**, 217 (1980)
98. J. Scheirs, G. Camino, W. Tumiatti, Eur. Polym. J. **37**, 933 (2001)
99. J.A. Lie, M.B. Hagg, J. Membr. Sci. **284**, 79 (2006)
100. X. Huang, Materials **2**, 2369 (2009)
101. C. Berrueco, P.Á lvarez, N. Díez, M. Granda, R. Menéndez, C. Blanco, R. Santamaria, M. Millan, *Fuel*, In Press (2011)
102. T. Matsumoto, Pure Appl. Chem. **57**, 1553 (1985)
103. D.M. Riggs, R.J. Shuford, R.W. Lewis, in *Graphite Fibers and Composites*, ed by G. Lubin, Handbook of Composites (Van Nostrand Reinhold Co., New York, 1982)
104. Kureha Co., U.S. Pat., 3,629,379 (1971)
105. C. Ungureanu, M. Onciu, D. Timpu, Mater. Plastice **33**, 57 (1996)
106. F.A. Smith, T.F. Eckle, R.J. Osterholm, R.M. Stichel, *Manufacture of Coal Tar and Pitches*, ed by A.J. Hoiberg, Bituminous Materials, vol III (RE Krieger, New York, 1966), p. 57
107. J.W. Smith, Fuel **45**, 233 (1966)
108. M. Guigon, A. Overlin, G. Desarmot, Fiber Sci. Tech. **20**, 177 (1984)
109. J.A. Monge, D.C. la Amoros, A.L. Solano, A. Oya, A. Sakamoto, K. Hoshi, Carbon **35**, 1079 (1997)
110. M.D. Kiser, D.C. Boyer, U.S. Pat., 0094454 (2004)
111. H. McGraw, *Concise Encyclopedia of Science and Technology*, 6th edn. (McGraw-Hill Inc., US, 1989)
112. D.C. Boyer, M.D. Kiser, S.E. See, U.S. Patent, 0187337 (2002)
113. M. Leavitt, *Technical Support Document for the 2004 Effluent Guidelines Program Plan*, (2004), United State Environmental Protection Agency, Chapter 6, p. 30
114. M. Sahoo, *Canadian Institute of Mining, Metallurgy and Petroleum*, ed. C. Fradet by Light Metals Section (Canadian Institute of Mining, Metallurgy and Petroleum 1998), p. 450
115. E. Mora, C. Blanco, V. Prada, R. Santamarıa, M. Granda, R. Menendez, Carbon **40**, 2719 (2002)
116. Committee on High Performance Synthetic Fibers for Composites, Commission on Engineering and Technical Systems, and National Research Council, *High-Performance Synthetic Fibers for Composites* (National Academy Press, Washington, DC, USA 1992), pp. 23, 56–64, 86
117. L.H. Peebles, *Carbon Fibers: Formation, Structure, and Properties* (CRC Press Inc., Boca Raton, 1995), p. 29
118. S.M. Zeng, T. Maeda, K. Tokumitsu, J. Mondori, Carbon **31**, 413 (1993)
119. Y.O. Choi, K.S. Yang, Fibers Polym. **2**, 178 (2001)
120. M.Z. Ozel, K.D. Bartle, Turk. J. Chem. **26**, 417 (2002)
121. I.C. Lewis, U.S. Pat., 4,032,430 (1977)
122. D.D. Edie, M.G. Dunham, Carbon **27**, 647 (1989)
123. H. Sasaki, T. Sawaki, US Pat., 4948574 (1990)
124. C.Q. Yang, Carbon **31**, 451 (1993)
125. A.A. Bright, L.S. Singer, Carbon **17**, 59 (1979)
126. J.B. Barr, S. Chwastiak, R. Didchenko, I.C. Lewis, R.T. Lewis, L.S. Singer, Appl. Polym. Symp. **29**, 161 (1976)
127. S. Otani, A. Oya, Status report on pitch-based carbon fiber in Japan, in *Composites '86: Recent Advances in Japan and the United States, Japan Society for Composite Materials*, ed. by K. Kawata, S. Umekawa, A. Kobayashi (Japan Society for Composite Materials, Tokyo, Japan, 1986), pp. 1–10

128. M. Majibur, R. Khan, Y. Gotoh, H. Morikawa, M. Miura, Y. Fujimori, M. Nagura, Carbon **45**, 1035 (2007)
129. M. Bengisu, E. Yilmaz, Carbohydr. Polym. **50**, 165 (2002)
130. V. Mottaghitalab, M. Farjad, J. Polym. Eng. **33**, 857 (2013)
131. E. Rojo, M.V. Alonso, J.C. Domínguez, B.D. Saz-Orozco, M. Oliet, F. Rodriguez, *J.Appl. Polym. Sci.* **130**, 2198 (2013)
132. N. Cordeiro, M. Faria, E. Abraham, L.A. Pothan, Carbohydr. Polym. **98**, 1065 (2013)
133. X. Zhao, X. Lu, W.T.Y. Tze, J. Kim, P. Wang, A.C.S. Appl, Mater. Interf. **5**, 8853 (2013)
134. X. He, M.B. Hägg, Chem. Eng. J. **215–216**, 440 (2013)
135. A.G. Dumanli, A.H. Windle, J. Mater. Sci. **47**, 4236 (2012)
136. X. He, M.B. Hägg, J. Membr. Sci. **378**, 1 (2011)
137. J.B. Donnet, R.C. Bansal, *Carbon Fibers*, 2nd edn. (Marcel Dekker, New York, NY, USA, 1990), pp. 1–145

Chapter 3
Matrices for Carbon Fiber Composites

Abstract Carbon fibers have become an important reinforcing material in advanced composites because of its extremely high strength, stiffness, heat resistance, and lightweight. Fiber-reinforced polymer composites have gained substantial interest owing to their very high strength-to-weight and stiffness-to-weight ratios, and are widely used in aerospace, engineering, marine, and automobile industries. In this chapter, we will present the various matrices and their characterization for carbon fibers' composites. It is described as classification into two types of thermosetting resins and thermoplastic resins.

3.1 Thermosetting Resins

3.1.1 Introduction

Thermosetting resins are polymer materials, which cure irreversibly, generally found in the liquid form. The cure may be affected by heat (e.g., cyanate ester resins), chemical reaction (e.g., epoxy resins), and irradiation (e.g., vinyl ester resins) such as an infrared, ultraviolet light, or electron beam. Once cured, the material cannot be reheated and melted back to its original liquid form.

Thermosetting resins are easy to process and laminate, do not necessarily need either pressure or heat to form, generally inexpensive, stronger than thermoplastics, and better suited for higher temperatures compared to thermoplastics. However, they are relatively more brittle.

Carbon fiber (CF) has become an important reinforcing material in advanced composites because of its extremely high strength, stiffness, heat resistance, and lightweight. Fiber-reinforced polymer composites have gained substantial interest owing to their very high strength-to-weight and stiffness-to-weight ratios, and are widely used in aerospace, engineering, marine, and automobile industries.

Among the thermosetting resins, cyanate ester, epoxy, phenolic, polyester, polyimide, and vinyl ester resins can be used as polymer matrices for CF-reinforced polymer composites [1–12].

S.-J. Park, *Carbon Fibers*, Springer Series in Materials Science 210,
https://doi.org/10.1007/978-981-13-0538-2_3

3.1.2 Cyanate Ester Resins

Cyanate ester (CE) resins belong to a class of high-performance thermosetting resins and contain at least two cyanate functional groups [13, 14]. Most commercial monomers can be represented by the structural model illustrated in Fig. 3.1.

CE resins can be cured by heating alone, either at elevated or lower temperatures in the presence of a suitable catalyst. The most common catalysts are transition metal complexes such as cobalt, copper, manganese, and zinc complexes. The chemistry of the cure reaction for CE resins involves the trimerization of three CN groups to form a triazine ring. The CE resins have two cyanate groups, which resulted in a three-dimensional polymer network structure, as shown in Fig. 3.2 [15].

CE resins will homopolymerize into a thermosetting material suitable for use in high-performance composites such as printed circuit boards, structural composites, and radomes. These resins have good processability, shelf-life, and compatibility with a variety of reinforcements.

CEs are a family of high-temperature thermosetting resins, which bridge the gap in the thermal performance between the engineering epoxy resins and high-temperature polyimides. In addition to their outstanding thermal performance, CE resins have several desirable characteristics, which justify their higher costs in many applications. CEs possess a unique balance of properties and are particularly notable for their low dielectric constant and dielectric loss, low moisture absorption, low shrinkage, and low outgassing characteristics. They can be used in high-performance adhesives, syntactic foams, honeycombs, and fiber-reinforced composites. They are often found in blends with other thermosetting resins such as epoxy, bismaleimide, and engineering thermoplastics [16, 17].

Fully cured bisphenol E CE has glass transition temperature (T_g) and decomposition temperature (T_d), at 5 wt% loss, of 274 and 438 °C, respectively [18, 19].

Marieta et al. [14] studied the effect of surface treatment on the interfacial shear strength (IFSS) of bisphenol A dicyanate (DCBA)/PAN-based CF composites. Figure 3.3 shows the apparent IFSS values of the composites with different treatments. The results indicated that the commercial sizing of epoxy was found to be effective in promoting adhesion because of the chemical reactions between epoxy sizing and DCBA during the curing process.

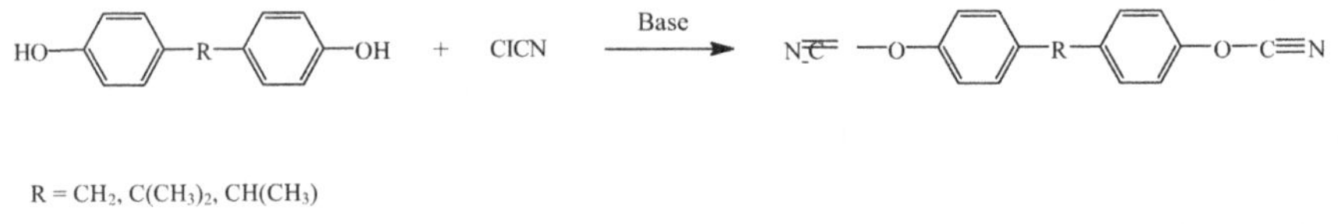

Fig. 3.1 Schematic of synthesis of CE monomers

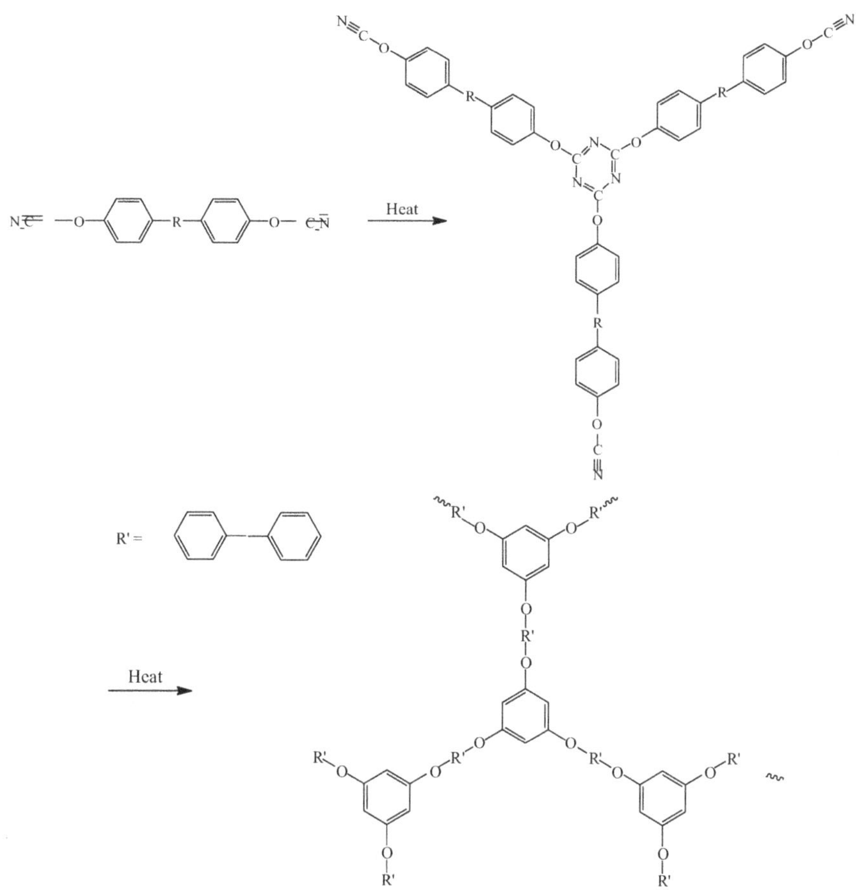

Fig. 3.2 Curing of CE resins via cyclotrimerization

Fig. 3.3 Apparent IFSS data of DCBA/CF composites subjected to different treatments using pull-out measurements [14]

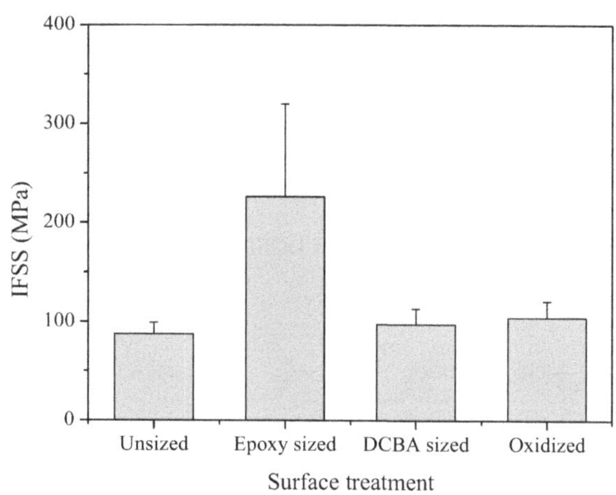

3.1.3 Epoxy Resins

Epoxy resins are molecules containing more than one epoxide group, as shown below:

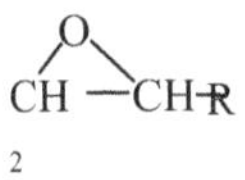

Epoxy resins are thermosetting polymers, which can be cured with a wide variety of curing agents via curing reactions. Epoxy resins have a wide range of applications, including fiber-reinforced materials, general-purpose adhesives, high-performance coatings, and encapsulating materials [20–26].

3.1.3.1 Bisphenol-A/F Epoxy Resins

Diglycidyl ether of bisphenol A (DGEBA) is produced by reacting epichlorohydrin with bisphenol A in the presence of a basic catalyst. Figure 3.4 depicts the chemical structure of DGEBA.

The properties of the DGEBA resins depend on the value of n, which is the number of repeating units commonly known as degree of polymerization. Low-molecular-weight molecules tend to be liquids while higher molecular weight molecules tend to be either viscous liquids or solids. DGEBA has been used commercially as a raw material and is the primary chemical building block for a broad spectrum of materials [27, 28].

Diglycidyl ether of bisphenol F (DGEBF) is synthesized by reacting epichlorohydrin with bisphenol F in the presence of a basic catalyst. Figure 3.5 shows the chemical structure of DGEBF.

DGEBF epoxy resins have low viscosity, and are widely used in solvent-free coatings, concrete reinforcements, adhesives, electrical insulation, and filament winding. The cured DGEBF epoxy resins provide improved resistance performance against solvents compared to that of DGEBA epoxy resins [29].

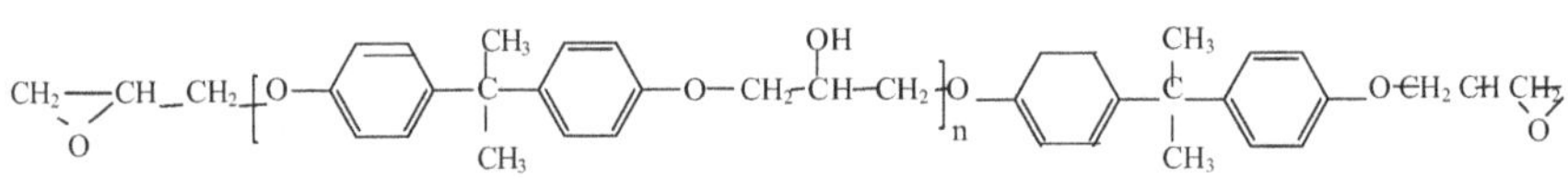

Fig. 3.4 Chemical structure of DGEBA

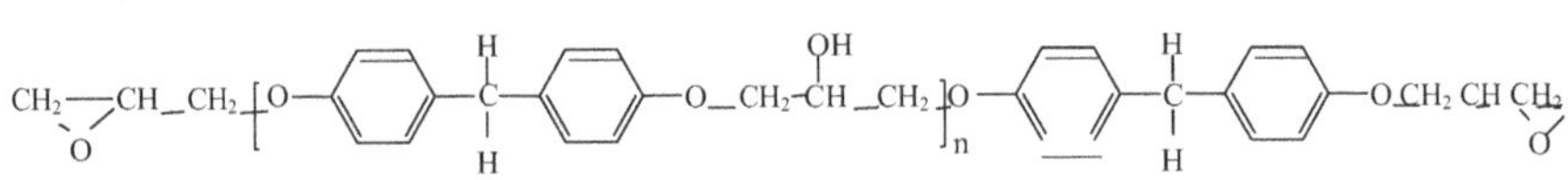

Fig. 3.5 Chemical structure of BFDGE

Diglycidyl ether of tetrabromobisphenol-A is prepared by reacting epichlorohydrin with tetrabromobisphenol A in the presence of a catalyst. Figure 3.6 shows the chemical structure of the diglycidyl ether of tetrabromobisphenol-A.

Diglycidyl ether of tetrabromobisphenol A with high T_g, heat resistance, good operation performance, fire resistance, and low moisture absorbance can be used in glass fiber laminates [30].

Park et al. [31] demonstrated the effect of fiber–polymer interactions on the fracture toughness behavior of CF-reinforced DGEBA epoxy matrix composites. Figure 3.7 shows the evolution of critical stress intensity factor (K_{IC}) of DGEBA/CF composites as a function of the electric current density. The K_{IC} of the composites continually increases with increasing current densities of the treatments up to 0.4 A/m^2, owing to the increased interfacial adhesion between the fibers and matrix.

3.1.3.2 Cycloaliphatic Epoxy Resins

A cycloaliphatic epoxy resin, 3′,4′-epoxycyclohexylmethyl-3,4-epoxycyclohexane-carboxylate, is synthesized by the reaction of 3′-cyclohexenylmethyl-3-cyclohexene-carboxylate with peracetic acid, as shown in Fig. 3.8.

The cycloaliphatic epoxy resin has an aliphatic backbone and fully saturated molecular structure, which shows excellent UV stability, good weatherability, and excellent electrical properties. The epoxy resin containing high epoxide content

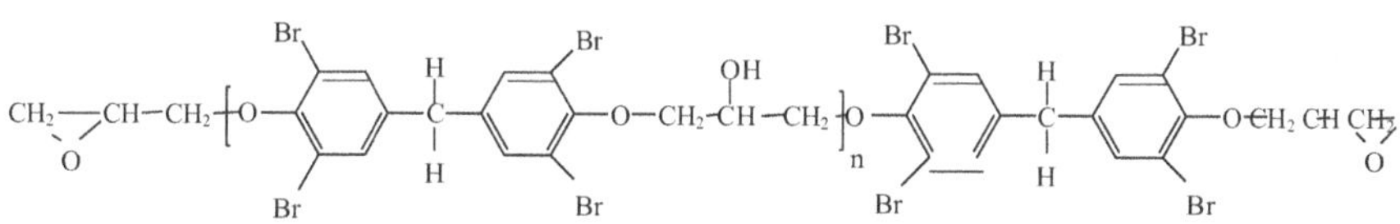

Fig. 3.6 Chemical structure of diglycidyl ether of tetrabromobisphenol-A

Fig. 3.7 Effect of current density on K_{IC} of DGEBA/CF composites [31]

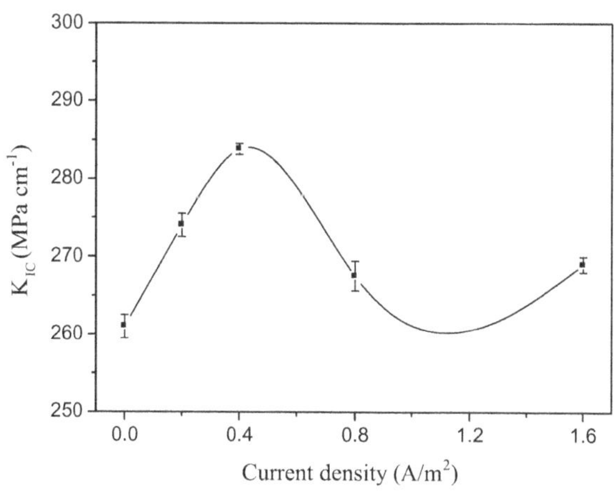

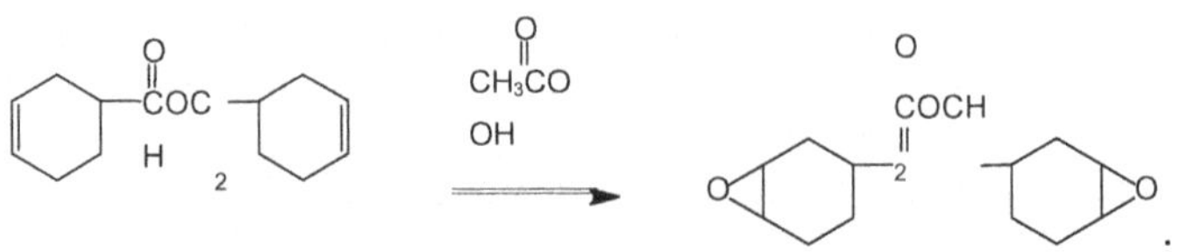

Fig. 3.8 Schematic of synthesis of 3′,4′-epoxycyclohexylmethyl-3,4-epoxycyclohexanecarboxylate

exhibits high cross-link density, T_g, and heat distortion temperature (HDT), which is used to fabricate structural components subjected to high temperatures. The epoxy resin containing short chains possesses low viscosity, which allows for the rapid fiber wet-out in commonly used processes, including filament winding, pultrusion, and resin transfer molding [32].

3.1.3.3 Trifunctional Epoxy Resins

A trifunctional epoxy resin, trimethylol propane-*N*-triglycidyl ether, is prepared by the reaction of trimethylol propane with epichlorohydrin, as shown in Fig. 3.9.

The epoxy resin has low viscosity, low-temperature curing, noncrystallinity, and plasticity [33, 34].

3.1.3.4 Tetrafunctional Epoxy Resins

Tetrafunctional epoxy resins are synthesized through the reaction of 1,3-diaminobenzene or 4,4′-aminodiphenyl methane with epichlorohydrin as shown in Fig. 3.10.

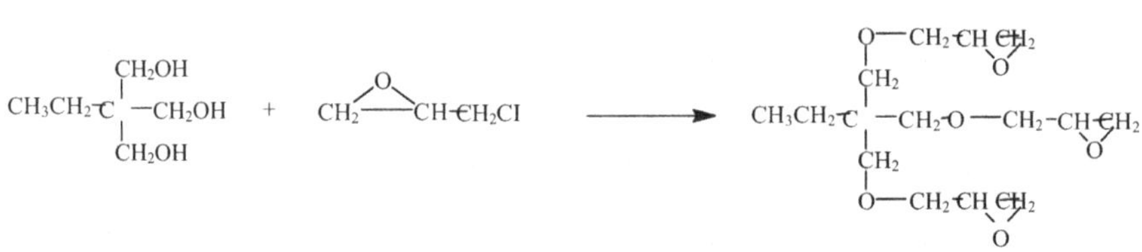

Fig. 3.9 Schematic of synthesis of trimethylol propane-*N*-triglycidyl ether

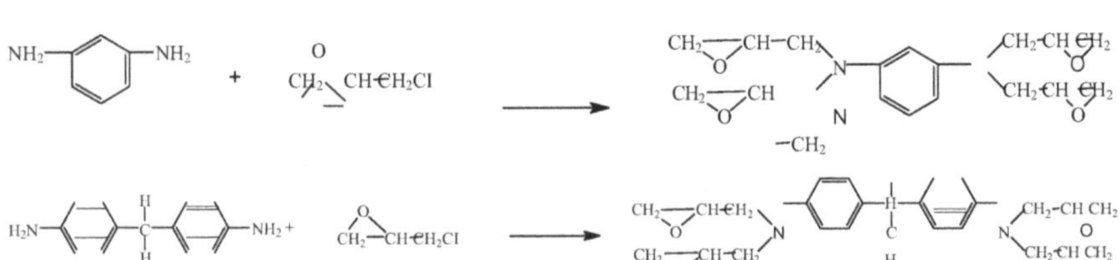

Fig. 3.10 Schematic of synthesis of tetrafunctional epoxy resins

Tetrafunctional epoxy resins with a large number of epoxy functionalities have high cross-linking densities, which are used in applications involving high-temperature resistance. The cured epoxy resins show excellent chemical resistance, high modulus, UV-blocking effect, and thermal stability [35, 36].

Chen et al. [18] studied the DGEBA/CF and 4,4′-tetradiglycidy diaminodiphenol methane (TGDDM)/CF filament wound composites. Table 3.1 lists the mechanical properties of the composites. It was observed that the two types of resins (DGEBA and TGDDM) containing either 4,4′-diaminodiphenyl methane (DDM) or DDM/diethyltoluene diamine (DETDA) as the hardener showed stronger tensile properties than those containing methyltetrahydrophthalic anhydride (MeTHPA) as the hardener. The TGDDM resin casts showed tensile strengths close to those of the DGEBA resin casts but with higher elongations.

3.1.3.5 Novolac Resins

Novolac resins are synthesized by reaction of a molar excess of phenol with formaldehyde in the presence of an acidic catalyst such as oxalic acid, hydrochloric acid, or sulfonate acid. Figure 3.11 shows the chemical structure of novolac resins.

Novolac resins are solids at room temperature, and soften and flow between 65 and 105 °C. The number average molecular weight (Mn) of a standard phenol novolac resin is between 250 and 900. Novolac resins are soluble in many polar organic solvents such as alcohols and acetone, but not in water [38].

Table 3.1 Tensile properties of [37] resin casts

Resin	Hardener	Tensile strength (MPa)	Elongation (%)	Tensile modulus (GPa)
	MeTHPA	60 ± 5	2.1 ± 0.2	2.2 ± 0.2
DGEBA	DDM	70 ± 5	2.3 ± 0.1	2.2 ± 0.1
	DDM/DETDA	72 ± 6	2.5 ± 0.2	2.0 ± 0.2
	MeTHPA	65 ± 6	3.0 ± 0.3	2.4 ± 0.2
TGDDM	DDM	71 ± 4	2.9 ± 0.3	2.7 ± 0.5
	DDM/DETDA	75 ± 6	3.1 ± 0.4	2.8 ± 0.4

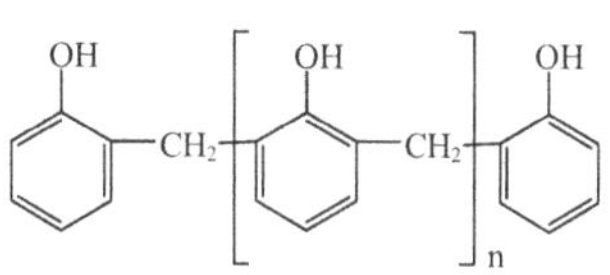

Fig. 3.11 Chemical structure of novolac resins

3.1.3.6 Novolac Epoxy Resins

Novolac epoxy resins are glycidyl ethers of phenolic novolac resins, which are synthesized by reacting phenolic novolac resin with epichlorohydrin. Figure 3.12 shows the chemical structure of novolac epoxy resins. Because an additional curing agent is required to complete the resin's cure, the industry commonly refers to novolac resins as "two-step" products.

The multiple epoxide groups in the novolac epoxy resins allow them to achieve high cross-linking density, thereby resulting in excellent resistance against temperature, chemicals, and solvents. Novolac epoxy resins are widely used to formulate the molding compounds for microelectronics packaging because of their superior performance at elevated temperatures, excellent moldability and mechanical properties, superior electrical properties, and heat and humidity resistance [39, 40].

3.1.3.7 Epoxy Diluents

The addition of epoxy diluents can reduce the viscosity of epoxy resins. The addition of reactive diluents might change the surface tension, reactivity, solvent resistance, physical strength, linear coefficient of expansion, flexibility/impact resistance, and abrasion resistance.

Epoxy diluents can be divided into nonreactive diluents and reactive diluents. Some examples of nonreactive diluents include benzyl alcohol, dibutylphthalate, hydrocarbon resins, and pine oil. Some examples of reactive diluents include *o*-cresyl glycidyl ether (EEW: 180 g/eq, viscosity: 7 mPa s at 25 °C), hexanediol diglycidyl ether (EEW: 142 g/eq, viscosity: 14 mPa s at 25 °C), and C12–C14 alkyl glycidyl ether (EEW: 280 g/eq, viscosity: 7 mPa s at 25 °C) [41, 42]. Figure 3.13 shows the chemical structures of *o*-cresyl glycidyl ether, hexanediol diglycidyl ether, and C12–C14 alkyl glycidyl ether.

3.1.3.8 Epoxy Curing Agents

For the curing of epoxy resins, a wide variety of curing agents such as amines, polyamides, phenolic resins, anhydrides, isocyanates, polymercaptans, and catalytic curing agents have been used. The cure kinetics and T_g of cured epoxy resins are dependent on the molecular structures of the curing agents. The stoichiometry of the epoxy/curing agent system also affects the properties of the cured epoxy resins.

Fig. 3.12 Chemical structure of novolac epoxy resins

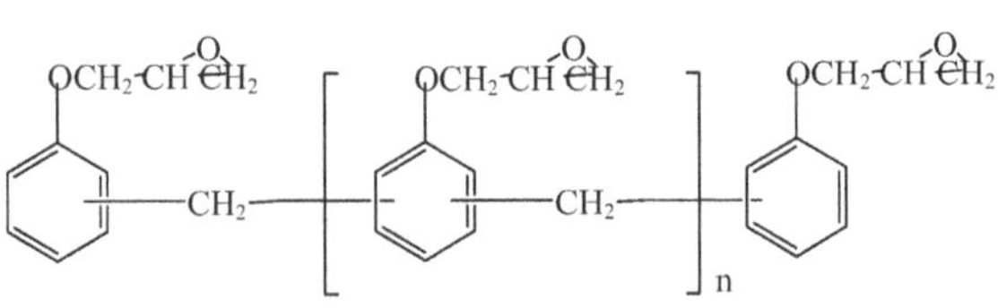

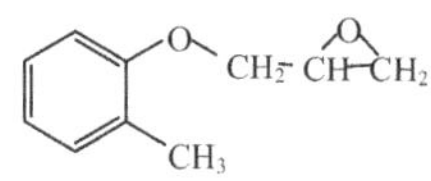

(a) *o*-Cresyl glycidyl ether

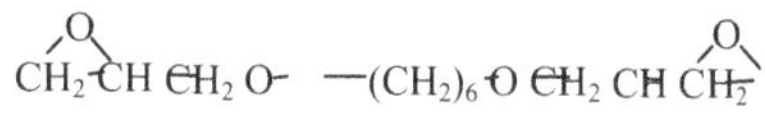

(b) Hexanediol diglycidyl ether

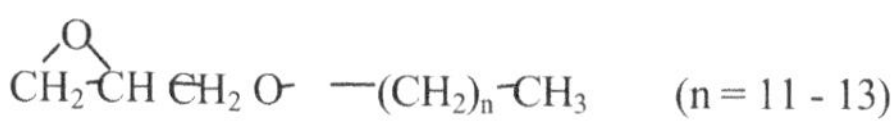

(c) C12–C14 alkyl glycidyl ether

Fig. 3.13 Chemical structures of reactive epoxy diluents

Employing different types and amounts of the curing agent determines the final network structure.

An agent which does not participate in the reaction is known as a catalytic curing agent. A reactive curing agent is generally used in much greater amounts compared to a catalytic curing agent. Also, it actually participates in the reaction. The curing agents are reacted with molecules and coupled directly into the cured system as a structural member of the polymer.

Epoxy curing agents can be divided into amine-type curing agents (triethylenetetramine and DDM), alkali curing agents (imidazole and tertiary amine), anhydrides (phthalic anhydride (PA) and nadic methyl anhydride), and catalytic curing agents [N-benzylpyrazinium hexafluoroantimonate (BPH) and N-benzylquinoxalinium hexafluoroantimonate (BQH)] [43–46]. The chemical structures of DDM, PA, BPH, and BQH are shown in Fig. 3.14.

3.1.3.9 Curing Process

Epoxy curing process is a chemical reaction in which the epoxide groups in the epoxy resins react with a curing agent to form a highly cross-linked, three-dimensional network [47, 48].

(1) Room temperature cure:

Epoxy resin cures with room temperature curing agent (such as aliphatic polyamine) at room temperature. A room temperature cure will provide lower T_g, higher flexibility, impact resistance, electrical and thermal shock resistance.

(2) Heat cure:

Generally, epoxy resins are cured with the curing agent at elevated temperatures, and the resulting cured epoxy resins yield higher glass transition temperature, tensile strength, heat resistance, and chemical resistance.

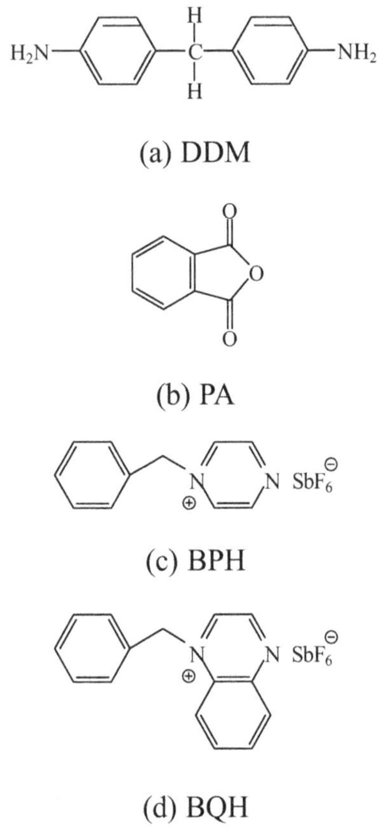

Fig. 3.14 Chemical structures of DDM, PA, BPH, and BQH

(3) Photocure

Epoxy resins are cured using photoinitiators via an addition cure reaction in the presence of an infrared, ultraviolet light, or electron beam.

3.1.3.10 Curing Mechanism

(1) Amine-based curing agents

Primary and secondary amines react through nucleophilic addition to the epoxide ring, as shown in Fig. 3.15. The reaction is exothermic. Proton donors such as tertiary amines are generally used as catalysts to accelerate the curing process with amines. The use of an excessive amount of catalyst achieves faster curing, but usually at the expense of the active life span and thermal stability [49].

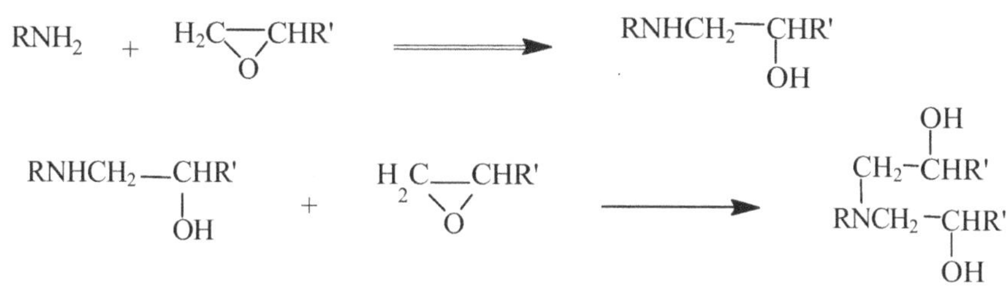

Fig. 3.15 Cure reaction mechanism of epoxide with amines

(2) Anhydride-based curing agents

Figure 3.16 shows the cure reaction mechanism of the epoxide with anhydrides. First, the anhydride reacts with the pendant hydroxyl groups in epoxy resins to produce ester acids, and then the carboxyl groups react with the epoxide groups to yield ester groups. The reaction with the anhydrides is usually catalyzed by tertiary amines and initiated by the activation of the anhydride with amine. Thus, the cure reaction is initiated at a relatively low temperature in the presence of tertiary amines [44, 50].

(3) Catalytic curing agents

Figure 3.17 shows the suggested mechanism of the polymerization of BPH-initiated epoxides. The BPH thermally decomposed to generate a benzyl cationic initiating species. In the initial propagation step (Fig. 3.17a), the epoxide group attacked a propagating cationic species. The reaction of the OH group with BPH generated a powerful protic acid ($H^+SbF_6^-$) (Fig. 3.17b). The propagation then continued through a subsequent attack on the epoxide groups, an alcohol attack, and proton transfer reactions (Fig. 3.17c). Also, the protic acid subsequently can attack the epoxide groups in the propagation step. Once the monomer concentration decreased, a pyrazine released from the salts can possibly attack, perhaps predominantly, the propagation species, thereby terminating the polymerization, as shown in Fig. 3.17d [51–54].

3.1.4 Phenolic Resins

Phenolic resins are normally prepared from phenol and formaldehyde. On the basis of the preparation methods, phenolic resins can be divided into resol and novolac resins [55–60].

3.1.4.1 Resol Resins

Resol resins are synthesized by the reaction of a molar excess of formaldehyde with phenol in the presence of a basic catalyst. Figure 3.18 shows the chemical structure of resol resins.

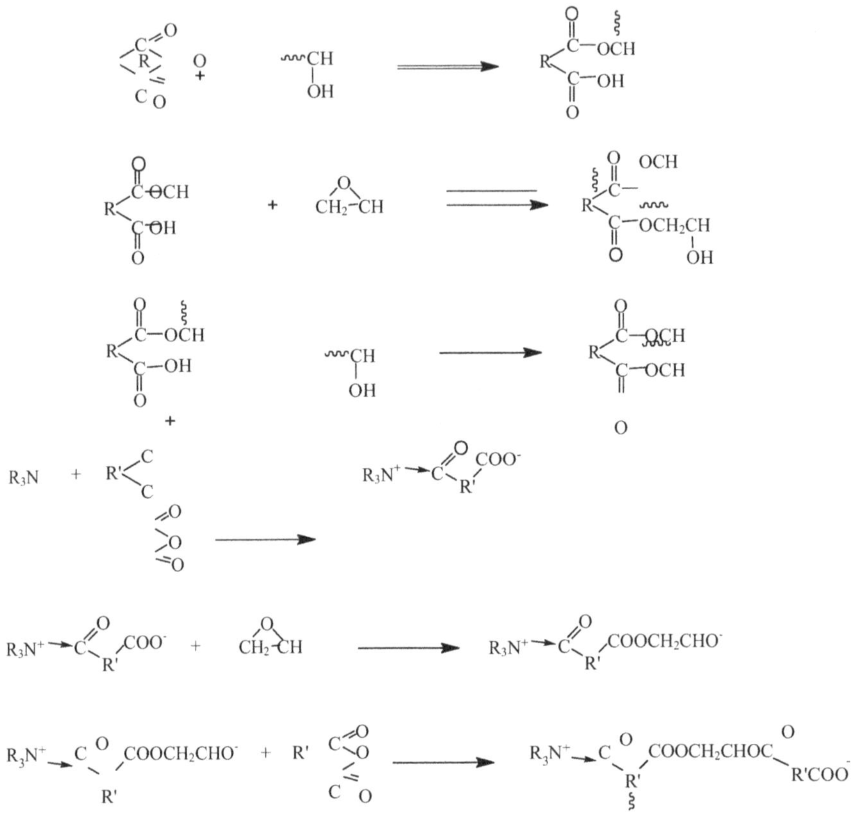

Fig. 3.16 Cure reaction mechanism of epoxide with anhydrides

When an excess of formaldehyde is used in the synthesis of resol resins, a sufficient number of methylol and dibenzyl ether groups remain reactive to complete the polymerization and cure the resin without the incorporation of a curing agent such as hexamethylenetetramine. For this reason, the industry commonly refers to the resol resins as "single-stage"-type products. The typical Mn of a straight phenol resol resin is between 200 and 450.

3.1.4.2 Novolac Resins

Novolac resins are synthesized by the reaction of a molar excess of phenol with formaldehyde in the presence of an acidic catalyst. Figure 3.19 shows the chemical structure of novolac resins.

The reaction creates a methylene bridge at either the ortho or para position of the phenolic aromatic rings. The final phenolic resins are unable to react further without the addition of a curing agent. Because an additional agent is required to complete

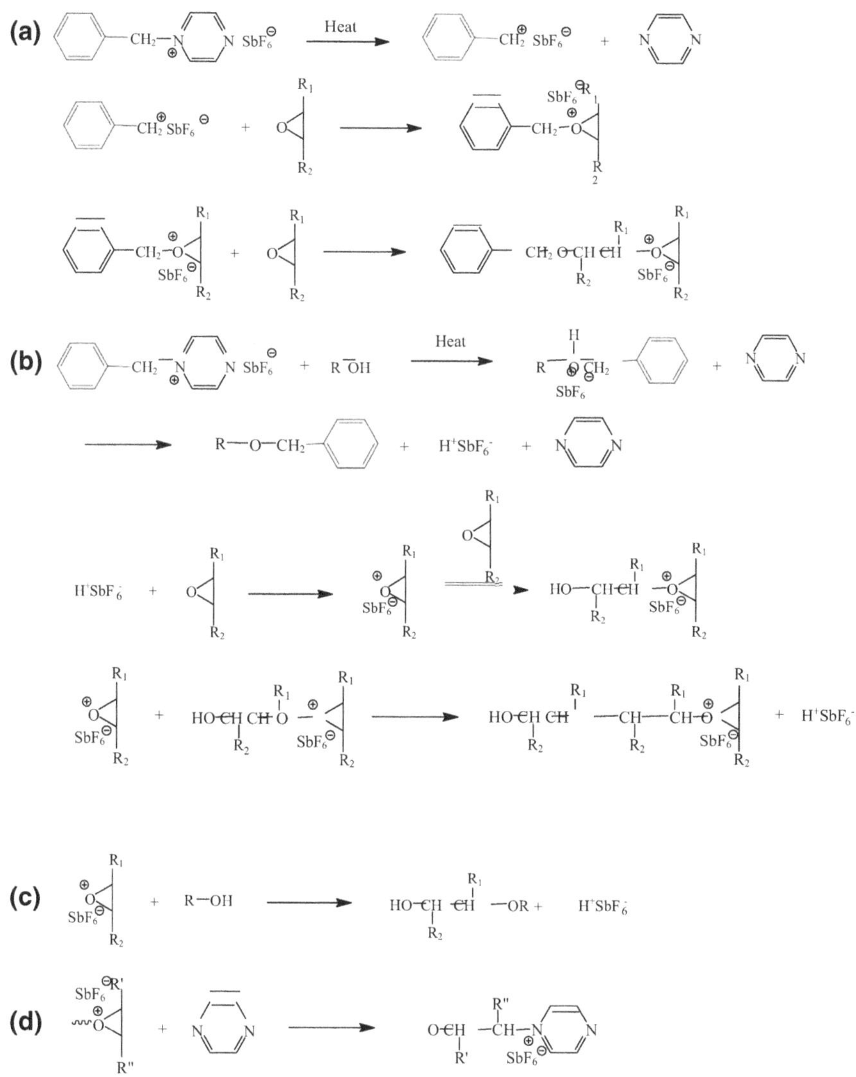

Fig. 3.17 Polymerization mechanism of BPH-initiated epoxides

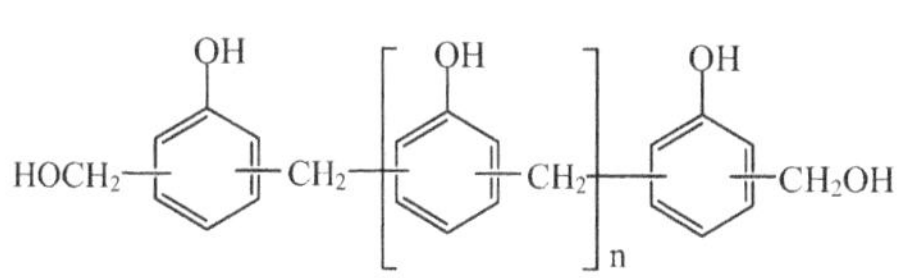

Fig. 3.18 Chemical structure of resol resins

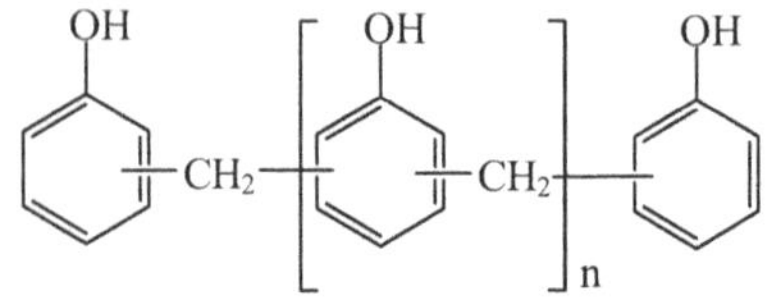

Fig. 3.19 Chemical structure of novolac resins

the curing of the resin, the industry commonly refers to novolac resins as "two-stage" products. The most common curing agent for novolac resins is hexamethylenetetramine.

Park et al. [56] demonstrated the interfacial characteristics and fracture toughness of the electrolytically Ni-plated CF-reinforced phenolic resin composites. Figure 3.20 shows the results of the K_{IC} tests of the composites as a function of the current density. The K_{IC} increased with the increasing current density up to 10 A/m^2, owing to the presence of nickelized functional groups and increased number of oxide functional groups on the CF.

He et al. [61] studied the effects of novolac resin modification on the mechanical properties of the epoxy/CF composites. Figure 3.21 shows the Charpy impact strength of the unmodified and modified composites on varying the content of the novolac resin. It was observed that on increasing the content of the novolac resin from 0 to 18 wt%, the impact strength first gradually increased and then decreased. The impact strength reached the maximum at 13 wt% of the novolac resin. The increased impact properties in epoxy/novolac resin/CF composites might be owing to the novolac resin transition layer, which was bonded to the CF surface and matrix.

3.1.5 Polyester Resins

Polyester resins are unsaturated resins formed by the reaction of dibasic organic acids and polyhydric alcohols, as shown in Fig. 3.22.

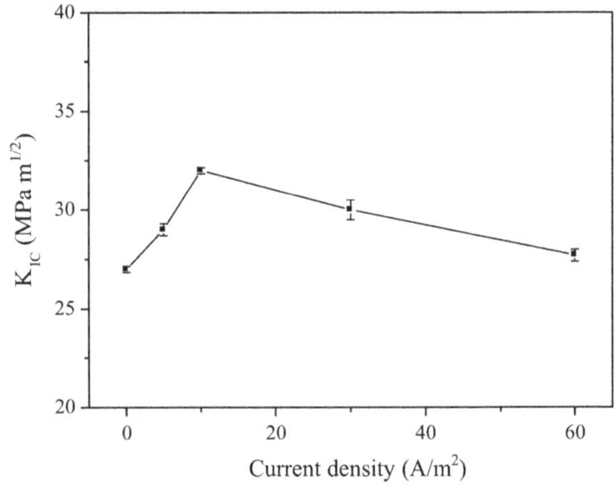

Fig. 3.20 Evolution of K_{IC} of phenolic resin/CF composites with current density [56]

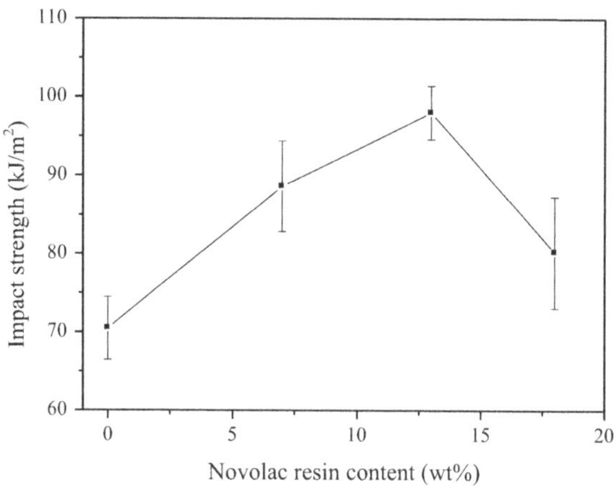

Fig. 3.21 Charpy impact strength of unmodified and modified composites with varying content of novolac resin [61]

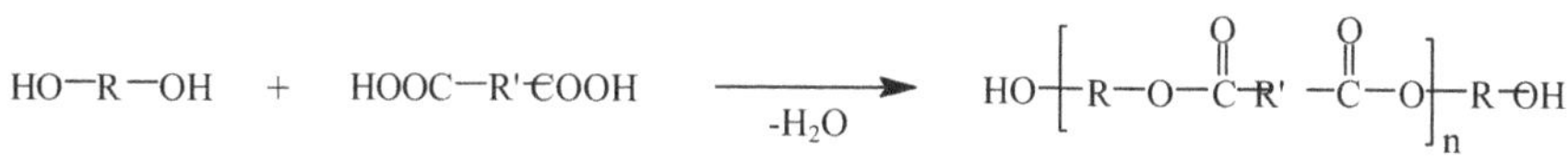

Fig. 3.22 Schematic of synthesis of polyester resins

Polyester resins have relatively good ultraviolet resistance, are long lasting, and show strong resistance to water. These resins are the most commonly used matrix in the marine and composite industries. The resins are used in sheet molding and bulk molding compounds and toners of laser printers. They are commonly used for auto repairs, casting, adhesives, and wood filling. These resins can be used with any type of fiberglass: CF or Kevlar [61–65].

Polyester resins are classified as thermosetting resins and will cure exothermically to a solid when the curing agent is added. Generally, organic peroxides such as benzoyl peroxide and methyl ethyl ketone peroxide are used as the curing agents.

Polyester resins are the most widely used resins in the composites industry. The resins are less expensive and more forgiving than the epoxy resins and offer corrosion resistance. The majority of all fiberglass parts are constructed using polyester resins because they are easy to use, fast curing, and resistance of the extremes of temperature and catalysts.

Some examples of commercially important linear polyesters are polyglycolic acid (PGA), polylactic acid (PLA), polycaprolactone (PCL), polyethylene terephthalate (PET), and polybutylene terephthalate (PBT).

PGA is prepared through the ring-opening polymerization of glycolide, as shown in Fig. 3.23.

PLA is synthesized through the ring-opening polymerization of lactide, as shown in Fig. 3.24.

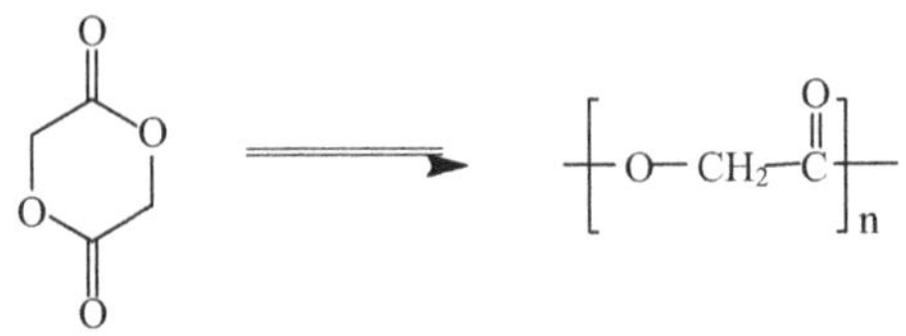

Fig. 3.23 Schematic of synthesis of PGA

Fig. 3.24 Schematic of synthesis of PLA

PCL is prepared through the ring-opening polymerization of ε-caprolactone using stannous octoate as a catalyst, as shown in Fig. 3.25.

PET is produced through the polycondensation of ethylene glycol and either dimethyl terephthalate or terephthalic acid. An excess ethylene glycol is reacted with the former in the melt in the presence of a basic catalyst, as shown in Fig. 3.26.

In the process involving terephthalic acid, the esterification of ethylene glycol and terephthalic acid is directly conducted at moderate pressure and high temperature, as shown in Fig. 3.27.

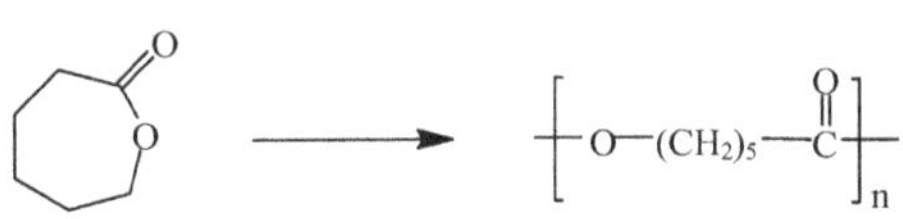

Fig. 3.25 Schematic of synthesis of PCL

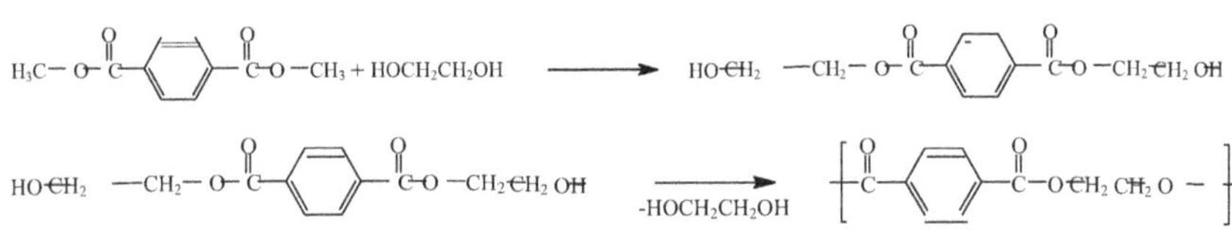

Fig. 3.26 Schematic of synthesis of PET by dimethyl terephthalate process

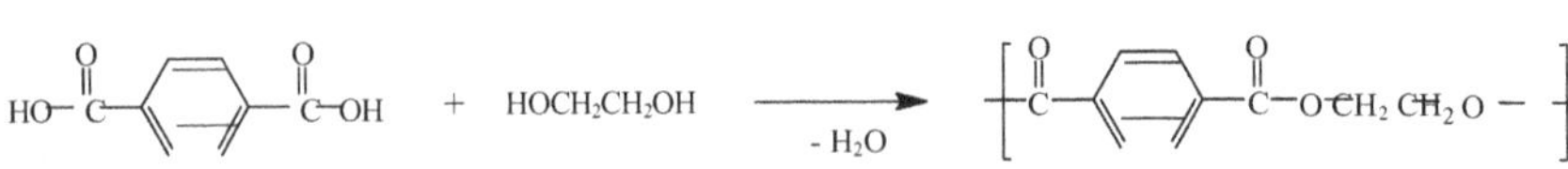

Fig. 3.27 Schematic of synthesis of PET using terephthalic acid

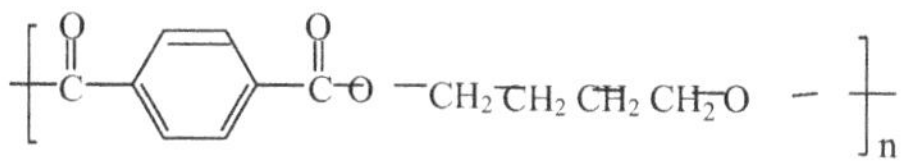

Fig. 3.28 Chemical structure of PBT

PBT is prepared through the polycondensation of terephthalic acid with 1,4-butanediol. Figure 3.28 shows the chemical structure of PBT.

Vilčáková et al. [66] demonstrated the electrical conductivities of polyester resin/CF composites in the percolation threshold region. Table 3.2 shows the average values of conductivity of the composites at laboratory temperature. A steep conductivity increase at a relatively low CF content (0.7 vol%) appeared, which was due to a special mechanism involved in the formation of the intrinsic fibrous structure of the composites.

3.1.6 *Polyimide Resins*

Polyimide (PI) is a polymer of imide monomers. Figure 3.29 shows the structure of PI [67–69].

On the basis of the composition of the main chain, the PIs can be divided into aliphatic PI, wherein the atoms of the imide group are part of a linear chain, and

Table 3.2 Conductivity of polyester/CF composites [66]

CF content (vol%)	Conductivity (S/cm)
0	$(4.09 \pm 0.00) \times 10^{-12}$
0.70	$(1.38 \pm 0.34) \times 10^{-11}$
0.77	$(1.41 \pm 1.51) \times 10^{-7}$
0.84	$(1.26 \pm 1.43) \times 10^{-6}$
0.91	$(2.38 \pm 1.15) \times 10^{-6}$
0.98	$(4.43 \pm 1.67) \times 10^{-6}$
1.05	$(3.70 \pm 1.03) \times 10^{-6}$
1.75	$(3.73 \pm 0.88) \times 10^{-6}$
2.10	$(1.11 \pm 0.24) \times 10^{-5}$
2.80	$(4.19 \pm 0.47) \times 10^{-5}$
3.85	$(2.42 \pm 0.05) \times 10^{-4}$
4.20	$(1.24 \pm 0.03) \times 10^{-4}$
4.55	$(1.71 \pm 0.02) \times 10^{-3}$
4.90	$(8.75 \pm 0.02) \times 10^{-3}$
5.25	$(2.64 \pm 0.03) \times 10^{-2}$
5.60	$(8.58 \pm 0.02) \times 10^{-2}$
6.00	$(1.32 \pm 0.00) \times 10^{-1}$
6.30	$(2.13 \pm 0.00) \times 10^{-1}$

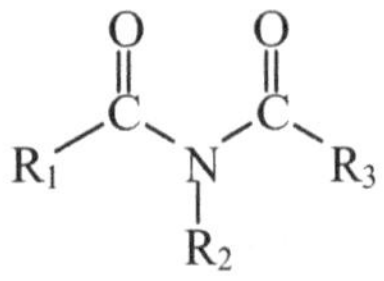

Fig. 3.29 Chemical structure of PI

aromatic PI, wherein the imide group is part of a cyclic unit in the polymer chain, as shown in Fig. 3.30.

PI possesses a greater resistance to heat compared to any other unfilled organic material. Unlike most plastics, the PIs are available in laminates, molded parts, and stock shapes. PI parts are fabricated using techniques ranging from powder-metallurgy to conventional injection, transfer, and compression molding, and extrusion.

PIs possess low dielectric constant, flexibility, polishability to angstrom-level surface finishes, excellent dimensional stability, low water absorption, high-temperature stability, excellent machinability, low outgassing and noncontaminating quality, exceptional mechanical strength, and low coefficient of thermal expansion. Consequently, PIs are used in place of metals and glass in many high-performance applications in the electronics, automotive, and even the aerospace industries.

PIs are solid, heat-resistant, incombustible substances having predominantly amorphous structures, with Mn of 50,000–150,000 and density of 1.35–1.48 g/cm^3 at 20 °C. Most PIs do not dissolve in organic solvents, are inert to the action of oils, and remain virtually unchanged under the action of dilute acids. However, they are hydrolyzed by alkalis and superheated steam. PIs are resistant to ozone, γ-rays, and fast electrons and neutrons, and resistant to heat.

Aromatic PIs with imide rings in the main chain have acquired practical significance because of their valuable physicochemical properties, which remain unchanged for long periods over a wide range of temperature (ranging from −270 to 300 °C). The most industrially valuable PI is polypyromellitimide, prepared

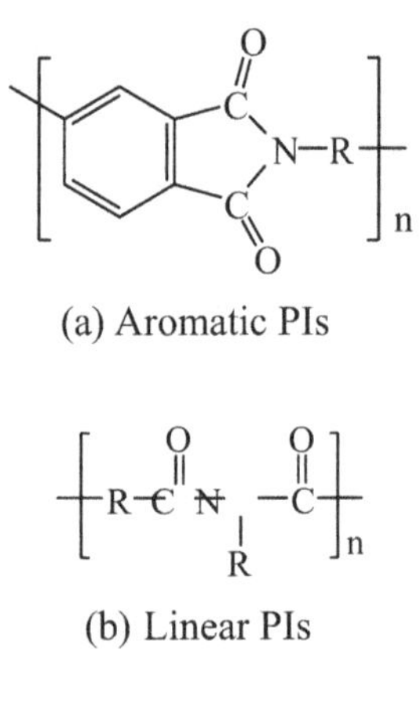

Fig. 3.30 Chemical structures of aromatic and linear PIs

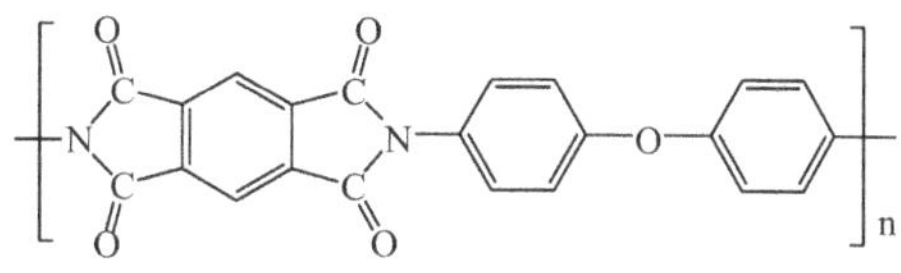

Fig. 3.31 Chemical structure of polypyromellitimide

through the dianhydride–diamine route, as shown in Fig. 3.31. The PI melts above 600 °C and exhibits thermal stability up to 500 °C in an inert atmosphere [70–73].

Li et al. [74] studied the friction and wear properties of surface-treated CF-reinforced PI composites under oil-lubricated conditions. As shown in Fig. 3.32, the friction coefficients of air-oxidized and rare-earth-solution-treated CF-reinforced PI composites are lower compared to those of untreated CF-reinforced PI composites, thereby indicating that the surface modification can improve the interfacial adhesion and compatibility between the CF and PI matrix.

3.1.7 *Vinyl Ester Resins*

Vinyl ester (VE) is a resin produced by the esterification of an epoxy resin with an unsaturated monocarboxylic acid, thereby combining the advantages of epoxy and polyester resins. The resins possess higher flexibility, analogous to the epoxy resins, and ease of processing, analogous to the polyesters [75].

VE resins are stronger than the polyester resins and cheaper than the epoxy resins. Vinyl ester has lower viscosity compared to the polyester and epoxy resins. VE resins offer better resistance to moisture absorption compared to the polyester resins. It is also known that vinyl ester resins bond well with fiberglass, but weakly so with Kevlar and CF owing to the nature of their exotic fibers [76].

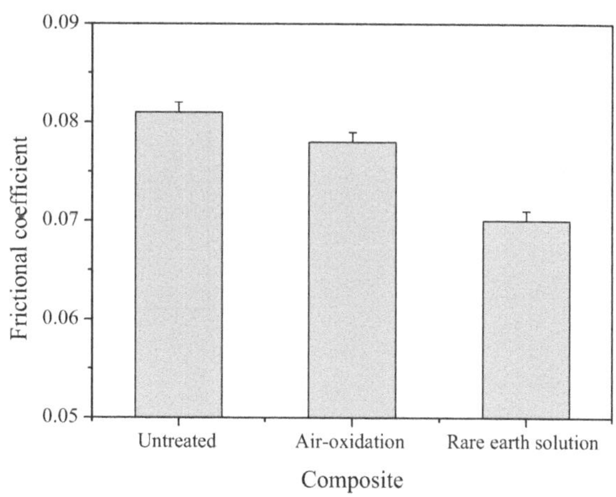

Fig. 3.32 Effect of surface treatment on friction of PI/CF composites [74]

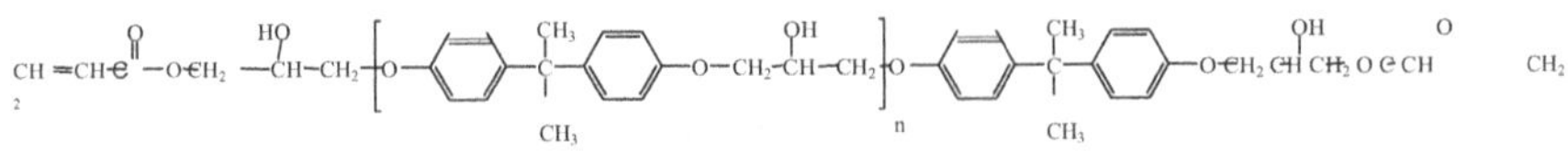

Fig. 3.33 Chemical structure of bisphenol A epoxy vinyl ester

Bisphenol A epoxy VE resin, as shown in Fig. 3.33, is based on bisphenol A epoxy resin, and provides resistance to a wide range of acids, alkalis, bleaches, and organic compounds for use in several chemical processing industry applications.

VE resins are exceptionally reactive and cure rapidly. Peroxides such as methyl ethyl ketone peroxide (MEKP), dicumyl peroxide, di(t-butylperoxy) cyclohexane, and t-butyl peroxybenzoate have been used as initiators for curing VE systems [77].

Cured VE is more flexible and possesses greater toughness compared to the polyester. VE can also withstand temperatures of up to approximately 200 °C without distortion.

Yamada et al. [78] demonstrated the plasma-graft polymerization of a monomer with double bonds onto the surface of CF and its adhesion to a VE resin. Figure 3.34 shows the pull-out force of yarns, which were grafted using poly(adipic acid divinyl ester) (AADE) chains and later embedded in the cured VE resin/ benzoyl peroxide/N,N-dimethylaniline mixture, as a function of the degree of grafting. The pull-out force increased with the increasing degree of grafting. The degree of grafting was proportional to the number of graft chains formed on the surface of the yarn because the average molecular weight of the graft chains did not depend on the degree of grafting.

Fig. 3.34 Pull-out force of grafted carbon yarn as a function of degree of grafting in AADE [78]

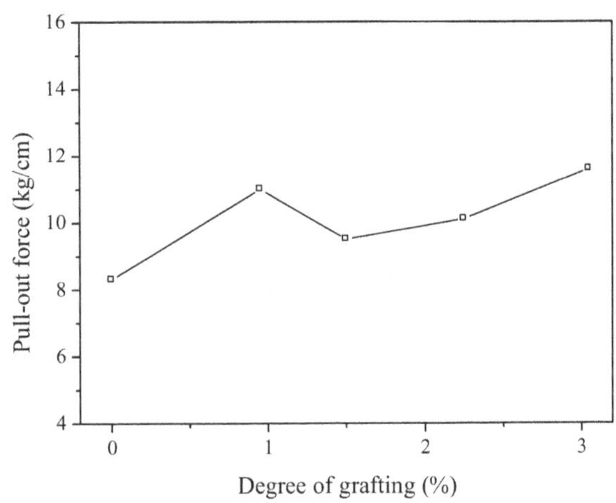

3.2 Thermoplastic Resins

3.2.1 Introduction

Thermoplastic resins consist of long polymer molecules, which are generally not cross-linked. The resin can be repeatedly melted and reused. Usually, no chemical change occurs when thermoplastic is cured.

Thermoplastic resins are often supplied as granules and heated to permit fabrication using conventional molding methods such as injection molding, rotational molding, extrusion, vacuum forming, and compression molding.

Thermoplastic resins have high impact strength, recyclability, and zero emissions. They can bond to other thermoplastics, and be molded or shaped with reheat.

Among the thermoplastic resins, acrylonitrile butadiene styrene, polyamide, polycarbonate, polyetheretherketone, polyetherimide, polyethersulfone, polyethylene, polyphenylene sulfide, and polypropylene resins can be used as polymer matrices for CF-reinforced polymer composites [79–81].

3.2.2 Acrylonitrile Butadiene Styrene Resins

Acrylonitrile butadiene styrene (ABS) is a terpolymer of acrylonitrile, butadiene, and styrene. Figure 3.35 shows the chemical structure of ABS. Many industrial methods such as bulk, bulk-suspension, emulsion, and emulsion graft-doped legitimate are used to prepare the ABS resins.

The typical compositions contain 50% styrene and the remaining 50% consisting of butadiene and acrylonitrile. The higher the content of acrylonitrile, the better the heat resistance, rigidity, and solvent resistance; however, the worse the flow character. Its heat resistance and solvent resistance are better compared to those of the high-impact polystyrene (HIPS). The impact strength, tensile strength, and surface hardness of ABS are better compared to those HIPS. ABS resin is tough, light, economic, heat- and stain-resistant, and an excellent engineering plastic [82, 83].

ABS materials can be processed using any of the standard thermoplastic processing methods. ABS resins have a relative density of 1.03–1.07, Izod impact strength of 186.2–215.8 J/m, tensile strength of 34.3–49 MPa, and Rockwell hardness of 62–118 [84].

ABS resins are widely used in fields such as automotive and electrical/electronic applications, housing, business equipment, and computer components.

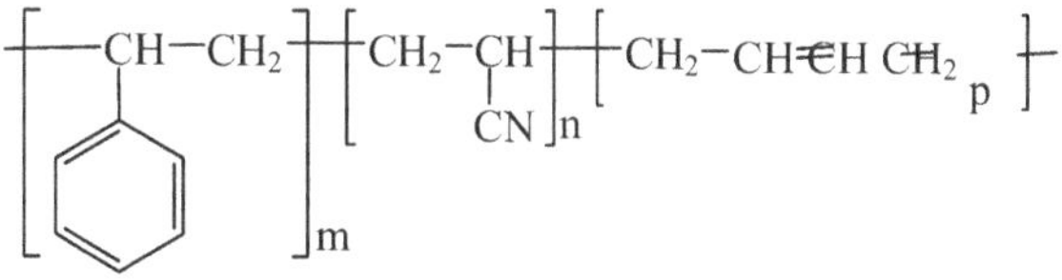

Fig. 3.35 Chemical structure of ABS

Li et al. [83] studied the CF surface treatment and addition of PA6 on the tensile properties of ABS composites. Figure 3.36 shows the effects of the shot carbon fiber (SCF) content on the tensile properties of ABS/SCF composites. Increasing the SCF content increased the tensile strength and modulus of the composites, in agreement with the well-known equation of Kelly and Tyson for mixing short fiber-reinforced composite materials.

3.2.3 Polyamide Resins

Polyamide (PA) is a polymer containing amide monomers joined by peptide bonds. PAs possess the CONH functional groups in the polymer backbone. They can be produced either by the interaction of an amine (NH_2) group and a carboxyl (COOH) group, or the polymerization of amino acids or amino-acid derivatives. PAs can occur both naturally and artificially. They are commonly used in textiles, automotives, carpets, and sportswear owing to their extreme durability and strength [85–87].

On the basis of the composition of the main chain, PAs are classified as aliphatic PAs, polyphthalamides, and aromatic PAs. On the basis of the number and types of repeating units, polyamides can be divided into homopolymers (PA 66, made from hexamethylenediamine and adipic acid) and copolymers (PA 6/66, made from caprolactam, hexamethylenediamine, and adipic acid). On the basis of their crystallinity, polyamides can be divided into semicrystalline and amorphous.

Poly(hexamethylenediamine adipamide) (Nylon 66, PA66) is one of the most widely used semicrystalline engineering thermoplastic resins. Nylon 66 is synthesized by reacting adipic acid with hexamethylene diamine, as shown in Fig. 3.37 [88, 89].

Nylon 66 is a semicrystalline polyamide, commonly used in fiber applications such as carpeting, clothing, and tire cords. It is also used as an engineering material in bearings and gears owing to its good abrasion resistance and self-lubricating

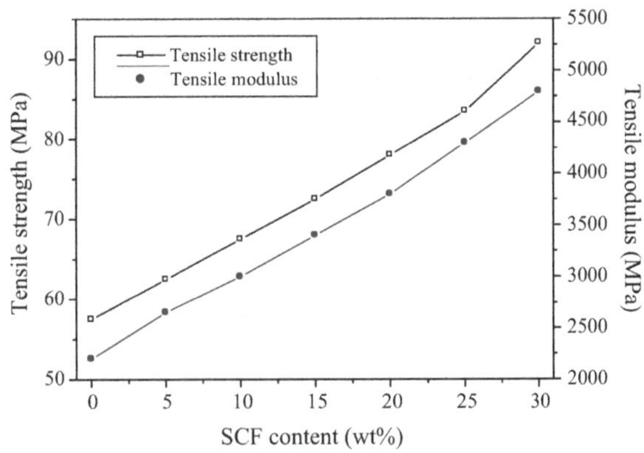

Fig. 3.36 Tensile properties of [83] ABS/SCF composites

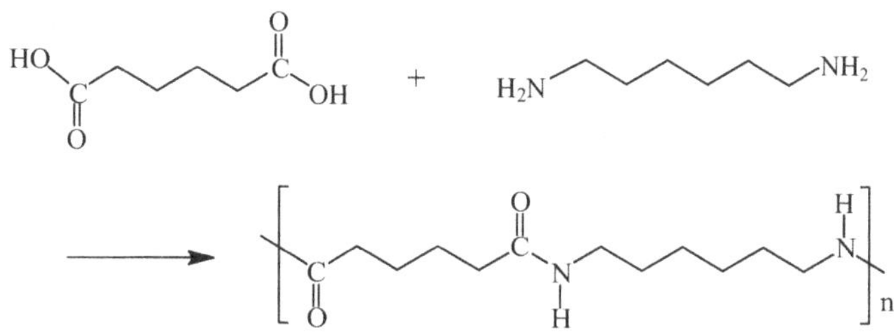

Fig. 3.37 Schematic of synthesis of nylon 66

properties [90]. Nylon 66 has good sliding properties, abrasion resistance, high tensile strength and melting point, and electrical insulation. It is widely used as gear wheels, friction strips, piston guides, impact plates, cam discs, etc. [91].

Nylon 66 has a melting point of 265 °C, which is high for a synthetic fiber, though not comparable to either polyesters or aramids such as Kevlar. Its long molecular chain results in more sites for hydrogen bonds, thereby creating chemical "springs," and making it very resilient.

The molten nylon 66 undergoes a spinning process, wherein nylon 66 is extruded and sent through a spinneret, which is a small metal plate with fine holes. Nylon 66 is then air-cooled to form a wide range of fiber types.

Botelho et al. [86] studied the mechanical behavior of CF-reinforced PA composites. Table 3.3 shows the tensile and compression properties of nylon 66/CF composites. An increase in the fiber volume fraction improved the mechanical properties of the composites.

3.2.4 Polycarbonate Resins

Polycarbonate (PC) is produced by the reaction of bisphenol A with phosgene $COCl_2$. The overall reaction is shown in Fig. 3.38.

PC resins are a particular group of thermoplastic polymers. PC is a novel, highly flexible, durable and nonbreakable, and cost-effective means of domestic and

Table 3.3 Tensile and compression properties of nylon 66/CF composites [86]

Properties	CF content (vol%)		
	40	50	60
Tensile strength (MPa)	408 ± 24	443 ± 31	512 ± 44.2
Tensile modulus (GPa)	35.4 ± 2.8	39 ± 1.6	51.1 ± 2.3
Compressive strength (MPa)	290 ± 29	352 ± 46	391 ± 45
Shear compression (MPa)	22.0 ± 3.0	25.5 ± 4.9	27.8 ± 2.3
Density (g/cm^3)	1.31	1.36	1.39

Fig. 3.38 Schematic of synthesis of PC

industrial usage. PC possesses high impact strength and exceptional clarity and is widely used in applications such as bulletproof windows, break-resistant lenses, and compact discs [92, 93].

PCs are used mainly as molding compounds. The commercially important PCs use bisphenol A and diphenyl carbonate. These polymers are clear plastics with a slight yellow coloration. They have excellent electrical properties and high impact strength.

PCs are easily worked, molded, and thermoformed. The plastics are widely used in the modern chemical industry. A balance of useful features, including temperature resistance, impact resistance, and optical properties, position the polycarbonates between commodity plastics and engineering plastics [94].

Choi et al. [95] demonstrated the production and characterization of PC composite sheets reinforced with vapor grown carbon fiber (VGCF). Figure 3.39 shows the Young's modulus of PC/VGCF cast and rolling composites reinforced with various amounts of VGCF. The Young's modulus of both the composites increased with the increasing VGCF content.

3.2.5 *Polyetheretherketone Resins*

Polyetheretherketone (PEEK) resins are synthesized using step-growth polymerization by the dialkylation of bisphenolate salts. A typical reaction involving 4,4′-difluorobenzophenone and disodium salt of hydroquinone, which is generated in situ by deprotonation with sodium carbonate, is shown in Fig. 3.40 [96].

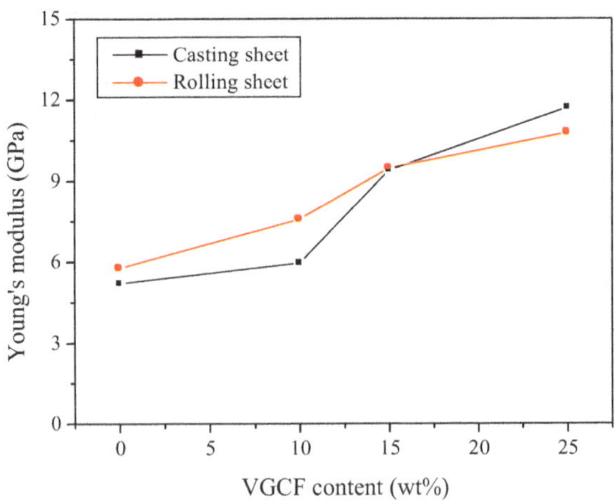

Fig. 3.39 Young's modulus of PC/VGCF cast and rolling composites reinforced with various amounts of VGCF [95]

Fig. 3.40 Schematic of synthesis of PEEK

PEEK is a semicrystalline thermoplastic with excellent mechanical and chemical resistance properties, which are retained at high temperatures. The Young's modulus is 3.6 GPa, while its tensile strength is over the range 90–100 MPa. PEEK has a T_g of approximately 143 °C and melts at approximately 343 °C. PEEK carries a V-0 flammability rating and exhibits low smoke and toxic gas emissions when exposed to flame. PEEK has excellent chemical resistance, low moisture absorption, inherently good wear and abrasion resistance, and remains unaffected by continuous exposure to either hot water or steam [97].

PEEK is generally used as an engineering material because it has high strength, low wear, and good resistance to corrosion, and operates effectively in an ultra-high vacuum environment [98, 99].

Gebhard et al. [100] studied the wear of aqueous lubricated short carbon-fiber (SCF)-reinforced PEEK composites. Figure 3.41 shows the fiber corrosion depths of the composites under different experimental and environmental conditions. The fiber corrosion depth of TGC (after a sliding experiment with simultaneous galvanic corrosion) showed the highest value in this study. This means that there is not only an increase in the wear rate owing to the occurrence of fiber corrosion, but also a supplementary effect of the tribological stress-enhanced CF corrosion.

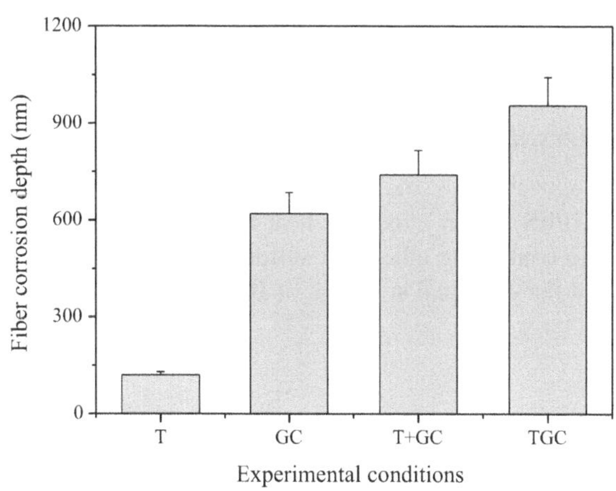

Fig. 3.41 Fiber corrosion depths of PEEK versus stainless steel pairing under different experimental and environmental conditions [100]. T: pure sliding, GC: galvanic corrosion, T + GC: mathematical addition, TGC: sliding experiment with simultaneous galvanic corrosion

3.2.6 Polyetherimide Resins

Polyetherimide (PEI) is an amorphous engineering thermoplastic resin. Figure 3.42 shows the chemical structure of PEI. PEI has high heat resistance, excellent electrical properties, high strength and modulus, and excellent processibility. T_g and amorphous density at 25 are 216 °C and 1.27 g/cc, respectively. Unmodified PEI resin is transparent and has inherent flame resistance and low-smoke evolution [101].

PEI can be processed on conventional thermoplastic molding equipment. PEI resin is available for general-purpose injection molding, blow molding, foam molding, and extrusion. PEI is extruded to produce profiles, coated wires, sheets, and films. The glass fiber reinforcement of the PEI provides it with both greater tensile strength and rigidity while simultaneously improving the dimensional stability [102].

PEI resists a broad range of chemicals under varied conditions of stress and temperatures. PEI is compatible with aliphatic hydrocarbons and alcohols such as gasoline and gasohol, mineral-salt solutions, dilute bases, and fully halogenated hydrocarbons [103].

PEI is often used instead of polysulfone and commonly specified for aerospace, automotive, medical, and packaging applications. PEI is resistant to UV and gamma radiation. This material can be easily machined and provides cost savings over ceramics in electrical applications [104].

Xian et al. [105] demonstrated the influence of SCF reinforcement on the sliding wear of PEI composites. Figure 3.43 shows the evolution of the specific wear rates with the SCF fraction. It was found that the addition of SCF reduced the specific wear rate of PEI by approximately 60 times at room temperature, and 80 times at 150 °C, respectively, owing to several functions of SCF, i.e., reinforcing, lubricating, and thermal conducting functions.

3.2.7 Polyethersulfone Resins

Polyethersulfone (PES) is an excellent heat-resistant, transparent, noncrystalline engineering plastic containing ether and sulfone groups in their backbone chains. Figure 3.44 shows the chemical structure of PES.

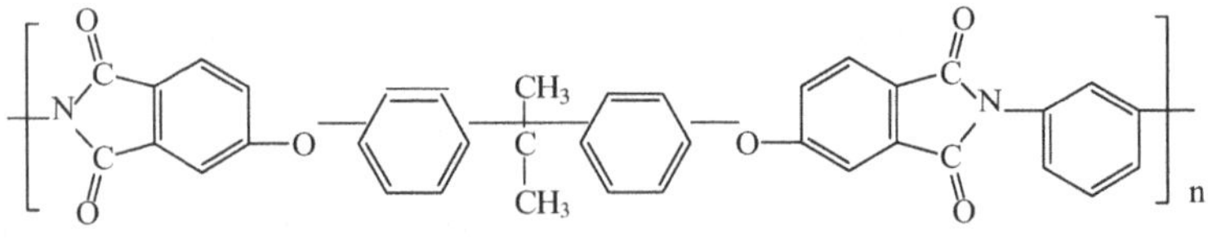

Fig. 3.42 Chemical structure of PEI

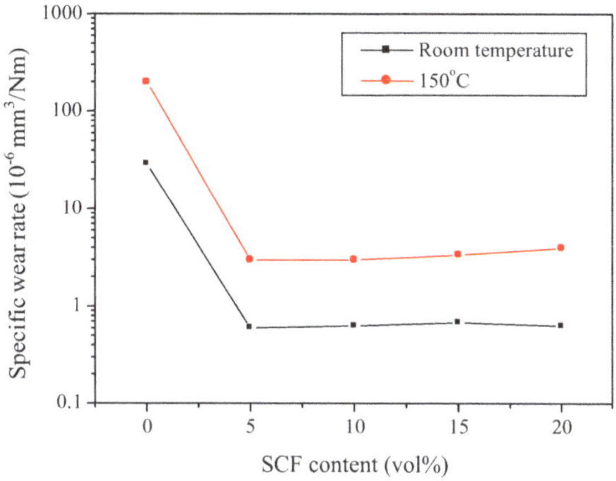

Fig. 3.43 Specific wear rate of neat PEI and its SCF composites at room temperature and 150 °C, measured with a pin-on-disc testing rig, 1 m/s, and 2 MPa [105]

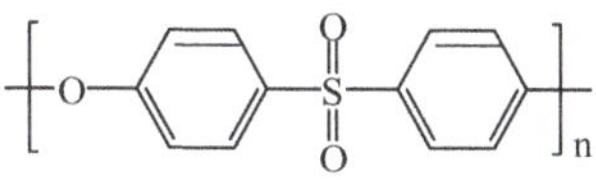

Fig. 3.44 Chemical structure of PES

PES resins exhibit high-temperature performance along with a high T_g, good dimensional stability, outstanding rigidity even at high temperatures, excellent insulation properties, biocompatibility, and inherent flame retardation [106].

PES can be molded using conventional plastics processing equipment such as injection molding, extrusion, compression molding, solution casting, and sintering. Because of its amorphous nature, PES has excellent dimensional stability and can be easily processed with highly polar solvents, which are suitable for applications requiring similar tolerances and little dimensional change over a wide temperature range. PES composites offer distinct advantages over thermosets, namely shorter processing times, better toughness, reduced storage and scrap, and reparability [107].

Wu et al. [108] studied the processing and properties of solution impregnated CF-reinforced PES composites. Table 3.4 lists the flexural test results of the composites using different molding cycles. A higher molding temperature and longer molding time increased the flexural strength from 44 to 71 MPa.

Table 3.4 Effects of molding cycle on transverse flexural properties of solution-processed CF-reinforced PES composites [104]

Sample	Molding temperature (°C)	Molding time t_1 (min)	Molding time t_2 (min)	Transverse flexural strength (MPa)	Transverse flexural modulus (GPa)
1	350	15	10	44	9
2	385	50	10	71	9

3.2.8 Polyethylene Resins

Polyethylene (PE) is a thermoplastic polymer consisting of long chains of the ethylene monomer, as shown in Fig. 3.45. PE is a light, versatile synthetic resin made by the polymerization of ethylene. PE can be produced by polymerization techniques, including radical, anionic addition, cationic addition, and ion coordination. Its simple basic structure can be either linear or branched to a greater or lesser degree.

PE is a member of the important family of polyolefin resins. It is the most widely used plastic worldwide; products ranging from clear food wraps and shopping bags to detergent bottles and automobile fuel tanks. It can also be slit or spun into either synthetic fibers or modified to take on the elastic properties of rubber [108–110].

PE is available over a range of flexibilities and other properties depending on the production process. PE can be formed using a wide variety of thermoplastic processing methods and is particularly useful when moisture resistance and low costs are necessary.

The mechanical properties of PE depend significantly on variables such as the extent and type of branching, crystal structure, and molecular weight. PE is classified into varieties such as low-density PE (LDPE), linear low-density PE (LLDPE), and high-density PE (HDPE) . LDPE, LLDPE, and HDPE typically have a density value ranging from 0.91 to 0.925 g/cm^3, 0.918 to 0.94 g/cm^3, and 0.935 to 0.96 g/cm^3, respectively [111, 112].

LDPE is prepared from gaseous ethylene under high pressures and temperatures in the presence of oxide initiators. LDPE has a high degree of short- and long-chain branching, and weak intermolecular forces as the attraction between the instantaneous dipole and induced dipole is less. This results in low tensile strength and increased ductility. The high degree of branches with long chains endows the molten LDPE with unique and desirable flow properties. LDPE is a flexible material with a melting point of approximately 110 °C. LDPE is used for both rigid containers and plastic film applications [113].

LLDPE is made through the copolymerization of ethylene with 1-butene and small amounts of 1-hexene and 1-octene, using either Ziegler-Natta or metallocene catalysts. LLDPE is a substantially linear polymer, with significant numbers of short branches. LLDPE exhibits higher tensile strength, impact, and puncture resistance compared to LDPE. LLDPE is used predominantly in the packaging film owing to its toughness, flexibility, and relative transparency [114].

HDPE can be produced at low temperatures and pressures, using Ziegler-Natta and metallocene catalysts or activated chromium oxide (Phillips catalyst). HDPE

Fig. 3.45 Chemical structure of PE

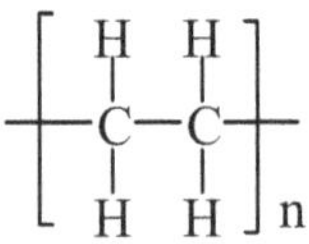

has a low degree of branching and, consequently, stronger intermolecular forces and tensile strength. HDPE is used in products and packaging such as milk jugs, detergent bottles, margarine tubs, garbage containers, and water pipes [115].

Zhang et al. [116] demonstrated the selective location and double percolation of CF-filled HDPE/isotactic polypropylene (iPP) blends. Figure 3.46 shows the dependence of the electrical conductivity on the VGCF content for HDPE, iPP, and HDPE/iPP blends. The HDPE samples exhibited a percolation threshold at 2.5 phr (parts per hundred parts resin) VGCF content, whereas for the iPP samples, the percolation threshold was 1.75 phr VGCF content. This was owing to the different filler dispersion states in the two polymers.

3.2.9 Polyphenylene Sulfide Resins

Polyphenylene sulfide (PPS) is formed by the reaction of sodium sulfide with *p*-dichlorobenzene, as shown in Fig. 3.47. PPS is an organic polymer having a symmetrical, rigid backbone chain consisting of recurring *p*-substituted benzene rings and sulfur atoms. PPS is a semicrystalline, high-performance engineering thermoplastic.

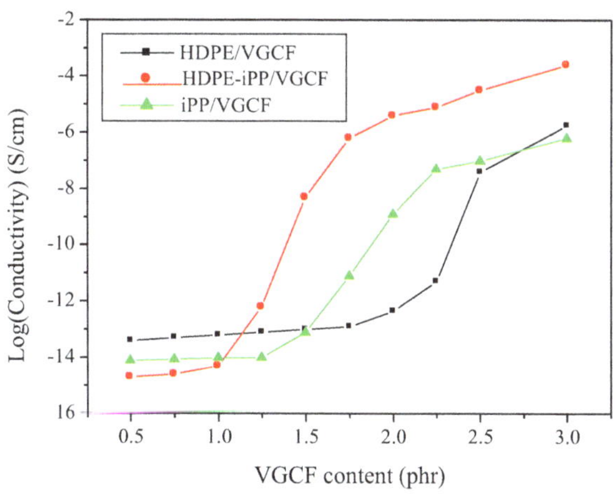

Fig. 3.46 Dependence of electrical conductivity on VGCF content for VGCF-filled HDPE, iPP, and HDPE/iPP (50/50) blends molded at 190 °C for 15 min [116]

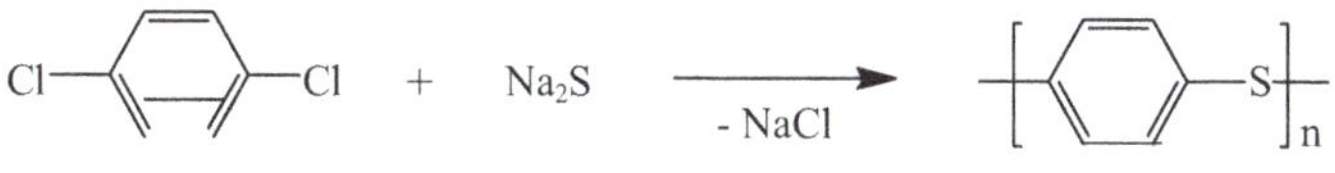

Fig. 3.47 Schematic of synthesis of PPS

PPS exhibits outstanding chemical resistance, thermal stability, dimensional stability, and fire resistance. The extreme inertness of PPS toward organic solvents, and inorganic salts and bases makes for an outstanding performance as a corrosion-resistant coating, suitable for contact with foods [117].

PPS can be molded, extruded, or machined to high tolerances. In its pure solid form, it may be opaque white to light tan in color. PPS resins are available as powders for slurry coating and electrostatic spraying. Because of its low viscosity, PPS can be molded with a high loading of fillers and reinforcements, which is needed to compensate for its inherent brittleness.

Fillers and reinforcements can also increase the strength, surface properties, dimensional stability, electrical properties, and overall cost. Glass fiber-reinforced PPS is used for the mechanical and electronic applications requiring high mechanical strength, impact resistance, and insulating characteristics [118].

Xu et al. [118] studied the tribological behavior of PPS/CF composite coating under dry sliding and water lubrication. Figure 3.48 shows the variations in the friction coefficient of the composite coating with the sliding content of CF at the rate of 0.43 m/s under 200 N. This showed that the friction coefficient of the composite coating appeared to slightly decrease with the increasing CF content, both under dry- and water-lubricated conditions. At the same time, the friction coefficient under water-lubricated conditions was much lower than that under dry sliding.

3.2.10 Polypropylene Resins

Polypropylene (PP) is a thermoplastic polymer with a linear structure made from the propylene monomer using Ziegler-Natta polymerization and metallocene catalysis polymerization, as shown in Fig. 3.49.

Fig. 3.48 Variations in friction coefficient of PPS/CF composite coating as a function of CF content [118]

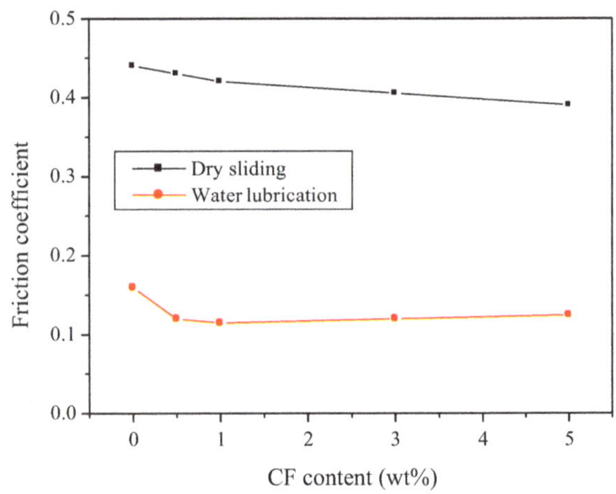

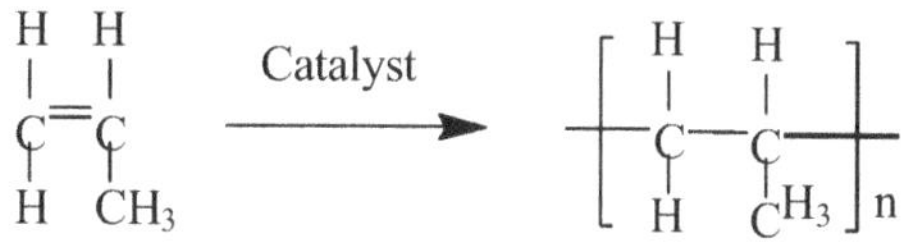

Fig. 3.49 Schematic of synthesis of PP

The short segments of PP show examples of isotactic and syndiotactic tacticity, as depicted in Fig. 3.50.

PP resins have cheap, lowest density (0.90–0.92 g/cm^3), high tensile and compressive strength, melting point of 160 °C, excellent dielectric properties, and nontoxicity, and are used in a wide variety of applications, including packaging and labeling, textiles, stationery, plastic parts and reusable containers of various types, laboratory equipment, loudspeakers, automotive components, and polymer banknotes. As an addition polymer made using the propylene monomer, it is unusually resistant to several chemical solvents, bases, and acids [119–121].

In general, homopolymers can be used for housing, housewares, packaging, cassette holders and fibers, monofilaments, and film tapes; copolymers are preferred for all applications exposed to cold environments and widely used for pipes, containers, boat hulls, seat shells, and automotive parts.

In order to improve a few properties, PP formulas may include additives such as pigments, carbon black, rubbers, antioxidants, and UV stabilizers. PP is available in the form of molding powder, extruded sheets, cast films, textile staples, and continuous filament yarns [122–124].

Li et al. [125] demonstrated the interfacial compatibility of PP composites filled with surface-treated CF. Figure 3.51 shows the three-point bending value of PP/CF composites as a function of treatment time. It was observed that the bending strength and modulus increased with the increasing treatment time, thereby indicating that the interfacial adhesion between CF and PP matrix increased.

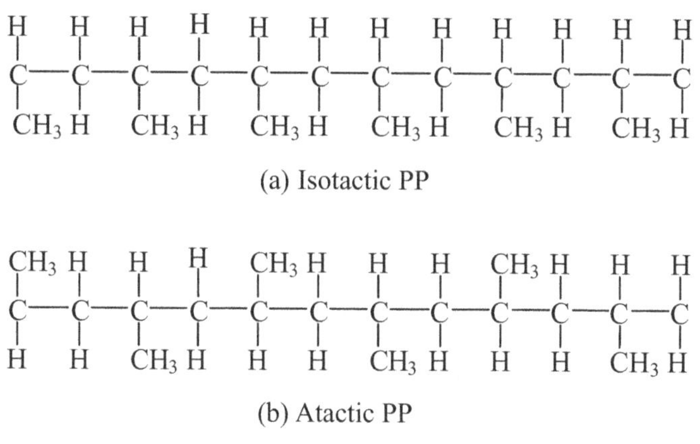

Fig. 3.50 Chemical structures of isotactic and atactic PPs

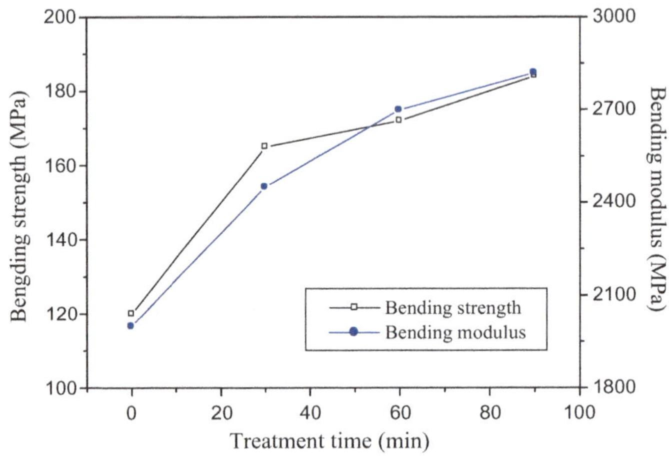

Fig. 3.51 Bending strength and modulus of PP/CF composites [125]

References

1. H.P.S.A. Khalil, M.A. Tehrani, Y. Davoudpour, A.H. Bhat, M. Jawaid, A. Hassan, J. Reinf. Plast. Comp. **32**, 330 (2013)
2. Y.H. Lee, M.S. Kim, H. Kim, J. Lee, D.J. Kim, J. Adhes. Sci. Technol. **27**, 508 (2013)
3. A. Molazemhosseini, H. Tourani, M.R. Naimi-Jamal, A. Khavandi, Polym. Test. **32**, 535 (2013)
4. S. Tiwari, J. Bijwe, Surf. Interface Anal. **45**, 1838 (2013)
5. R. Hamzaoui, S. Guessasma, B. Mecheri, A.M. Eshtiaghi, A. Bennabi, Mater. Des. **56**, 60 (2014)
6. K. Bilge, S. Venkataraman, Y.Z. Menceloglu, M. Papila, Comp. Part A: Appl. Sci. Manuf. **58**, 73 (2014)
7. J. Han, Y. Liang, Appl. Mech. Mater. **446–447**, 1405 (2014)
8. A.S. Yudin, D.I. Buyaev, O.V. Afonicheva, I.G. Goryacheva, A.P. Krasnov, J. Fric, Wear **34**, 245 (2013)
9. G. Russo, M. Pauletta, A. Cortesia, Eng. Struct. **52**, 422 (2013)
10. F.L. Jin, S.J. Park, Polym. Degrad. Stabil. **97**, 2148 (2012)
11. A.A. Azeez, K.Y. Rhee, S.J. Park, D. Hui, Comp. Part B: Eng. **45**, 308 (2013)
12. W. Dong, H.C. Liu, S.J. Park, F.L. Jin, J. Ind. Eng. Chem., Article in press (2013)
13. J.M. Barton, I. Hamerton, J.R. Jones, J.C. Stedman, Polymer **37**, 4519 (1996)
14. C. Marieta, E. Schulz, I. Mondragon, Compos. Sci. Technol. **62**, 299 (2002)
15. P. Ren, G. Liang, Z. Zhang, Polym. Compos. **27**, 591 (2006)
16. K. Chung, J.C. Seferis, Polym. Degrad. Stabil. **71**, 425 (2001)
17. P. Ren, G. Liang, Z. Zhang, Polym. Compos. **27**, 402 (2006)
18. X. Sheng, M. Akinc, M.R. Kessler, J. Therm. Anal. Calorim. **93**, 77 (2008)
19. M. Thunga, W.Y. Lio, M. Akinc, R. Kessler, Compos. Sci. Technol. **71**, 239 (2011)
20. E. Moaseri, M. Maghrebi, M. Baniadam, Mater. Des. **55**, 644 (2014)
21. E. Moaseri, M. Karimi, M. Maghrebi, M. Baniadam, Int. J. Solid Struct. **51**, 774 (2014)
22. M.A.A. Bakar, S. Ahmad, W. Kuntjoro, S. Kasolang, Appl. Mech. Mater. **393**, 136 (2013)
23. G. Liu, X.I. Hu, P. Zhang, R.L. Yu, J.W. Bao, M.H. Chen, Q.W. Li, X.S. Yi, Acta Polym. Sinica **10**, 1334 (2013)
24. I.D.G. Ary Subagia, Y. Kim, J. Mech. Sci. Technol. **27**, 987 (2013)

25. X. Jia, G. Li, B. Liu, Y. Luo, G. Yang, X. Yang, Comp. Part A: Appl. Sci. Manuf. **48**, 101 (2013)
26. M. Li, Y. Gu, Y. Liu, Y. Li, Z. Zhang, Carbon **52**, 109 (2013)
27. F.L. Jin, C.J. Ma, S.J. Park, Mater. Sci. Eng., A **528**, 8517 (2011)
28. L. Zhu, F.L. Jin, S.J. Park, Bull. Korean Chem. Soc. **33**, 2513 (2012)
29. K. Satoh, K. Ohyama, N. Aoki, M. Iida, F. Nagai, Food Chem. Toxicol. **42**, 983 (2004)
30. A.I. Balabanovich, A. Hornung, D. Merz, H. Seifert, Polym. Degrad. Stabil. **85**, 713 (2004)
31. S.J. Park, M.H. Kim, J.R. Lee, S. Choi, J. Colloid Interf. Sci. **228**, 287 (2000)
32. M.J. Yoo, S.H. Kim, S.D. Park, W.S. Lee, J.W. Sun, J.H. Choi, S. Nahm, Eur. Polym. J. **46**, 1158 (2010)
33. S.J. Park, T.J. Kim, J.R. Lee, J. Polym. Sci., Part B: Polym. Phys. **38**, 2114 (2000)
34. G.H. Kwak, S.J. Park, J.R. Lee, J. Appl. Polym. Sci. **78**, 290 (2000)
35. S.J. Park, F.L. Jin, J.R. Lee, Mater. Sci. Eng., A **374**, 109 (2004)
36. M.C. Lee, T.H. Ho, C.S. Wang, J. Appl. Polym. Sci. **62**, 217 (1996)
37. W. Chen, Y. Yu, P. Li, C. Wang, T. Zhou, X. Yang, Compos. Sci. Technol. **67**, 2261 (2007)
38. S. Miloshev, P. Novakov, V. Dimitrov, I. Gitsov, Polymer **32**, 3067 (1991)
39. S.J. Park, M.K. Seo, J.R. Lee, J. Polym. Sci., Part A: Polym. Chem. **38**, 2945 (2000)
40. B. Guo, D. Jia, W. Fu, Q. Qiu, Polym. Degrad. Stabil. **79**, 521 (2003)
41. F. Mustata, D. Rosu, C.N. Cascaval, Polym. Test. **19**, 927 (2000)
42. S. Montserrat, J. Málek, P. Colomer, Thermochim. Acta **336**, 65 (1999)
43. H.J. Xu, F.L. Jin, S.J. Park, Bull. Korean Chem. Soc. **30**, 2643 (2009)
44. S.J. Park, F.L. Jin, Polym. Degrad. Stabil. **86**, 515 (2004)
45. S.J. Park, F.L. Jin, J.R. Lee, Macromol. Rapid Comm. **25**, 724 (2004)
46. S.J. Park, F.L. Jin, J.R. Lee, J.S. Shin, Eur. Polym. J. **41**, 231 (2005)
47. S.J. Park, F.L. Jin, J.S. Shin, Mater. Sci. Eng., A **390**, 240 (2005)
48. S.J. Park, F.L. Jin, Polym. Int. **54**, 705 (2005)
49. F.L. Jin, S.J. Park, Bull. Korean Chem. Soc. **30**, 334 (2009)
50. J. Rocks, L. Rintoul, F. Vohwinkel, G. George, Polymer **45**, 6799 (2004)
51. F.L. Jin, S.J. Park, J. Ind. Eng. Chem. **13**, 808 (2007)
52. F.L. Jin, S.J. Park, Polym. Int. **57**, 577 (2008)
53. F.L. Jin, S.J. Park, Mater. Sci. Eng., A **478**, 402 (2008)
54. F.L. Jin, S.J. Park, J. Ind. Eng. Chem. **14**, 564 (2008)
55. M.H. Choi, B.H. Jeon, I.J. Chung, Polymer **41**, 3243 (2000)
56. S.J. Park, Y.S. Jang, J. Colloid Interf. Sci. **237**, 91 (2001)
57. H. An, B. Feng, S. Su, Carbon **47**, 2396 (2009)
58. J. Fei, H.J. Li, Y.W. Fu, L.H. Qi, Y.L. Zhang, Wear **269**, 534 (2010)
59. J.M. Park, D.J. Kwon, Z.J. Wang, G.Y. Gu, K.L. Devries, Comp. Part A: Appl. Sci. Manuf. **47**, 156 (2014)
60. Z.J. Wang, D.J. Kwon, G.Y. Gu, W.I. Lee, J.K. Park, K.L. DevVries, J.M. Park, Comp. Part A: Appl. Sci. Manuf. **52**, 151 (2013)
61. H. He, K. Li, J. Wang, J. Wang, J. Gu, R. Li, Polym. Compos. **32**, 227 (2011)
62. K.T. Hsiao, Compos. A **39**, 834 (2008)
63. V. Cecen, M. Sarikanat, Y. Seki, T. Govsa, H. Yildiz, I.H. Tavman, J. Appl. Polym. Sci. **102**, 4554 (2006)
64. M. Monti, M. Natali, R. Petrucci, J.M. Kenny, L. Torre, Polym. Compos. **32**, 766 (2011)
65. Z. Wu, L. Meng, L. Liu, Z. Jiang, L. Xing, D. Jiang, Y. Huang, J. Adhesion Sci. Technol. **28**, 444 (2014)
66. J. Vilčáková, P. Sáha, O. Quadrat, Eur. Polym. J. **38**, 2343 (2002)
67. X.R. Zhang, X.Q. Pei, Q.H. Wang, Mater. Design **30**, 4414 (2009)
68. P. Samyn, G. Schoukens, Mater. Chem. Phys. **115**, 185 (2009)
69. L. Mascia, Z. Zhang, S.J. Shaw, Compos. A **27A**, 1211 (1996)
70. T. Ogasawara, Y. Ishida, R. Yokota, T. Watanabe, T. Aoi, J. Goto, Compos. A **38**, 1296 (2007)
71. X. Zhang, X. Pei, Q. Wang, Mater. Chem. Phys. **115**, 825 (2009)

72. V.E. Yudin, M.Y. Goykhman, K. Balik, P. Glogar, G.N. Gubanova, V.V. Kudriavtsev, Carbon **38**, 5 (2000)
73. M.P. Stevens, *Polymer Chemistry* (Oxford University Press, New York, 1999)
74. J. Li, X.H. Cheng, Mater. Chem. Phys. **108**, 67 (2008)
75. N.S. Broyles, K.N.E. Verghese, S.V. Davis, H. Li, R.M. Davis, J.J. Lesko, J.S. Riffle, Polymer **39**, 3417 (1998)
76. F. Vautard, S. Ozcan, H. Meyer, Compos. A **43**, 1120 (2012)
77. F.L. Jin, K.Y. Rhee, S.J. Park, Mater. Sci. Eng., A **435–436**, 429 (2006)
78. K. Yamada, H. Yamane, K. Kumada, S. Tanabe, T. Kajiyama, J. Appl. Polym. Sci. **90**, 2415 (2003)
79. G. Yan, X. Wang, D. Wu, J. Appl. Polym. Sci. **129**, 3502 (2013)
80. Y. Chen, X. Wang, D. Wu, Polym. Adv. Techol. **24**, 364 (2013)
81. X. Yang, Z. Wang, M. Xu, R. Zhao, X. Liu, Mater. Des. **44**, 74 (2013)
82. C.Y. Huang, C.C. Wu, Eur. Polym. J. **36**, 2729 (2000)
83. J. Li, C.L. Cai, Curr. Appl. Phys. **11**, 50 (2011)
84. J. Li, Y.F. Zhang, Surf. Interface Anal. **41**, 610 (2009)
85. S.H. Wu, F.Y. Wang, C.C.M. Ma, W.C. Chang, C.T. Kuo, H.C. Kuan, W.J. Chen, Mater. Lett. **49**, 327 (2001)
86. E.C. Botelho, L. Figiel, M.C. Rezende, B. Lauke, Compos. Sci. Technol. **63**, 1843 (2003)
87. Y. Li, Z. Wan, G. Gao, X. Xiong, C. Wang, H. Wan, Luo. Mater. Des. **51**, 257 (2013)
88. N. Feng, X. Wang, D. Wu, Curr. Appl. Phys. **13**, 2038 (2013)
89. S. Motozuka, M. Tagaya, Y. Hotta, M. Morinaga, T. Ikoma, T. Honma, T. Daimon, J. Tanaka, Ind. Eng. Chem. Res. **52**, 2182 (2013)
90. A.Y. Feldman, M.F. Gonzalez, E. Wachtel, M.P. Moret, G. Marom, *Polymer*, 7239 (2004)
91. S. Senthilvelan, R. Gnanamoorthy, Polym. Test. **25**, 56 (2006)
92. M.A. Montes-Morán, A. Martínez-Alonso, J.M.D. Tascón, M.C. Paiva, C.A. Bernardo, Carbon **39**, 1057 (2001)
93. J.A. Carneiro, C.A. Covas, G. Bernardo, F.W.J.V. Caldeira, J.M. Hattum, R.L. Ting, M.L. Alig, Lake. Compos. Sci. Technol. **58**, 401 (1998)
94. J.M. Park, J. Colloid Interf. Sci. **225**, 384 (2000)
95. Y.K. Choi, K. Sugimoto, S.M. Song, M. Endo, Compos. A **37**, 1944 (2006)
96. S.M. Kurtz, in *Chapter 2 - Synthesis and Processing of PEEK for Surgical Implants,* PEEK Biomaterials Handbook (2012), pp. 9–22
97. C.C.M. Ma, N.H. Tai, S.H. Wu, S.H. Lin, J.F. Wu, J.M. Lin, Compos. B **28B**, 407 (1997)
98. G.Y. Xie, G.X. Sui, R. Yang, Compos. Sci. Technol. **71**, 828 (2011)
99. A. Gebhard, T. Bayerl, A.K. Schlarb, K. Friedrich, Corros. Sci. **51**, 2524 (2009)
100. A. Gebhard, T. Bayerl, A.K. Schlarb, K. Friedrich, Wear **268**, 871 (2010)
101. G.M. Wu, J.M. Schultz, Polym. Compos. **21**, 223 (2000)
102. M. Hou, L. Ye, H.J. Lee, Y.W. Mai, Compos. Sci. Technol. **58**, 181 (1998)
103. K.Y. Kim, L. Ye, Compos. A **35**, 477 (2004)
104. R. Akkerman, P.E. Reed, K.Y. Huang, L. Warnet, Eur. Struct. Integr. Soc. **27**, 3 (2000)
105. G. Xian, Z. Zhang, Wear **258**, 776 (2005)
106. A. Fernández, E. Arbelaiz, I. Diaz, Polym. Compos. **25**, 480 (2004)
107. G.M. Wu, J.M. Schultz, Polym. Compos. **21**, 223 (2000)
108. A.A.J.M. Peijs, J.M.M.D. Kok, Composites **24**, 19 (1993)
109. I. Turku, T. Kärki, Eur. J. Wood Wood Prod., Article in press (2013)
110. A. May-Pat, A. Valadez-González, P.J. Herrera-Franco, Polym. Test. **32**, 1114 (2013)
111. D. Hertel, R. Valette, H. Münstedt, J. Non-Newton, Fluid Mech. **153**, 82 (2008)
112. M.W. Spencer, L. Cui, Y. Yoo, D.R. Paul, Polymer **51**, 1056 (2010)
113. N. Chand, S. Kreuzberger, G. Hinrichsen, Composites **25**, 878 (1994)
114. Q. Ling, J. Sun, Q. Zhao, Q. Zhou, Mater. Sci. Eng., B **162**, 162 (2009)
115. N. Cohen, T. Rae, Biomaterials **5**, 352 (1984)
116. X.S. Zhang, H. Yi, S. Yui, M. Asai, Sumita, Mater. Letter **36**, 186 (1998)

117. Z. Jiang, L.A. Gyurova, A.K. Schlarb, K. Friedrich, Z. Zhang, Compos. Sci. Technol. **68**, 734 (2008)
118. H. Xu, Z. Feng, J. Chen, H. Zhou, Mater. Sci. Eng., A **416**, 66 (2006)
119. N.G. Karsli, A. Aytac, M. Akbulut, V. Deniz, O. Güven, Radiat. Phys. Chem. In Press
120. R.C.L. Dutra, B.G. Soares, E.A. Campos, J.L.G. Silva, Polymer **41**, 3841 (2000)
121. N.G. Karsli, A. Aytac, Mater. Design **32**, 4069 (2011)
122. J. Taketa, L. Ustarroz, S.V. Gorbatikh, I. Lomov, Verpoest, Composites: Part A **41**, 927 (2010)
123. R. Rezaei, N.A. Yunus, Ibrahim, Mater. Design **30**, 260 (2009)
124. S.Y. Fu, B. Lauke, E. Mäder, X. Hu, C.Y. Yue, J. Mater. Process. Tech. **89–90**, 501 (1999)
125. J. Li, Appl. Surf. Sci. **255**, 8682 (2009)

Chapter 4
Surface Treatment and Sizing of Carbon Fibers

Abstract Surface treatment and sizing for carbon fibers are discussed primarily to identify the mechanism of interphase formation. The objective of this chapter is to review the surface treatment and sizing of carbon fibers used in the manufacturing composites. The surface properties of the carbon fibers were experimentally characterized, including the chemical compositions, surface free energy, acid–base interactions, and dispersive and polar components. It will be also presented that a number of oxidizing agents and gases have been used to surface treat the carbons, such as the plasma, electrochemical oxidation, and wet chemical and thermal treatments.

4.1 Introduction

Carbon fibers (CFs) have received much attention lately for their many potential applications in different matrix materials owing to their properties, processability, and recyclability. However, raw CFs need to be treated and/or sized as part of the manufacturing process. Designing a suitable surface-treatment method is a requisite to ensure that the high strength of the CFs is maintained during handling and composite manufacture. The surface treatment or sizing method is also equally important to ensure the formation of CF–matrix interface [1].

It is well known that the surface treatment of CFs is essential for improving the adhesion between the fibers and matrix. The interfacial and physical or chemical properties of CF-reinforced matrix composites are significantly influenced by the interfacial characteristics existing between the reinforcing CFs and matrix. The characteristics of CF–matrix interface could be either chemically or physically altered by both fiber surface treatment and sizing [1–20].

Select suitable surface-treatment methods might modify the CF surface by increasing the surface area, surface functional groups, electro properties components on the surface for providing the CF–matrix interactions, and enhancing other properties of the CFs. For example, the mechanical properties of CFs/polymer composites could be improved through the interface between the CFs and polymer

S.-J. Park, *Carbon Fibers*, Springer Series in Materials Science 210,
https://doi.org/10.1007/978-981-13-0538-2_4

using certain methods for the modification of the reinforcing CF surface. The bonding strength between the fibers and matrix could be enhanced through the fiber surface treatment [15–22]. Until now, the commercially available CF surfaces were normally coated with a sizing layer, which was usually either a solution or an emulsion consisting of polymeric components. The sizing could alter the handling of the CF, which included fiber protection, fiber alignment, and fiber wettability [23].

CF surface treatment and sizing are discussed primarily to identify the mechanism of interphase formation. The objective of this chapter is to review the surface treatment and sizing of CFs used in the manufacturing composites. Also, the surface properties of the CFs were experimentally characterized, including the chemical compositions, surface free energy, acid–base interactions, and dispersive and polar components. A number of oxidizing agents and gases have been used to surface treat the CFs, such as the plasma, electrochemical oxidation, and wet chemical and thermal treatments [24–70].

4.2 Oxidation of Carbon Fibers

The effects of oxidation on the CFs have been thoroughly investigated because of its importance in the fields of nuclear, medical, and materials science. Basically, the effect of the oxidation on CFs manifests as the displacement of carbon atoms from their graphitic structures. A number of oxidizing agents and gases have been used to surface treat the CFs, such as HNO_3 [24, 25], HSO_4 [26], H_2O_2 [27], NaOCl [28], $KMnO_4$ [28], RuO_4 [28–30], $NaClO_3$ [31, 32], $Na_2Cr_2O_7$ [32], $NaIO_4$ [33], air [34, 35], CO_2, and H_2O [17, 36].

4.2.1 Gaseous Oxidants

Dry oxidation normally employs air, oxygen (O_2) or ozone (O_3) as the oxidizing gas. Gas-phase oxidation (such as air, O_2, O_3, and CO_2) of CFs has been pursued in order to enhance the fiber–matrix bond. The advantages of oxidation in air included low cost, simple operation, lack of pollution, and excellent homogeneity, which make it one of the best approaches [38–47].

1. Air oxidation

When exposed to air at high temperatures, the CFs oxidizes, the properties of composites change, and mechanical properties degrade. Tong et al. treated polyacrylonitrile-based CFs with air at temperatures over the range 400–700 °C [39]. With increasing oxidation temperature, the crystallite height decreased gradually, while the d_{002} spacing increased presumably owing to the reaction of oxygen with carbon. The tensile strength decreased significantly, while the tensile modulus was nearly unchanged. An increase in the CF density after oxidation suggested that the glass-like carbon in the fiber was easier to burn off compared to the more graphitic carbon. The spacing d_{002} of the graphite layers increased, while the graphite size in the fiber decreased. The larger d_{002} after high-temperature oxidation might have been because of the reaction of oxygen with some C atoms in the graphite layers to produce functional groups such as –COOH, C–O–C, and C=O [39, 40, 45]. The gradual decrease in Lc(002) with increasing oxidation temperature was in agreement with the increasing mass loss. The size of the graphite crystal in the outer layer of the fiber became smaller owing to the burning of C atoms during oxidation, resulting in the smaller average crystal size Lc(002). The higher the oxidation temperature, the greater was the mass being burned off, and the smaller was the average crystal size in the resultant CFs [39].

Wang et al. found changes in the microporous structure and surface properties of pitch-based activated CFs upon air oxidation at 300, 400, 500, and 600 °C [41]. The adsorption isotherms on *P* and oxT-P with *T* below 400 °C were actually similar to each other except for a slight enhancement in the adsorption by oxidation, which appeared in or began from the sharply rising range at low P/P_0. On the other hand, the oxidation of *P* above 500 °C changed the N_2-adsorption isotherms from the I_a type, having risen sharply, to a round I_b type, indicating the formation of wider micropores. Air oxidation at 600 °C greatly increased the N_2-adsorption content, thereby manifesting in either further opening of the pores or micropore formation by air activation under this condition [41].

Wang and Wu prepared a three-dimensionally braided CF-reinforced carbon/epoxy resin (C/Ep)Carbon composites and studied the effect of air oxidation of the fibers on the properties of the composites [42]. They found that the specific surface area and surface roughness of the oxidized CFs increased, while their tensile strength decreased with increasing oxidation temperature or time. The flexural strength, flexural modulus, and shear strength of the C/Ep composites under the optimum treatment at 450 °C for 1 h increased by 149, 91, and 29%, respectively, compared to those of the untreated samples, but the impact strength decreased by 23% [42].

Rong et al. investigated that the effect of air oxidization on the adsorption of formaldehyde by the rayon-based activated carbon fibers (ACFs) [45]. Because pure formaldehyde was not stable in air, the adsorption of its vapor was simultaneously investigated with that of water vapor. ACF0 (pure ACFs) and A-350-1 h (oxidized at 350 °C for 1 h) possess a reasonable number of micropores, with a small fraction of mesopores, but without any macropores. A-420-1 h (oxidized at 420 °C for 1 h) is also microporous and possesses a reasonable number of mesopores with relatively high pore volumes. A-450-1 h (oxidized at 450 °C for 1 h) is microporous with some proportion of mesopores, but both the micropore and mesopore volumes of A-450-1 h are obviously lower than those of A-420-1 h and A-350-1 h. This might be owing to the violent reaction between the carbon and air at 450 °C, as mentioned earlier herein. This showed that the oxidation temperature obviously affected the pore size distributions of ACFs. Both A-420-2 h (oxidized at

420 °C for 2 h) and A-420-3 h (oxidized at 420 °C for 3 h) are microporous, containing a small proportion of mesopores and but without any macropores. The pore volumes, especially of micropores, are much lower than those of A-420-1 h and ACF0, illustrating that prolonged air oxidation time (in this experiment, oxidation time >1 h) could destroy the micropore skeleton of ACFs, thereby decreasing the pore volume. Therefore, it could be inferred that the air oxidation time could affect the pore size distributions of ACFs dramatically [45].

Seo and Park prepared the air-oxidized multiwalled carbon nanotube (MWCNT) electrodes from catalytically grown MWCNTs of high purity and narrow diameter distribution. The experimental results showed that the air oxidation modified the intrinsic structures of individual MWCNTs and slightly improved the dispersity of the MWCNTs. The TEM images of pristine and air-oxidized MWCNTs at 600 °C are shown in Fig. 4.1. The pristine MWCNTs were approximately 30–40 nm in diameter, several micrometers in length and randomly entangled, while the air-oxidized MWCNTs were detached mutually, and their lengths were <1 μm. Comparing the two transmission electron microscopy (TEM) images, it was evident that the lengths of the air-oxidized MWCNTs were shorter than those of the pristine MWCNTs, and the dispersity of the air-oxidized MWCNTs was enhanced compared to that of the pristine MWCNTs. This was good for the enlargement of the BET specific surface area and pore volume. Moreover, the specific area increased compared to that of the pristine MWCNTs. The MWCNT powder was oxidized by air, and the amorphous carbonaceous matter jammed in the MWCNT aggregates could be etched away or stripped off. Moreover, oxygen attacked the curvatures of the MWCNTs, resulting in the breaking of the MWCNT walls. The binding between the entangled MWCNTs was weakened by the broken MWCNTs and removal of the amorphous carbon. Many MWCNTs were detached from the aggregates, resulting in either partial or complete disassembly of the aggregates, and an increase in the surface area. This increase was the result of two factors: Firstly, the MWCNT surfaces were exposed after the entangled MWCNTs were

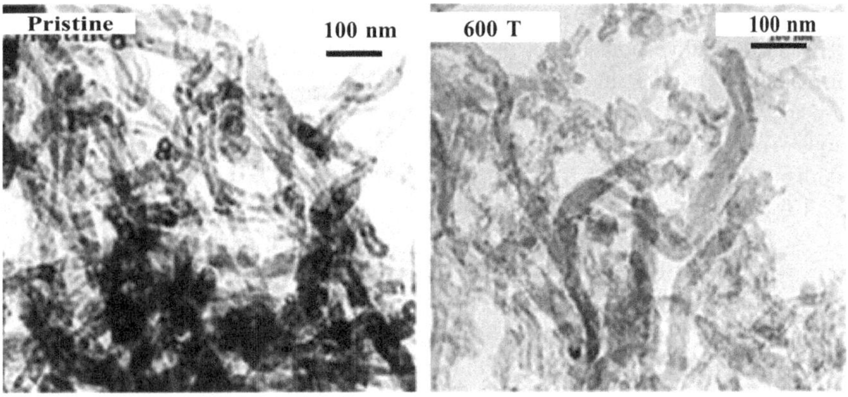

Fig. 4.1 TEM images of pristine and air-oxidized MWCNTs at 600 °C [47]

Table 4.1 Characteristics of pristine and air-oxidized MCWCNTs at various temperatures [47]

Samples	Specific capacitance (F/g)	Specific surface area (m^2/g)	Pore volume (cm^3/g)	Rospores (<2 nm, %)
Pristine	25	180	0.60	12.5
400 *T*	32	230	0.72	8.2
500 *T*	36	236	0.73	7.1
600 *T*	50	290	0.84	4.6

dispersed; secondly, the internal cavities of the MWCNTs were opened owing to the caps being destroyed and cylindrical walls being broken. As the annealing temperature increased, more of the MWCNTs were broken and uncapped MWCNT aggregates were disassembled. Moreover, oxygen attacked the cylindrical walls of the MWCNTs. As shown in the TEM image of air-oxidized MWCNTs at 600 °C, some outer graphitic layers were etched off, yielding rough "convex–concave" walls. This resulted in the further enhancement of specific area, as listed in Table 4.1. Subsequently, the specific surface area decreased slightly with the increasing air oxidation temperature up to a maximum of 600 °C [47].

2. O_2, O_3 oxidation

To study the influence of O_2 oxidation kinetics on the surfaces of the CFs, Poila, and Serra demonstrated that T-300 CF oxidation appeared to be gas-phase diffusion, which was limited over the temperature range 816–1538 °C [48]. CF oxidation kinetics was consistent with the predicted weak dependence on the temperature and concentration for the transport through a laminar boundary layer. Under most conditions, the C/SiC composite oxidation kinetics was dominated by C fiber oxidation rather than SiC oxidation. The weight losses in the composites were observed at all temperatures between 816 and 1538 °C and all oxygen partial pressures between 50 and 0.1% O_2. The weight loss rates for the composites were highest at 816 °C and decreased at higher temperatures [48].

The results of Lamouroux et al. for temperatures <700 °C showed that there was a large activation energy that reflected the energetics of the carbon–oxygen reaction ($C + 1/2O_2\ (g) = CO(g)$) [49]. At the higher temperatures in this study ($T > 800$ °C), the weaker temperature dependence suggested a change in the rate-controlling mechanism. Based on the discussion of Walker et al., it was proposed that the oxygen contacting the surface reacted with the CF rapidly relative to the transport rate of oxygen to the surface through the gas boundary layer. Thus, the reaction at high temperatures was limited by the gas-phase diffusion of O_2 to the surface [50].

Hosokai et al. found that O_2 influenced the resulting product distribution in completely different ways at 700 and 800 °C [51]. O_2 was consumed at 700 °C mainly by the oxidation of tar and light oxygenates, forming CO, CO_2, and H_2O, unless the ratio of O_2/carbon involved in the nascent volatiles (O/C) exceeded 0.7 mol of O/mol of C. On the other hand, the oxidation at 800 °C selectively consumed H_2 and lower hydrocarbons, leaving the residual tar yield nearly unchanged. Such different behaviors of O_2 were mainly because of the difference in the characteristics of the thermal cracking of the tar and light oxygenates between 700 and 800 °C. The thermal cracking at 700 °C left the tar with a molecular mass (MM) range of up to >1000, which was decomposed in the presence of O_2 but incompletely. At 800 °C, the tar was decomposed quickly even in the presence of O_2, forming refractory aromatics with a MM range of up to 400 together with soot, while the lighter gases were oxidized. A portion of tar, obtained by cracking at 700 °C, was deposited inside a ceramic—fiber filter downstream of the reactor, whose temperature was 350–500 °C; however, the deposit could not be removed mechanically. The entire tar, obtained from the cracking at 800 °C, was allowed to pass through the filter, regardless of O/C. It was thus found that the gas-phase temperature range 700–800 °C was critical for the tar property, which was relevant for dust removal over the range 350–500 °C [51].

The ozone method was also used to modify the surface activity of CFs, which improved the compressive and flexural strengths of carbon/carbon composites [52]. Jin et al. prepared the carbon/carbon composites with the untreated and ozone-treated CFs. The ozone treatment increased the compressive and flexural strengths of the carbon/carbon composites. This was probably because of the chemical interaction between the CFs and pitch in the matrix and increasing surface roughness of CFs conduced to the interfaces of CFs and the carbon matrix strengthened. It has been previously shown that the chemical bonding was important for increasing the adhesion between the CFs and carbon matrix. It could be ascertained that the adhesion of carbon/carbon composites showed better improvement with ozone treatment at 120 °C for 6 min, compared to that observed in the other methods. At 160 °C, the ozone rapidly decomposed to oxidize the CFs. Therefore, the compressive and flexural strengths of the carbon/carbon composites decreased. In conclusion, the optimum conditions for the ozone treatment were the temperature of 120 °C and duration of 6 min [52].

Ozone is a powerful oxidant with a high selectivity for unsaturated bonds. Its reaction with saturated bonds proceeds slowly. It is known that ozone produces hydroperoxy (HOO) and superoxide (O_2^-) radicals as intermediates during decomposition in the liquid phase [59]. Usually, they react again with ozone and produce hydroxyl (OH) with ozonide (O_3^-) radicals as intermediates. The hydroxyl radical has greater oxidizing power (*E*o: +3.06 V) compared to ozone, and its reactions with organics are rapid. Park et al. have previously reported the influence of ozone surface treatment on the activated carbons [58, 61]. The purpose of that study was to determine the effect of the ozone treatment on the adsorption of Cr(VI) and Cu(II) on ACFs [58]. The ozone was added to either NaOH or H_2O_2 initiator solutions to produce the hydroxyl radicals. The structural and functional groups of ozone-treated ACFs were analyzed by varying the initiator. Table 4.2 shows the structural characteristics of the ACFs before and after ozone treatments for 1 h in each solution. The average pore diameter was not largely affected by the ozone treatment. However, the specific surface area of the ozonized ACFs in 1 M NaOH solution decreased because of the pore-blocking phenomena by the sodium ions of NaOH. On the other hand, despite the pore-blocking phenomena, the specific

Table 4.2 Structural characteristics of ACFs before and after ozone treatments [58]

Specimen	Smic (m^2/g)	Sex (m^2/g)	Vmic (cm^3/g)	Vtotal (cm^3/g)	d (nm)
Nontreated ACFs	1516	1508	8.2	0.68	1.91
Ozone-treated ACFs in 1-M NaOH	1364	1354	9.8	0.54	1.76
Ozone-treated ACFs in 3-M H_2O_2	1706	1697	9.6	0.76	1.88

S_{mic} Micropore surface area
S_{Ex} External surface area
V_{mic} Micropore volume
V_{total} Total pore volume
d Average pore diameter

surface area of the ACFs ozonized with the H_2O_2 initiator increased. This was because the carbon atoms on the surfaces of the ACFs reacted with the hydroxyl radicals and were directly oxidized to carbon dioxide and converted into soluble decomposition products. These were then oxidized to produce carbon dioxide continuously, resulting in an increase in the specific surface area. The production of carbon dioxide was verified by the conversion of the lactonic groups to phenolic groups, as shown in Fig. 4.2 [58].

Activated carbons were modified by the ozone treatment to enhance the efficient removal of ammonia gas over the activated carbons. The variation in the adsorption of ammonia against the reaction time of each specimen is shown in Fig. 4.3 [61]. It shows that the ozone-treated activated carbons reveal much higher removal efficiency compared to that of the pristine activated carbons. This was because of the acid–base interaction of each functional group containing oxygen on the surfaces of activated carbons, which played a key role in the determination of the gas-adsorption capacity.

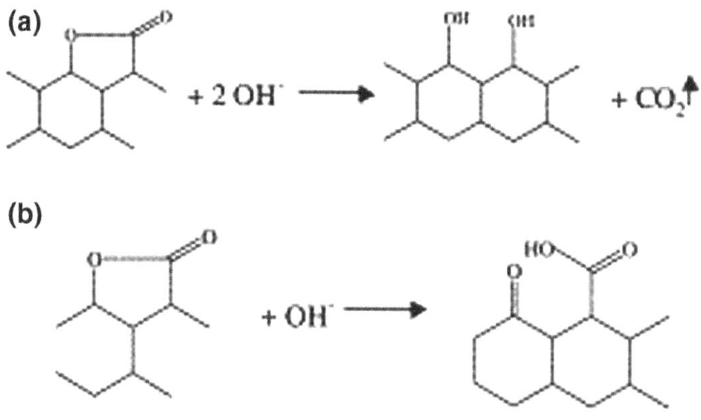

Fig. 4.2 Deformation of lactonic groups by hydroxyl (OH) radicals; formation of **a** phenolic and **b** carboxylic groups [58]

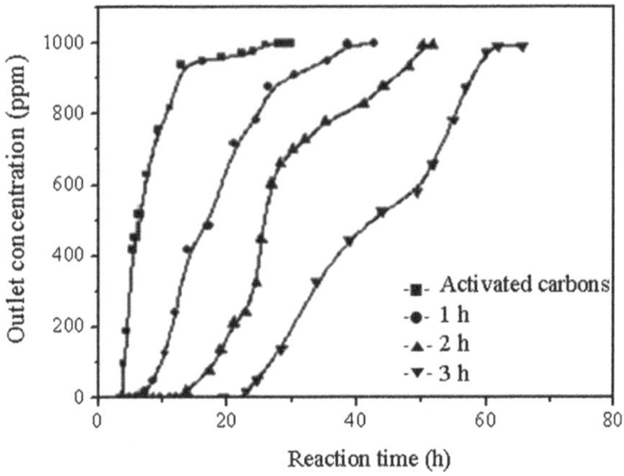

Fig. 4.3 Ammonia-removal efficiency of activated carbons as a function of ozone-treatment time [61]

3. CO_2, H_2O oxidation

Sun et al. reported that polyacrylonitrile (PAN)-based carbon hollow fibers (PAN-ACHFs) can be activated using carbon dioxide [36]. The burn-off and shrinkage ratios were determined from a change in the weight and length before and after activation. The burn-off of PAN-ACHF slowly increased with the increasing oxidation time, from 1 to 5 h, and then hardly changed. The shrinkage of PAN-ACHF almost did not change with the increasing oxidation time. The BET surface area of the PAN-ACHF and surface area of mesopores gradually increased on extending the activation time and reached the maximum values, 780 and 180 m^2 g^{-1}, respectively, when the fibers were activated at 800 °C for 100 min. The number of pores on the surface of PAN-ACHF increased with the extending activation time. The number of mesopores in the PAN-ACHF, composed of fibers activated for different durations, increases with the extending activation time, while the dominant pore sizes of the mesopores in the PAN-ACHF were over the range 2–5 nm [36].

Kelemen and Freund used XPS to characterize the glassy carbon surfaces following exposure to O_2 and CO_2. The clean glassy carbon samples were oxidized in O_2 and CO_2 and characterized using XPS [53]. The O(1s) signal was on the same intensity scale. Each sample was oxidized for 300 s under specified conditions and cooled in the reactant gas. A negligible loss of carbon to the gaseous products occurred at 300 °C. The oxidations in O_2 and CO_2 at 700 °C corresponded to the carbon weight losses of 20 and <0.1%, respectively. They found multiple O(1s) components in the observed peak, which will be discussed later. It was striking that the same amount of oxygen along with the same O(1s) line shape was observed after CO_2 exposure at 300 and 700 °C. The areas of the O(1s) and C(1s) signals were used in conjunction with the standard XPS sensitivity factors to determine the oxygen concentration in the sampling region. The amount following the oxidation by CO_2 corresponded to 15 oxygen atoms per 100 carbon atoms [53].

Zaini et al. produced ACFs derived from PAN by using steam and CO_2 as activating agents and examined their ability to remove heavy metals from the aqueous solution [54]. The drawback was that missing delocalized π-electrons were produced during the activation of oxidized fibers initially treated at 300–350 °C. This definitely resulted in the formation of the imperfect graphitic structure. Conversely, a near-ideal graphitization could be anticipated during the activation of the polymer ladder structures of oxidized fibers previously treated at 400–450 °C. The activated carbons derived from PAN were highly microporous with the pores centered in the supermicropore region. Steam was found to be better than CO_2 in generating a higher surface area, but the resulting yield was relatively lower [54].

4.2.2 Acid Oxidation

After carbonization, the CFs had a surface that did not bond well with the epoxies and other materials used in composite materials. The main aim of oxidation of the CF surfaces is to obtain more hydrophilicity. The addition of oxygen atoms to the surface provides better chemical bonding properties. In general, the oxygen-containing groups increase the adhesive interfacial force in the polymeric matrix by behaving as either acids or bases, which possess ion-exchange properties. Among the oxidation treatments, acid oxidation is the most widely used method to increase the total acidity during the wet oxidation treatment [55–77].

1. Inorganic acid oxidation

Various densities of the functional groups have been produced on CF surfaces as a function of the reaction time. Oxidative treatments produce active sites by forming carbon–oxygen functional groups, but oxidation also increases the surface roughness and surface area by etching [55–73].

Park et al. studied the effect of nitric-acid treatment on the multimetal adsorption by basic ACFs from the interfacial and textural points of view [57]. One molar nitric acid at its boiling point was used for the oxidation of functional groups, without causing damage to the fiber surfaces. The results of 1-mM adsorptions of both copper and nickel on ACFs are shown in Fig. 4.4a. From the results, it was evident that the copper ions were more favorably adsorbed on the ACFs than the nickel ions over the pH range investigated. Such selective adsorption resulted from the interaction between the adsorbate and adsorbent and could be ascribed to the characteristics such as pore size and distribution and specific polarity. Also, the size of the hydrated copper ion was much smaller than that of the nickel ion; consequently, the hydrated copper ion could more easily penetrate into the pores of the ACFs. The amount of the adsorbed metal ions on the ACFs increased, reaching a maximum at pH 6, followed by a drastic decrease. This was because of the isoelectric point of the ACFs, which was quantitatively changed by the amount of the oxygen-containing surface functional groups. Figure 4.4b shows the removal of the multimetal ions from the acidic ACFs as a function of pH. The amounts of both adsorbed copper and nickel ions relatively increased at the low end of the pH range

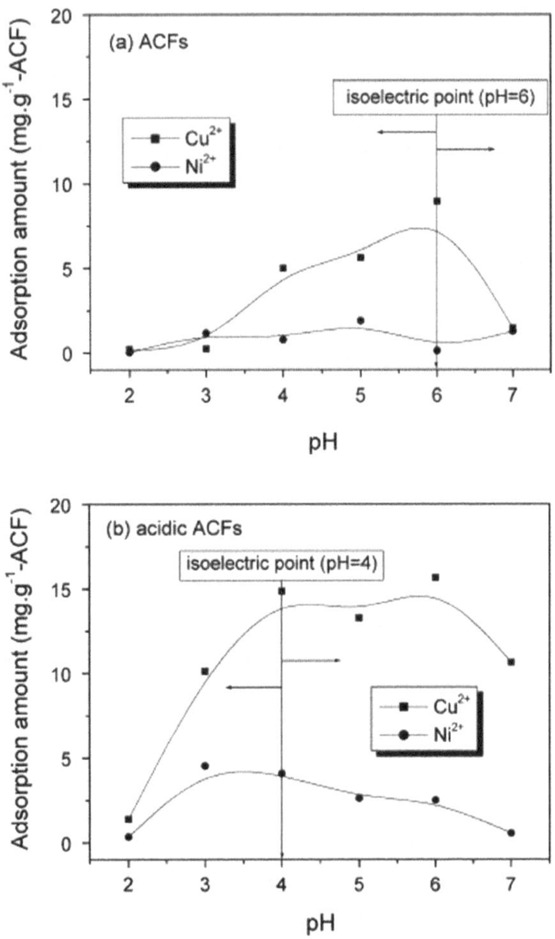

Fig. 4.4 Removal of multimetal ions from **a** ACFs and **b** acidic ACFs as a function of pH at 25 °C, $[Cu^{2+}] = [Ni^{2+}] = 1$ mM [57]

when compared to the results of the ACFs. This was mostly ascribed to the fact that the oxygen-containing surface functional groups increased from the modification with nitric acid, thereby resulting in the shift of the isoelectric point of the ACFs to the lower pH. This increase in the amount of the adsorbed metal ions was mostly because of the increase in the total acidity of the ACFs. This demonstrated that there were certain types of carbons that exhibited amphoteric behavior owing to the presence of the weakly acidic functional groups. For relatively high pH values, the carbon behaved as an acid by adsorbing OH^-, whereas at low pH values it behaved as a base; i.e., it adsorbed H^+ ions. Therefore, the adsorption capacity of the metal ions was mainly influenced by the weakly acidic functional groups (e.g., lactones) on the carbon surface at pH < p*I* and the strongly acidic functional groups (e.g., carboxyl) at pH > p*I* [57].

Perrard et al. investigated the oxidation kinetics of an excellulose-activated carbon cloth using an aqueous solution of NaOCl at ambient temperature [75]. The oxidation resulted from a coupling between the chemical kinetics and physical transport kinetics. During the first stage (0–7 min), the molar ratio CO/CO_2 sharply decreased from 5.1 to 2.2, while the selectivity ratio S^{III}/S^{I} rapidly increased. Expectedly, with NaOCl being a strong oxidizing agent, the oxidation process led to the formation of carboxylic groups, though a large number of quinone groups remained. In contrast, the loss of matter dominated the second stage (7–60 min) because the samples lost weight during the oxidation and a decrease in the fiber diameter was observed. At the end of the second stage, 21 wt% oxygen remained in the fibers. Such high oxygen content involving a large proportion of carboxylic groups, whose creation was accompanied by C–C bond rupture, indicated a profound modification of the fiber structure. Further oxidation during the third stage (60–90 min) led to severe aromatic ring cleavages, for instance the oxidation of triphenylene to mellitic acid. The amount of oxygen fixed by the solid per gram of the sample decreased, while the ratio CO/CO_2 slightly increased. This occurred because the intensive attack on the fibers yielded strongly oxidized organic compounds, which proceeded to the liquid phase. Consequently, the CFs lost their mechanical strength, and the fabric was deeply damaged under the stirring conditions of the solution in the reactor. The change in the mean diameter of the fibers had been measured using amplified SEM micrographs [75]. It was observed that there was no change in this diameter for oxidation times smaller than 15 min. For oxidation times over the range 15–60 min, the diameter of the fibers slightly decreased from 10.2 to 9.4 μm. A decrease in the mean diameter was steeper for oxidation times larger than one hour (third stage). During this stage, the transfer of carbon-containing compounds or matter to the aqueous medium largely dominated. The production of a small quantity of carbon dioxide also possibly occurred [75].

Fathi et al. reported the surface chemistry and surface roughness of CFs after the oxidation treatment with sulfuric acid and the effect on fiber–matrix adhesion and thus the performance of the composite after the treatment [76]. The values of stiffness, flexural modulus, and flexural strength are listed in Table 4.3. It can be seen that the flexural modulus of the treated CF composites slightly increased. This increase, as stated previously for the storage modulus, could have been because of the better compatibility with the polyester resin compared to the untreated fibers. Also, the flexural strength of the composite rods was significantly improved compared to the untreated CF composites, especially CFS10.

2. Organic acid oxidation

Table 4.3 Mechanical properties of composite samples (CF samples treated with various concentrations of sulfuric acid) [76]

Type of fiber	Flexural modulus (GPa)	Flexural strength (MPa)	Stiffness (N/m)
CF (untreated)	61.1	596	28,800
CFS20	63.2	637	29,800
CFS10	61.5	716	29,000

Tang and coauthors have reported the PA (polyamide) 66 composites filled with arylboronic acid-treated CF. They studied the influence of the treatment of the CF surface with arylboronic acid on the mechanical and tribological behavior of the PA66 composites [77]. To study the changes of the carbon surfaces, the specific ratio of each oxygen functional group in the C_{1s} regime is shown in Table 4.4. We found that the number of oxygen functional groups, including C–O (BE = 286.1–286.3 eV) and O–C=O (BE = 288.8–289.1 eV), increased after the treatment, while the number of C=C groups decreased. This indicated that the surface treatment probably broke the C=C (BE = 284.6 eV) bond and bound the oxygen functional groups to the fiber surfaces. The SEM results showed that PA66 peeled off the fibers on the worn surfaces of the composites with the untreated fibers, where there are fibers pulling out from the composites. The surfaces of the composites were severely worn out, while the PA66 matrices and treated fibers had good adhesion to the worn surfaces of the composites with treated fibers, containing scarce CF [77]. The CFs on the worn surfaces of the composites bore most of the load between the contact surfaces and reduced the real contact area of the counterpart and shear evolution on the adhesive point. Also, the self-lubrication effect of the CF reduced the friction coefficient. The peeling CF could cause abrasion on the contact surfaces of the composites, which could accelerate the peeling of the matrix [77].

Xu et al. studied the oxidation–reduction and preirradiation-induced methods to study the effect of the modification by acrylic acid on the wetting and adsorption abilities of CFs in the epoxy solution and the interfacial properties of the CFs/epoxy [78]. The chemical bonding, interdiffusion, and hydrogen bonding in the interface between the CFs and matrix resins might contribute to an improvement in the interfacial adhesion. The increasing surface energy of CFs benefited the increase in the surface energy difference between the CFs and epoxy resins and thereby improved the wettability of the fibers with epoxy resins. As a result, the interfacial adhesion of CF/epoxy composites was strengthened and ILSS increased. The ILSS of the modified CF/epoxy composites increased by more than 15% compared to that of the untreated CF/epoxy composites. Also, the preirradiated CF/epoxy composites outperformed the oxidation–reduction-grafted CF/epoxy composites in the strength value of ILSS [78].

Table 4.4 Relative percentages of functional groups on [77] surface-treated CFs

Relative atomic percentage of functional groups			
Type of fiber	C=C (284.6 eV)	C–O (286.1–286.3 eV)	O–C¼ O (288.8–289.1 eV)
Untreated	95.8	2.9	1.3
Arylboronic acid treatment	66.5	11.5	22

4.2.3 Electrochemical Oxidation

Recently, several studies dealing with the environmental applications of electrochemistry have demonstrated that many compounds could be treated with the electrochemical oxidation instead of traditional methods [78–85]. This electrochemical process requires neither chemical pretesting nor chemical adjustment of the wastewater, little space, and produces fewer by-products and less sludge. The electrochemical oxidation yielded either oxygenated products or electrophilic species readily reacting with unsaturated compounds, depending on the experimental conditions. The oxidized carbons possessed certain unique properties owing to the oxygen-containing functional groups such as carboxylic, lactonic, and phenolic groups on their surfaces [86–90].

The fibers were subjected to the electrochemical oxidation under different conditions (Fig. 4.5). Sodium hydroxide (NaOH), ammonium hydrogen carbonate (NH_4HCO_3), ammonium carbonate ($(NH_4)_2CO_3$), sulfuric acid (H_2SO_4), and nitric acid (HNO_3) were used as electrolytes at 3, 5, 10, and 20 wt%, respectively. The voltage of the oxidation changed over the range 0.5–5.0 V (0.5, 1, 2, and 5 V). The oxidation was carried out in a continuous process; the treatment time changed linearly with the passage of the fiber through the electrolyte; two passage rates, 20

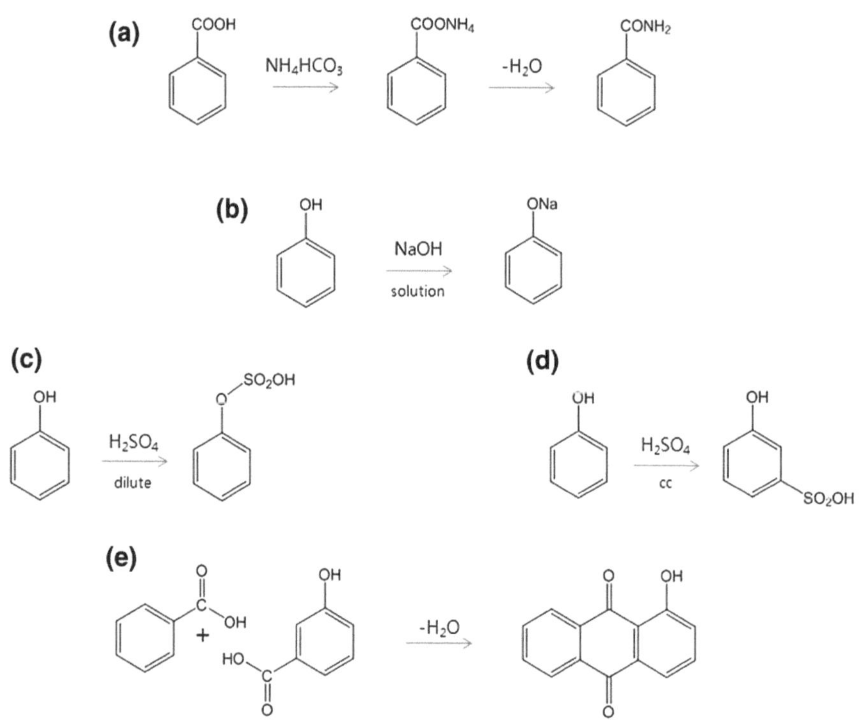

Fig. 4.5 Schematic of electrochemical oxidation of CFs under different conditions [86]

and 30 cm/min, were used in the experiments. The oxidized fibers were washed with the distilled water in a continuous process [86].

The fragmentation experiments revealed considerable differences in the behavior of the fibers oxidized under different conditions. The interaction between the fiber and epoxy matrix was characterized by the IFSS changes over a wide range depending on the type and concentration of the electrolyte as well as voltage used. The IFSS values determined under different conditions are also listed in Table 4.5. The average standard deviation of IFSS was approximately 20%, which corresponded to the values previously published in the literature [42–94]. The oxidation in 20 wt% ammonium carbonate and ammonium hydrogen carbonate solutions at 0.5 V led only to an insignificant increase in IFSS. Under the same conditions, the effect of sulfuric acid was also moderate, while the oxidation in nitric acid and sodium hydroxide resulted in an enhancement of the IFSS values, thereby indicating strong interfacial interaction. At low voltage (0.5 V), oxidation in HNO_3 led to an increase in IFSS with increasing concentration; also, increasing voltage had the same effect at low electrolyte concentration (3 wt%). At high concentration and voltage, IFSS somewhat decreased. This result emphasized the importance of optimization and that appropriate oxidation conditions must be used in order to achieve maximum efficiency.

Table 4.5 Interfacial shear stress measured in an epoxy matrix on CFs oxidized under different [86] conditions

Electrolyte		Voltage (V)	IFSS (MPa)
Type	Concentration (wt%)		
Neat	–	–	17.3 ± 2.8
$(NH_4)_2CO_3$	20	0.5	18.8 ± 2.5
NH_4HCO_3	20	0.5	17.1 ± 2.1
HNO_3	3	0.5	23.1 ± 4.6
	3	5.0	29.3 ± 8.6
	20	0.5	36.0 ± 14.2
	20	5.0	25.8 ± 8.8
NaOH	3	5.0	23.7 ± 5.7
	5	5.0	23.8 ± 4.9
	10	5.0	25.7 ± 3.5
	20	0.5	37.7 ± 8.6
	20	1.0	32.6 ± 7.8
	20	2.0	29.0 ± 3.8
	20	5.0	27.5 ± 5.9
H_2SO_4	3	5.0	29.0 ± 4.8
	5	5.0	25.6 ± 5.1
	10	5.0	30.1 ± 4.2
	20	0.5	20.6 ± 7.6
	20	1.0	23.0 ± 3.0
	20	2.0	27.8 ± 3.7
	20	5.0	36.0 ± 9.6

Park et al. studied the anodic oxidation of ACFs in order to introduce the surface functional groups onto the ACF surfaces [81]. Their research indicated that the extent of adsorption and adsorption rate of Cr(VI) increased with increasing number of the surface oxide groups of ACFs, even though both the specific surface area and micropore volume did not significantly change, as was the case with the anodic oxidation treatments on ACFs. The phenol-based ACFs (mat type, Kuraray) were subjected to an electrolytic reaction in an aqueous solution of sodium hydroxide (35 wt%), wherein the negative ions were attracted to the surfaces of the ACFs acting as anodes, thereby modifying the ACF surfaces. In the FT-IR spectra, as seen in Fig. 4.6, the O–H stretching vibrations were observed over the range 3600–3200 cm^{-1} owing to the existence of surface hydroxyl groups and chemisorbed water after the anodic oxidation [81]. The absorption bands in the 1640–1500 cm^{-1} region suggested the overlapping of the aromatic ring bands and double band (C=C) vibrations with the bands of C=O moieties **13** and **14**. The oxidation of the ACFs also enhanced the absorption over the range 1250–1000 cm^{-1}, which suggested an increase in the presence of ether (symmetric stretching vibrations), hydroxylic, and phenolic structures (vibrations at 1180 cm^{-1}). The amount of adsorption and the adsorption rate of Cr(VI) from the aqueous Cr(VI) solution increased when the surface chemistry of the ACFs was modified through the anodic oxidation, as shown in Fig. 4.7, which then led to an increase in the number of surface functional groups without significantly changing the surface area. As a result, the amount of adsorption and adsorption rate of Cr(VI) from an aqueous solution increased owing to a large amount of the surface functional groups on the ACFs.

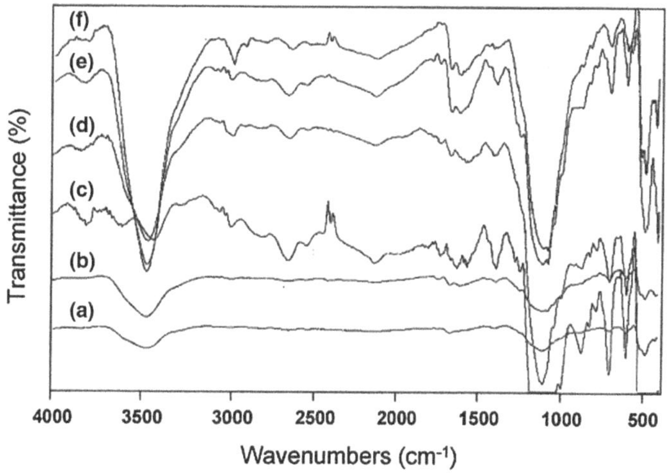

Fig. 4.6 FT-IR spectra of ACFs studied (mA m^{-2}): **a** 0; **b** 5; **c** 15; **d** 45; **e** 150; **f** 450 (oxidation rate 1 m min^{-1}) [81]

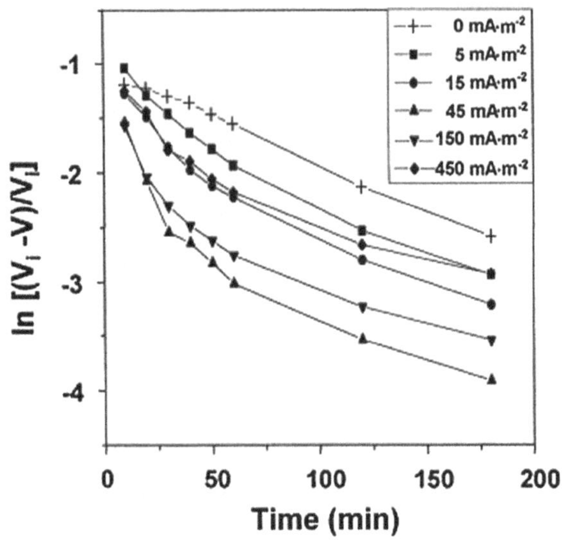

Fig. 4.7 Relationship between $\ln[(V_i - V)/V_i]$ and time on ACFs for Cr(VI) at 298 K. V_i, initial concentration of Cr(VI) (mol/l); V, amounts of Cr(VI) adsorption in a given time (mol/L) $[(V_i - V)/V_i]$ represents residue extent of Cr (VI) in aqueous solution after adsorption in a given time [81]

4.2.4 Treatment with Nonoxidative Agents

Ehlert and Sodano reported that the surfaces of CFs were functionalized by the reaction of the existing surface defects with isopropylidene malonate [91]. The reaction used the existing defects rather than generating new ones to graft the terminal malonic esters, thereby creating a carboxyl-functionalized surface. The reaction did not employ oxidation and preserves the tensile strength of the fiber. Also, the reaction was validated using X-ray photoelectron studies and Fourier transform infrared spectroscopy, whereby the expected malonic ester structure was confirmed. The usable surface coverage of the carboxylic acid groups doubled to 8.4%. The effects of solvent, temperature, concentration, and reaction time were studied. This functionalization could improve the composite interfaces without sacrificing the fiber strength and potentially open up a new paradigm in the functionalization of the carbon surfaces [91].

4.3 Plasma Treatment

Plasma is a partially or fully ionized gas containing electrons, ions, and neutral atoms or molecules; also called the fourth state of matter [92–110]. Gas plasma can be produced by introducing the desired gas into a vacuum chamber, typically 0.1–10 Torr, and subsequently exciting the gas using radiofrequency energy. The free radicals and electrons created in the plasma collide with the exposed surface of the material, rupturing covalent bonds and creating free radicals. Plasma composed of inorganic gases such as argon, helium, hydrogen, nitrogen, and oxygen leads to the implantation of atoms, radical generation, and etching reactions while plasma composed of organic gases such as hydrocarbons and alkylsilanes leads to polymer-forming reactions (Table 4.6) [110]. The schematic of the plasma device is shown in Fig. 4.8.

Montes-Morán et al. treated the ultrahigh-modulus (UHM) CFs using oxygen plasma [94]. The "cleaning" effect of the plasma oxidation was depicted using the scanning tunneling microscopy (STM). The SEM results showed that all the P120 fibers examined showed surfaces with long granular patterns as well as scratched areas, most likely associated with the skin flakes mentioned previously. After the surface treatment, the new surfaces created by the plasma oxidation consisted primarily of homogeneously stacked patches, with rare occurrences of skin fragments. These patches, which also confirmed the relatively flat areas observed on the untreated P120 fibers, might consist of graphene structures whose atomic resolution could be eventually achieved. Additionally, no ballistic sputtering was detected on the treated surfaces.

Park and Kim carried out the atmospheric-pressure plasma treatment containing oxygen on ACFs in order to enhance the surface functionality of the ACFs [95]. It was found that the number of oxygen functional groups such as C_6H_5OH and O–C=O increased after the plasma treatment. However, a slight decrease in the specific surface area was observed. The HCl-removal efficiency of all the ACFs

Table 4.6 Selected systems used in plasma treatment [94–109]

Plasma treatment	References
O_2, Air	[94, 95, 97, 101–105, 108]
Ar/O_2	[96, 99]
F_2, CF_3CH_2F	[98, 109]
NH_3	[106, 107]
CO_2, Ar	[108]
He	[100]

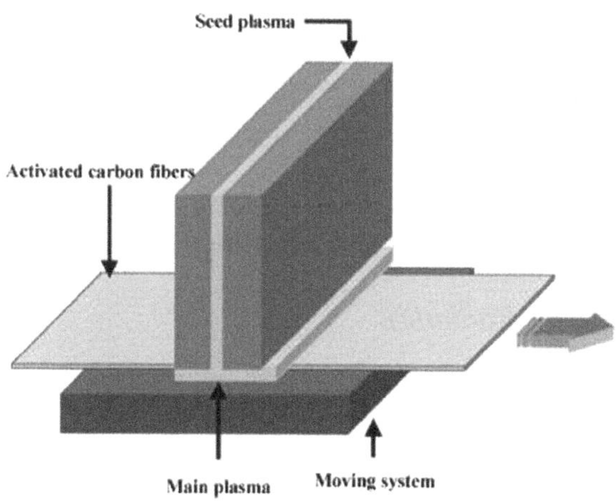

Fig. 4.8 Schematic of atmospheric-pressure plasma device [95]

Table 4.7 Chemical composition of ACFs as a function of treatment time [95]

	C_{1s} (at.%)	O_{1s} (at.%)	N_{1s} (at.%)	O_{1s}/C_{1s} (%)
As-received	90.4	7.8	1.7	8.6
P-O-1	88.5	9.8	1.6	11.1
P-O-2	86.3	11.5	1.7	13.3
P-O-3	83.2	14.7	1.7	17.7
P-O-4	80.7	17.2	1.8	21.3

modified by plasma treatment was higher than that of the untreated samples. The efficiency increased up to P-O-3 and then decreased at P-O-4. This indicated that the HCl-removal efficiency was basically dependent on the surface functionality; however, the control of the specific surface area was also important. In conclusion, if the plasma treatment is accomplished with high oxygen content and well-developed micropores, this technique could be a potentially useful and economic method to remove the highly toxic hydrochloride gases from carbon surfaces. The surface atomic compositions obtained using XPS are presented in Table 4.7. It was observed that the surface of the ACFs was mainly composed of carbon, oxygen, and nitrogen. The carbon and oxygen contents of the as-received sample were 90.4% and 7.8 at.%, respectively; however, those of P-O-4 were 80.7 and 17.2 at.%. The O_{1s}/C_{1s} ratio of plasma-treated ACFs increased with the treatment time owing to the conformation of the oxygen-containing surface functional groups on the carbon surfaces.

Kim and Park have described the preparation of ACFs coated with graphite nanofibers (GNFs) [96]. Low-pressure plasma–mixed-gas (Ar/O_2) treatment of the ACFs led to the growth of the GNFs on their surfaces. Figure 4.9 shows the GNF-coated ACFs prepared in this study. The ACFs treated at 300 W were fully covered by the GNFs, but the diameter sizes of the GNFs produced were not homogeneous. Figures 4.9b, c shows the magnified images of the GNFs produced. As evident in the images, the GNFs were fairly straight. Figure 4.9d depicts the TEM image of a single GNF fiber. It shows that the structure of the GNFs is in the form of stacked graphite layers of the herringbone type. As shown in Fig. 4.9, we successfully produced the GNF-coated ACFs under certain conditions (300 W, 13.56 MHz, 50 cm^3 min^{-1} of Ar/O_2 feeding) of low-pressure plasma. The diameters of the GNFs varied over the range 50–100 nm.

4.4 Other Surface Modification Methods

4.4.1 Radiation

Methods used for the oxidation of such nonpolar CF surfaces are mentioned in part 4.3. Each of the methods mentioned above might cause a loss of fiber strength if the treatment is excessively extended. Similarly, gamma-ray or neutron radiation

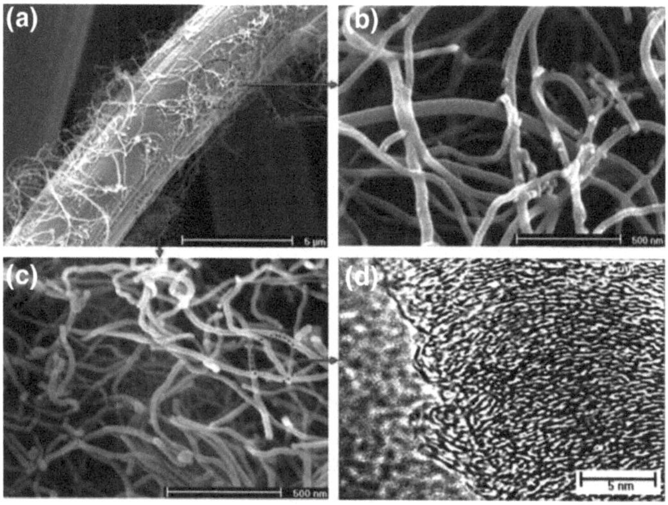

Fig. 4.9 **a**, **b**, **c** SEM, and **d** TEM images of graphite nanofiber (GNF)-coated activated carbon fibers (ACFs) produced at 300 W: **a** GNF-coated ACFs, **b**, **c** GNFs coated on ACFs, and **d** a TEM image of GNFs [96]

change the microstructure of the carbon materials and cause losses in the mechanical properties at high radiation doses [97, 111–114]. It is known that the radiation affects the crystal lattice by either the displacement of atoms within the lattice or electronic excitation [112–114]. The electrons stripped from the atoms are believed to cause dimensional (topographical) change in the CFs and create active sites on the fiber surfaces, which may bind with the functional groups of the bulk polymers [112].

In particular, gamma-ray radiation has been more widely used for radiation grafting through which reactive groups on the surfaces of polymers could be created [97, 112, 114]. Xu et al. investigated the effect of γ-ray radiation on the microstructural and mechanical properties of CFs, which were irradiated by a ^{60}Co source. The interlayer spacing d_{002} of the CFs decreased after irradiation [111]. Tiwari et al. employed the γ-ray radiation technique with varying doses (100–300 kGy) on the carbon fabric to develop composites with the polyetherimide (PEI) matrix through the impregnation method followed by compression molding [114]. Any improvement in the friction and wear properties was correlated with that in ILSS because of the carbon fabric treatment. The higher the dosage, the higher the enhancement in ILSS and tribo-performance of the composites. Both these factors were thought to be responsible for enhancing the fiber–matrix interface. For in depth analysis, various techniques such as fiber tension test, adhesion test, and Raman spectroscopy were carried out [111].

Wan et al. treated the PAN-based carbon fibers (PAN-CFs) with gamma-ray radiation [97]. The efficiency of the gamma-ray radiation was compared with air oxidation in terms of the variations in the surface structures of the CFs and the mechanical performance of their composites. The observed higher content of the carboxyl group on the surfaces of the gamma-irradiated CFs was likely to be responsible for the relatively stronger fiber–matrix bonding. Thus, it was concluded that the gamma-ray irradiation was effective in tailoring the surface properties of the CFs. The data pertaining to the mechanical properties (including flexural strength and modulus and shear and impact strengths) of the 3D MC nylon-based composites with various CFs demonstrated that after gamma-ray irradiation, the flexural strength and modulus and shear strength increased by 36.2, 19.0, and 14.7%, respectively.

The corresponding values for the composite samples reinforced with the air-treated CFs were 31.7, 11.4, and 6.2%, respectively. The stronger mechanical properties of the gamma-irradiated fiber composites compared to those of the air-treated fiber composites were attributed to their higher interfacial adhesion (32 ± 2.9 Pa vs. 24 ± 1.1 MPa) tested using an interlaminar shear test. This suggested that the gamma-ray irradiation was more efficient in improving the fiber–matrix adhesion compared to air oxidation [97].

Park and Seo et al. also reported that the surface modification of the CFs via ion-beam irradiation had a tremendous influence on the improvement of the interfacial adhesive strength between the CFs and matrix resins in CF-reinforced composites [101]. They studied the effects of the Ar^+-ion-beam irradiation on the physicochemical properties such as surface composition and mechanical interfacial properties of PAN-based high-strength CFs irradiated by Ar^+-ion-assisted reaction (IAR) method to form a strong CF/epoxy matrix interface in the final composite system. The FT-IR spectra of the as-received and 0.8-keV Ar^+-irradiated CFs are shown in Fig. 4.10a. A new carbonyl peak was formed at 1710 cm^{-1}, and the range of the –OH peak broadened. This affected the surface characteristics of the fibers, thereby resulting in an improvement in the interfacial adhesion between the fibers and matrix owing to an increase in the specific polarity and formation of H-bonding by the CFs [102]. Figure 4.10b shows the experimental τ_{max} values of the Ar^+-ion-beam-irradiated single CF in the epoxy resins. It shows that the Ar^+-ion-beam irradiation increases the τ_{max} of the single CF up to 1.0 keV of Ar^+-ion-beam energy followed by a small decrease. This was in good agreement with the results of the surface analyses. In other words, the acidic functional groups on the CF surfaces were more efficient and reactive in undergoing an interfacial reaction with the matrix materials [101].

4.4.2 Fluorination

Recently, several studies pertaining to the fluorination of CFs have been published. The degree of fluorination depends on the nature of the fibers and fluorination method used [115]. A large amount of the intercalated fluorine could be obtained using either well-graphitized structure or more extensive fluorine treatment at higher temperature and pressure. The reaction of fluorine and graphite at low

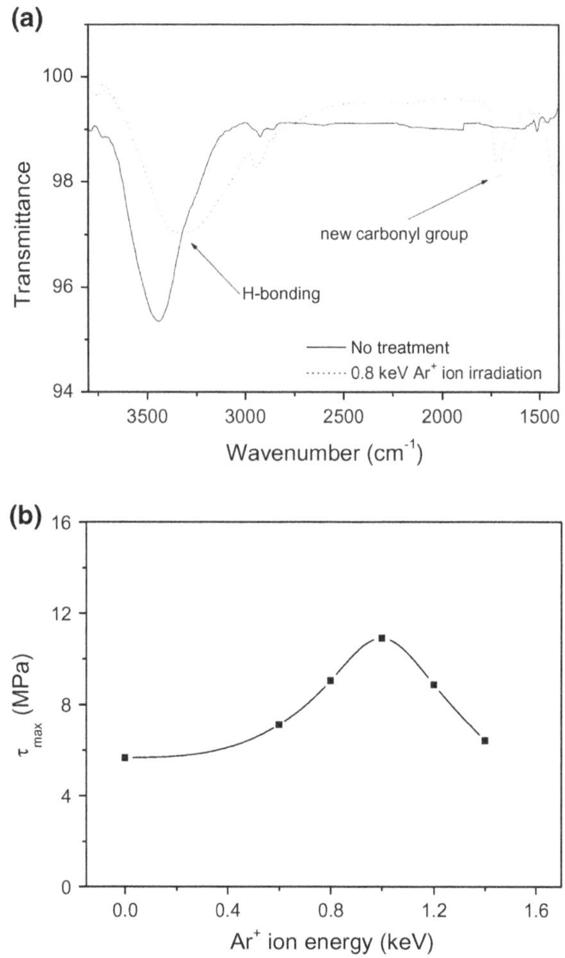

Fig. 4.10 **a** FT-IR transmittance spectra of as-received and 0.8-keV Ar^+-ion-beam-irradiated CFs; **b** maximum interfacial shear strength of Ar^+-ion-beam-irradiated CFs as a function of Ar^+-ion-beam intensity

temperature is kinetically hindered by the formation of covalent C–F bonds on the outer surface, which prevents further migration of fluorine into the interplanar space, and thus further fluorination [115–123]. The fluorination of the CFs, in particular the formation of graphite intercalation compounds or covalent and insulating graphite fluorides, $(CF)_n$ or $(C_2F)_n$, strongly influences the physical properties such as layer structure and tensile and interfacial strength [115].

Park et al. used different fluorination methods to modify the surface properties of the CFs. The relationship between the degree of fluorination and physicochemical properties of CFs was studied [115]. The fluorination reaction was performed in a batch reactor made of nickel with an outer electric furnace, as shown in Fig. 4.11a. After evacuation, fluorine (F_2) or fluorine and oxygen mixture (F_2/O_2 mixture) were introduced into the reactor at room temperature, and then, the reactor was heated to the treatment temperature. After the reaction, the sample was cooled down to the

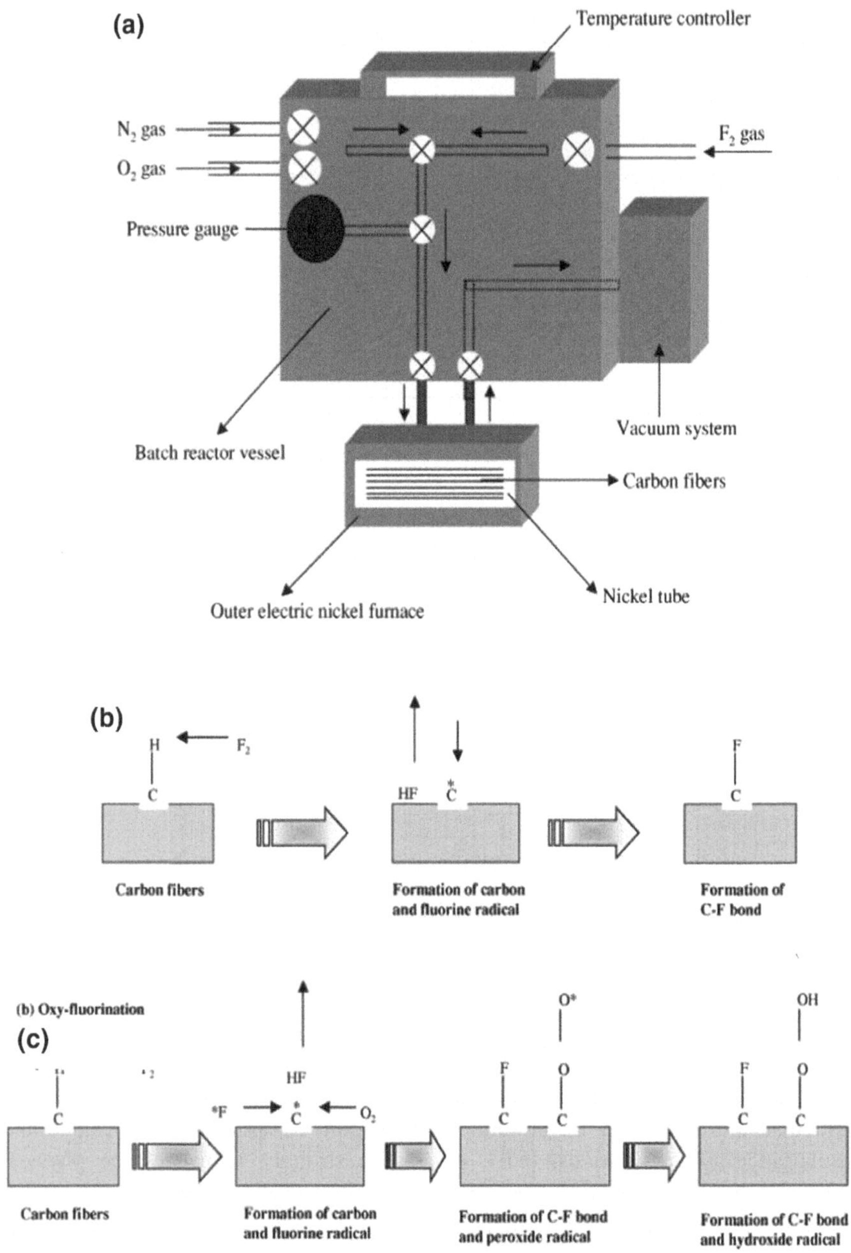

Fig. 4.11 **a** Schematic of fluorination reactor; **b**, **c** suggested mechanism of direct fluorination of CFs [115]

room temperature, and then, the reactive gases were purged from the reactor with nitrogen. In the case of the reaction at room temperature, the reactor was cooled and evacuated in a cooling bath prior to charging fluorine. The reactor was removed from the cooling bath after purging fluorine with nitrogen. The fluorine pressure was 0.1 MPa, and the nominal reaction time was 2 h at the treatment temperature. Generally, highly oriented graphite reacts with fluorine at high temperatures (above 700 °C), with fluorination of all the carbon layers, to yield carbon tetrafluoride. Depending on the temperature, fluorination leads to either graphite fluoride $(CF)x$ or various other graphite fluorides in the ideal case (addition of F_2 to all carbon–carbon double bonds) [115, 121]. Park et al. investigated the direct fluorination of the CFs using pure F_2 and F_2/O_2 mixtures at different heat-treatment temperatures [115]. A suggested mechanism of direct fluorination of the CFs is shown in Fig. 4.11b, c. Figure 4.12 shows the morphological properties of the fluorinated

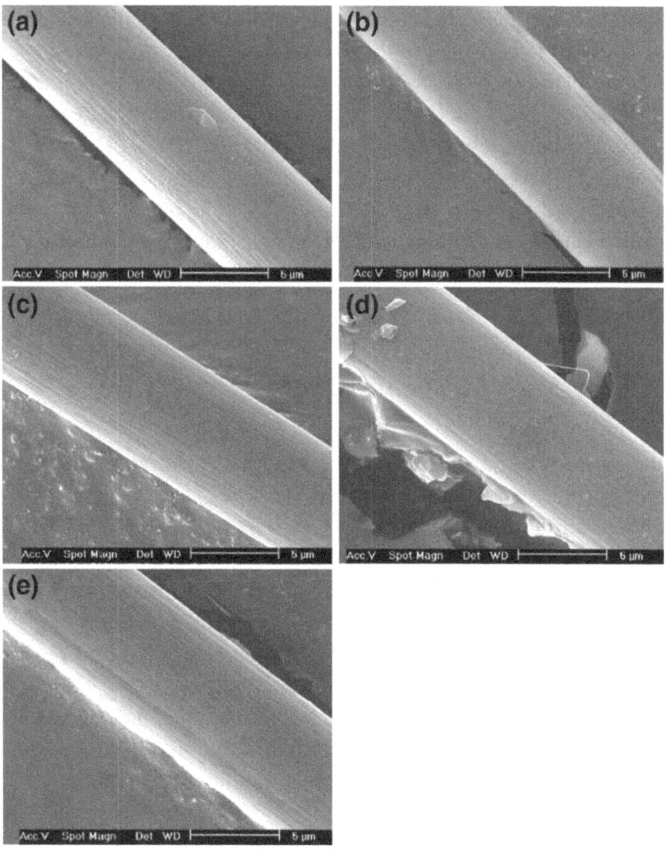

Fig. 4.12 SEM images of fluorinated CF surfaces. **a** C-0, **b** CF-20, **c** CF-125, **d** CFO-20, **e** CFO-125 (nonfluorinated (C-0)) and fluorinated fibers (CF-20, CF-125, CFO-20, and CFO-125) [115]

CFs. Mild fluorination (F_2 only) as well as aggressive fluorination (F_2/O_2 mixture) smoothened and changed the surface of the fibers, which indicated that the surface modification by fluorination took place, resulting in an enhancement of the mechanical properties of the fluorinated CFs.

4.4.3 Polymer Coating

Wei et al. reported the grafting of polyesters onto the surfaces of the vapor-grown carbon nanofibers (VGCF) by the anionic ring-opening alternating copolymerization of epoxides with cyclic acid anhydrides initiated by COOK groups introduced onto the VGCF surface [124]. In addition, the electrical properties of the conductive composites prepared from the polyester-grafted VGCF were investigated. They investigated the stability of the dispersion of poly(SO-alt-PAn)-grafted HDVGCF in THF. The untreated HDVGCF precipitated within 1 d as mentioned above. On the contrary, poly(SO-alt-PAn)-grafted HDVGCF yielded a stable dispersion in a solvent such as THF, suitable for grafted polyesters. The precipitation of poly (SO-alt-PAn)-grafted HDVGCF was scarcely observed even after 4 weeks. The grafted polyester chains were mainly grafted onto the edges of VGCF because of the presence of the carboxyl groups on the edges. However, it seemed that the grafting of the polyester onto the wall of the VGCF had also proceeded. Both polyester chains on the edges and wall of VGCF played an important role in the destruction the entanglement [123].

4.4.4 Grafting with Inorganic Materials

Recently, the surfaces of the CFs have been modified with inorganic materials such as silica, metals, and carbon nanomaterials. [124–127]. Deng et al. prepared the two-dimensional ultrathin mesoporous $NiCo_2O_4$ nanosheets on the CF paper (CFP) through a facile solvothermal method combined with postthermal treatment [124]. The well-interconnected ultrathin $NiCo_2O_4$ nanosheets directly grown on the carbon nanofibers would allow for easy diffusion of the electrolyte, shorten the transport path of ions and electrons, and accommodate the strain during cycling. The Ni–Co precursor nanosheets were uniformly grown on the carbon nanofibers forming a large-scale conformal coating. Moreover, the nanosheets were interconnected with each other to form a wall-like structure, which possessed good mechanical strength. After annealing at a relatively low temperature of 350 °C for 3 h with a slow heating rate of 1 °C min^{-1}, these Ni–Co precursor nanosheets could be easily converted to crystalline $NiCO_2O_4$. Importantly, the well-interconnected ultrathin nanosheet morphology was perfectly retained after the annealing treatment, which could be attributed to the robust support of the carbon nanofibers and the slow heating rate applied during the annealing [124].

Waller et al. reported a low-temperature hydrothermal process for the growth of the conformal coatings of highly crystalline $LiMn_2O_4$ directly onto a CF current collector, thereby completely eliminating the process steps and materials associated with the conventional tape casting approach (binders, solvents, and metal foils) [125]. The resulting electrodes tested at a rate of 1 °C showed an initial discharge capacity of 125 mAh g^{-1} and an average energy efficiency of 92.4% over 100 cycles. At the hydrothermal temperature of 100 °C, a conformal coating with a flower-like morphology was observed using SEM. The "flowers" had "petals," which were a few tens of nanometers thick. The coatings produced at 140 °C and above were coated with a layer of discrete multifaceted crystallites, over the size range 50–500 nm. These coatings showed thickness of approximately 2 μm, with nodules protruding beyond this threshold in certain parts, which were also observed in the flower-like coatings [125].

Du et al. prepared a novel and sensitive CF electrode (CFE) modified with graphene flowers and used to simultaneously determine ascorbic acid (AA), dopamine (DA) and uric acid (UA) [126]. They found that the layered-petal graphene flowers homogeneously bloomed on the surface of the CFE. Moreover, sharp and obvious oxidation peaks were found at the resultant electrode, in contrast with the CFE and glassy carbon electrode (GCE), for the oxidation of AA, DA, and UA. Also, the linear calibration plots for AA, DA, and UA were observed, respectively, over the ranges 45.4–1489.23, 0.7–45.21, and 3.78–183.87 μM. By simultaneously changing the concentrations of AA, DA, and UA, their oxidation peaks appeared at −0.05, 0.16, and 2.6 V, and the good linear response ranges were found to be 73.52–2305.53, 1.36–125.69, and 3.98–371.49 μM, respectively. In addition, the resultant electrode showed satisfactory results when employed for the determination of AA, DA, and UA in urine and serum samples [126].

Shazed et al. prepared the growing carbon nanotubes (CNTs) on the surface of the high-performance carbon fibers (CFs) [127]. The growth of the CNT on CF was conducted via floating catalyst chemical vapor deposition (CVD). CNT was vigorously grown on the surface of the CF at 800 and 700 °C. The high density of CNT was desirable owing to the purpose of the production of high-performance fiber-reinforced composites. However, the coatings were found to be more uniform with minimal clumping and agglomeration at 700 °C. It was deduced that under the given experimental conditions, higher temperature led to a higher degree of catalyst agglomeration, which affected the homogeneous growth of the CNT. Furthermore, a lower synthesis temperature was desirable to minimize the heating energy and protect the CF from being damaged by unnecessarily high heat levels. The growth CNTs were of the multiwalled type. After 30 min of reaction time, it could be seen that a coating with an apparent thickness of 3 μm had been established [127].

4.5 Sizing

The commercially available CFs are normally coated by a sizing layer on the surface, which usually is either a solution or emulsion consisting of polymeric components. The presence of sizing in the CFs/sizing and sizing/matrix interfaces plays an important role in controlling certain properties of the composites [128–130]. Conventional sizing such as film formers, emulsifiers, antistatic, and coupling agents is generally designed to protect the fiber surface and promote the adhesion between the fiber and matrix [131–137]. The sizing could alter the handling of the CFs, which includes protection, alignment, and wettability of fibers. The typical sizing materials include epoxy, polyester, nylon, urethane, and others. Table 4.8 lists the major components involved in typical sizing and, in brief terms, the major purpose of each component.

The application of sizing is normally termed either size or finish and could be achieved by:

- deposition from solution of a polymer
- deposition of a polymer onto fiber surface by electrodeposition
- deposition of a polymer onto fiber surface by electropolymerization
- plasma polymerization.

Zhang et al. prepared the CFs sized with three different amounts of the emulsifier in the sizing agent in order to improve the performances of CFs and interfaces of CF composites [128]. The sizing agents, E-1, E-2, and E-3, based on the three different amounts of the emulsifier contained therein, 10, 15, and 20 wt%, respectively, were used [128, 129]. The content of the sizing agent used for the CFs in the research is 1.5%. The sizing agent phase between the carbon fiber and resin was important; based on the results of the present study, the emulsifier content of the sizing agent affected the IFSS of the composites. The IFSS of the sizing agents, E-2- and E-3-sized carbon fiber composites, were enhanced by 4.6 and 8.6%, respectively, compared to the E-1-sized carbon fiber composites. For the E-3-sized carbon fiber/epoxy matrix system, with the same resin matrix and technological conditions, high emulsifier content of the sizing agent E-3 could appreciably help enhance the interface bonding strength. This could result in better mechanical performance than those of the other two composite systems [128].

Table 4.8 Components of typical sizing for fiberglass [133]

Component	Major function
Water	Carrier for sizing mixture
Coupling agent	Improves bond between matrix and reinforcement
Lubricant	Prevents mechanical damage to fibers
Film former	Holds fiber bundles together and coat fibers for protection
Wetting agent	Assists in wetting fibers through resin matrix and film former
Cross-linking agent	Cross-links to resin matrix or film former or both
Antistatic agent	Dissipates static charge that builds on fibers

Recently, sizing agents containing CNTs were used to size the glass fibers or fabrics in order to improve the mechanical or electrical properties of glass–fiber composites [134, 138]. Liu et al. disclosed a sizing approach to fabricate the CNT/CF hybrid fiber in detail by employing a sizing agent containing MWCNTs [138]. MWCNTs were chemically modified with an amine monomer and mixed with a thermoplastic polymer sizing agent. After the modification, the MWCNTs were uniformly dispersed in the sizing agent. Unsized CF tow was sized using the sizing agents containing MWCNT, while the CNT/CF hybrid fiber was fabricated. The MWCNT–COOH was reacted with phthalazinone-containing diamine (DHPZDA) in dimethyl formamide (DMF) at 130 °C followed by the addition of dicyclohexylcarbodiimide over 6 h under uninterrupted stirring to introduce the amine monomer on the surfaces of the MWCNTs (marked as MWCNT–DHPZDA). It was found that the CNTs were attached to the fiber surface along the outer edges of the grooves and a hierarchical structure was observed in all the images. It was also observed that there were wide and deep grooves along the axial direction of the CFs before and after sizing. This indicated that the PPEK absorption did not fill the grooves on the fiber surface because of the nanosized thickness of the sizing-agent layer [138].

In manufacturing companies, CFs are generally sized or coated with various polymers in order to protect the fibers and make handling easier after the surface oxidation treatment. Remarkably, it has been found that the sizing also plays an important role in improving the interfacial adhesion between the fibers and matrix [139]. Yuan et al. prepared an organic solvent-free polyamic acid (PAA) nanoemulsion by direct ionization of the solid PAA in deionized water, with an average particle size of 261 nm and zeta potential of −55.1 mV. The nanoemulsion was used for CF sizing to improve the interfacial adhesion between the CF and polyethersulfone (PES) [140]. A continuous and uniform PAA sizing layer was formed on the surfaces of CFs, and the surface energy of the CFs increased from 42.91 to 54.55 mN/m after the sizing treatment. The single-fiber pull-out testing was also performed, which showed a 47.9% increase in the interfacial shear strength (IFSS) of the CF/PES composites, from 33.6 to 49.7 MPa. The primary reasons for the improved interfacial adhesion were the increased van der Waals forces between the PES matrix and sizing layer, and the chemical bonding between the sizing layer and CF surface. Furthermore, the PAA sizing positively affected the interfacial adhesion of the CF/PES composites under hydrothermal conditions [140].

Cao investigated a heat-resistant sizing agent composed of thermoplastic polyimide and thermosetting epoxy for PAN-based CFs [141]. The sized fibers were reported to possess an improved wear resistance and 97% improvement in the interfacial shear strength compared to the unsized fibers. However, not all types of sizing could improve the interfacial adhesion between the fibers and matrix resins. Dilsiz found that the polyimide- and polyurethane-sized fibers in an epoxy matrix showed lower interfacial shear strength compared to the unsized fibers, from the single-fiber fragmentation test. This was attributed to the effect of the surface chemical changes on the fiber/matrix adhesion [142]. Interfacial adhesion was expected to be dependent on the surface properties of the fiber, including roughness, porosity, and functional groups; it is also dependent on the chemical characteristics of the matrix [143]. Dai et al. evaluated the influence of sizing on the interfacial adhesion of the CF/matrix [144]. Desized CFs (T300B and T700SC) were prepared by extracting the sizing on the fiber surface and comparing the CFs with those sized commercially. Also, the surface properties of the CFs, including the chemical compositions, surface free energy, acid–base interactions, dispersive components, and polar components, were experimentally characterized. Besides, the adhesion between the CFs and solid epoxy resins was also evaluated with respect to the IFSS, in order to determine the primary factors responsible for the fiber/matrix interfacial adhesion [141].

Recently, the use of nanoparticles such as nano-SiO_2, nano-Al_2O_3, and nano-TiO_2 for modifying the mechanical interfacial properties of composites has been reviewed by several researchers [145–150]. On the basis of the significant effects of sizing on the interfacial properties of the composites, we selected inorganic nano-SiO_2 to modify the emulsion sizing of CFs so as to increase the extent of the interfacial adhesion of the CFs/epoxy composites. Yang et al. used nano-SiO_2 particles to modify the epoxy emulsion sizing of the CFs in order to improve the interfacial properties of the CF-reinforced epoxy composites [150]. The results indicated that the modified sizing with nano-SiO_2 slightly increased the surface energy, number of hydroxyl functional groups, and surface roughness of CFs compared to the unmodified sizing. Consequently, the interfacial shear strength (IFSS) of the single-fiber composites and interlaminar shear strength (ILSS) of composites were enhanced. The SEM images of the fractured sections of the composites confirmed that the interfacial adhesion between the fibers and matrix improved after the sizing treatment with the nano-SiO_2-modified emulsion.

Toray Industries, Inc., of Japan is recognized as the worldwide leader in the CF industry, having pioneered the development of PAN-based CFs over 30 years ago. Toray's TORAYCA® brand of CFs is globally recognized for its outstanding performance, quality, and processing consistency, thereby making the TORAYCA® brand the preferred choice of designers, engineers, and fabricators worldwide. TORAYCA® CFs are critical for modern aviation and aerospace applications, including for the primary and secondary structures for Boeing and Airbus commercial aircrafts, International Space Station, satellites, rocket motor casings, and expendable launch vehicles such as Boeing Delta programs. TORAYCA® T800 CF is the only CF that has attained the certification from the FAA for the critical flight-path components for Boeing 777 [151]. TORAYCA fibers were treated with various sizing agents to enhance the handling and bonding characteristics with various resin systems. Listed below are the sizing types developed for the TORAYCA fibers. Not all the sizings were available with every fiber. The data available for the sizing of a particular fiber are listed in Tables 4.9 and 4.10 [151].

The importance of CF-based composites has rapidly increased since the 1960s. The CFs are primarily designed for high-strength structural applications and widely used in the aircraft, defense, catalyst, cell, and sports industries owing to their excellent properties. The wettability of the pristine CFs might negatively affect the

Table 4.9 Sizing agents of TORAYCA fibers [151]

Sizing type	Resin system compatibility
1	Epoxy
3	Epoxy
4	Epoxy, phenolic, BMI
5	General purpose: epoxy, phenolic, polyester, vinyl ester
6	Epoxy
F	Designed for vinyl ester, compatible with epoxy
9	No sizing fiber designation

Table 4.10 Designation of TORAYCA fibers [151]

Designation	Fiber length (mm)	Sizing	Sizing content	Packaging weight (kg)
T008-3	3	Epoxy	1	5
T008-6	6	Epoxy	1	5
T008-12	12	Epoxy	1	5
T010-3	3	No	0	5
T010-6	6	No	0	5
T010-12	12	No	0	5
T010-24	24	No	0	5

physical and mechanical properties of the resulting composites because of the surface incompatibility between the hydrophilic inorganic fillers of the carbon fiber and nonpolar resin [152–159]. Therefore, the main purpose of this chapter was to review the surface treatment and surface wettability of CF surfaces. Accordingly, a variety of surface-treatment methods have been applied to modify the surfaces of CFs to promote the interfacial adhesion and improve the properties of the composites. These methods include oxidation, electrochemical and plasma treatments, fluorination, and electron-beam-irradiation pretreatment.

References

1. W.Z. Nie, J. Li, Mech. Compos. Mater. **46**, 251 (2010)
2. S.W. Lee, H.Y. Lee, S.Y. Jang, S. Jo, H.S. Lee, W.H. Choe, S. Lee, Carbon **55**, 361 (2013)
3. H.M. Lee, K.M. Bae, H.R. Kang, K.H. An, H.G. Kim, B.J. Kim, Appl. Chem. Eng. **24**, 370 (2013)
4. J.D. Atkinson, Z. Zhang, Z. Yan, M.J. Rood, Carbon **54**, 444 (2013)
5. J.S. Marcuzzo, C. Otani, H.A. Polidoro, S. Otani, Mater. Res. **16**, 137 (2013)
6. H. Zhang, S. Wang, Y. Liu, Ceramics Int. (2013), Article in press
7. R.M. Alway-Cooper, D.P. Anderson, A.A. Ogale, Carbon **59**, 40 (2013)
8. W. Gao, G.J. Zhao, App. Mecha. Mater. **433–435**, 2003 (2013)
9. Z. Sun, Y. Yu, S. Pang, D. Du, Appl. Surf. Sci. **284**, 100 (2013)

10. Y. Yao, V. Velpari, J. Economy, J. Mater. Chem. A **1**, 12103 (2013)
11. U. Świetlik, B. Grzyb, K. Torchała, G. Gryglewicz, J. Machnikowski, Fuel Process. Technol. **119**, 211 (2014)
12. B.V. Kaludjerović, V.M. Jovanović, S.I. Stevanović, Ž.D. Bogdanov, Ultrasonics Sonochem. **21**, 782 (2014)
13. U.K. Fatema, F. Okino, Y. Gotoh, J. Mater. Sci. **49**, 1049 (2014)
14. M.C. Hung, S.Y. Yuan, S.I. Chang, J.W. Liao, T.H. Ko, C.L. Cheng, Carbon **68**, 628 (2014)
15. C.M. Du, D.W. Huang, H.X. Li, M.D. Xiao, K. Wang, L. Zhang, Z.Y. Li, T.F. Chen, J.M. Mo, D. Gao, Y.H. Huang, S.K. Liu, L. Yu, C.R. Zhang, Plasma Chem. Plasma Process. **33**, 65 (2013)
16. B. Jiang, J. Zheng, Q. Liu, M. Wu, Z. Yan, S. Qiu, Q. Xue, Z. Wei, H. Xiao, M. Liu, Chem. Eng. J. **215–216**, 969 (2013)
17. H. Ge, S. Li, H. Liu, D. Wang, J. Chen, J. Appl. Polym. Sci. **131**, 39843 (2014)
18. W. Lee, J.U. Lee, H.J. Cha, J.H. Byun, RSC Adv. **3**, 25609 (2013)
19. S. Zhang, W.B. Liu, L.F. Hao, W.C. Jiao, F. Yang, R.G. Wang, Comp. Sci. Technol. **88**, 120 (2013)
20. I. Giraud, S. Franceschi-Messant, E. Perez, C. Lacabanne, E. Dantras, Appl. Surf. Sci. **266**, 94 (2013)
21. L. Wenbo, Z. Shu, H. Lifeng, J. Weicheng, Y. Fan, L. Xiaofei, W. Rongguo, Polym. Comp. **34**, 1921 (2013)
22. F.R. Jones, J. Adhesion Sci. Technol. **24**, 171 (2010)
23. Z. Dai, F. Shi, B. Zhang, M. Li, Z. Zhang, Appl. Surf. Sci. **257**, 6980 (2011)
24. J. Li, C.L. Cai, Curr. Appl. Phys. **11**, 50 (2011)
25. L. Li, Surf. Interface Anal. **41**, 759 (2009)
26. B. Fathi, M. Esfandeh, A.K. Soltani, I.A. Amraei, Polym. Plastics Technol. Eng. **50**, 564 (2011)
27. Y. Lu, J. Chen, H. Cui, H. Zhou, Compos. Sci. Technol. **68**, 3278 (2008)
28. A. Rasheed, J.Y. Howe, M.D. Dadmun, P.F. Britt, Carbon **45**, 1072 (2007)
29. P. Morgan, in *CFs and Their Composites* (CRC Press, 2005)
30. D.D.L. Chung, in *CF Composites* (Butterworth-Heinemann, 1994)
31. E.G. Rakov, Russian. Chem. Rev. **70**, 827 (2001)
32. J.C. Goan, L.A. Joo, G.E. Sharpe, surface treatment for graphite fibres. 27th Ann. Tech. Conf. Reinf. Plast. Compos. Inst. **27**(21-E) (1972)
33. J.B. Donnet, P. Ehrburger, Carbon **15**, 143 (1977)
34. W. Shen, H. Wang, R. Guan, Z. Li, Colloids Surf. A **331**, 263 (2008)
35. Y.G. Wang, G.S. Wu, X.T. Cailiao, New Carbon Mater. **22**, 88 (2007)
36. J. Sun, C. He, S. Zhu, Q. Wang, J. Appl. Polym. Sci. **106**, 470 (2007)
37. Y.Z. Wan, Y.L. Wang, Y. Huang, H.L. Luo, J. Mater. Sci. **40**, 3355 (2005)
38. I.M.K. Ismail, W.C. Hurley, Carbon **30**, 419 (1992)
39. Y. Tong, X. Wang, H. Su, L.X. Corros, Science **53**, 2484 (2011)
40. Z.L. Liu, Y.W. Jin, H. Su, L.H. Xu, Y.J. Tong, J. Beijing Univ. Chem. Technol. **37**, 78 (2010)
41. Z.M. Wang, N. Yamashita, Z.X. Wang, K. Hoshinoo, H. Kanoh, J. Colloid Interface Sci. **276**, 143 (2004)
42. Y.G. Wang, G.S. Wu, New Carbon Mater. **45**, 1380 (2007)
43. L.Y. Zheng, L.X. Zhao, J.J. Zhang, Wear **262**, 1026 (2007)
44. H.J. Wang, P.Z. Gao, Z.H. Jin, Mater. Lett. **59**, 486 (2005)
45. H. Rong, Z. Ryu, J. Zheng, Y. Zhang, Carbon **40**, 2291 (2002)
46. Y. Takahashi, H. Fujita, A. Sakoda, Adsorption **19**, 143 (2013)
47. M.K. Seo, S. J. Park Curr. Appl. Phys. **10**, 241 (2010)
48. E.J. Poila, J.L. Serra, J. Am. Cer. Soc. **94**, 2185 (2011)
49. F. Lamouroux, X. Bourrat, R. Naslain, J. Sevely, Carbon **31**, 1273 (1993)
50. P.L. Walker Jr., F. Rusinko Jr., L.G. Austin, Adv. Catal. **XI**, 133 (1959)
51. S. Hosokai, K. Kishimoto, K. Norinaga, C.Z. Li, J.I. Hayashi, Ener. Fuels **24**, 2900 (2010)

52. Z. Jin, Z. Zhang, L. Meng, Mater. Chem. Phys. **97**, 167 (2006)
53. S.R. Kelemen, H. Freund, Energy Fuels **2**, 11 (1988)
54. M.A.A. Zaini, Y. Amano, M. Machida, J. Hazardous Mater. **180**, 552 (2010)
55. M. Li, H. Liu, Y. Gu, Y. Li, Z. Zhang, Appl. Surf. Sci. **288**, 666 (2014)
56. B.K. Kim, S.K. Ryu, B.J. Kim, S.J. Park, J. Colloid Interface Sci. **302**, 695 (2006)
57. S.J. Park, J.S. Shin, J.W. Shim, S.K. Ryu, J. Colloid Interface Sci. **275**, 342 (2006)
58. K.R. Ko, S.K. Ryu, S.J. Park, Carbon **42**, 1864 (2004)
59. S.J. Park, Y.S. Jang, J.W. Shim, S.K. Ryu, J. Colloid Interface Sci. **260**, 259 (2003)
60. S.J. Park, B.J. Kim, Mater. Sci. Eng., A **408**, 269 (2005)
61. S.J. Park, S.Y. Jin, J. Colloid Interface Sci. **286**, 417 (2005)
62. S.H. Han, H.J. Oh, S.S. Kim, Comp. Part B Eng. **60**, 98 (2014)
63. S.Y. Lee, S.J. Park, J. Colloid Interface Sci. **389**, 230 (2013)
64. B.J. Kim, W.K. Choi, M.K. Um, S.J. Park, Surf. Coat. Technol. **205**, 3416 (2011)
65. B.J. Kim, S.J. Park, J. Colloid Interface Sci. **325**, 297 (2008)
66. S.J. Park, Y.S. Jang, J. Colloid Interface Sci. **261**, 238 (2003)
67. S.J. Park, J.S. Shim, J. Colloid Interface Sci. **264**, 39 (2003)
68. S.J. Park, B.J. Kim, J. Colloid Interface Sci. **282**, 124 (2005)
69. S.Y. Lee, S.J. Park, Int. J. Hydrogen Energy **35**, 6757 (2010)
70. S.J. Park, S.Y. Lee, Int. J. Hydrogen Energy **35**, 13048 (2010)
71. J.R. Choi, Y.S. Lee, S.J. Park. J. Ind. Eng. Chem. (2013), Article in press
72. S.J. Park, M.K. Seo, Chapter 10—Comprehension of Nanocomposites, Interface Sci. Technol. **18**, 777 (2011)
73. S.Y. Lee, S.J. Park, J. Solid State Chem. **194**, 307 (2012)
74. N. Uddin, R.R. Kalyankar, Inter. J. Polym. Sci. **201**, 963549 (2011)
75. A. Perrard, L. Retailleau, R. Berjoan, J. Joly, Carbon **50**, 2226 (2012)
76. B. Fathi, M. Esfandeh, A.K. Soltani, I.A. Amraei, Polym. Plast. Technol. Eng. **50**, 564 (2011)
77. G. Tang, D. Chang, G. Wei, H. Lu, W. Yan, B. Liu, Polym. Plast. Technol. Eng. **51**, 329 (2012)
78. Z. Xu, L. Chen, Y. Huang, J. Li, X. Wu, X. Li, Y. Jiao, Eur. Polym. J. **44**, 494 (2008)
79. H.K. Shin, M. Park, P.H. Kang, H.S. Choi, S.J. Park. J. Ind. Eng. Chem. (2014), Article in press
80. S.J. Park, J.S. Kim, Carbon **39**, 2011 (2001)
81. S.J. Park, B.J. Park, S.K. Ryu, Carbon **37**, 1223 (1999)
82. D.I. Jang, S.J. Park, Fuel **102**, 439 (2012)
83. S.J. Park, Y.M. Kim, Mater. Sci. Eng., A **391**, 121 (2005)
84. S.J. Park, J.B. Donnet, J. Colloid Interface Sci. **206**, 29 (1998)
85. B.J. Kim, S.J. Park, Int. J. Hydrogen Energy **36**, 648 (2011)
86. J. Gulyás, E. Földes, A. Lázár, B. Pukánszky, Compos. Part A **32**, 353 (2001)
87. K.G. Gallagher, T.F. Fuller, Phys. Chem. Chem. Phys. **11**, 11557 (2009)
88. S.J. Park, Y.M. Kim, J. Colloid Interface Sci. **278**, 276 (2004)
89. J.W. Shim, S.J. Park, S.K. Ryu, Carbon **39**, 1635 (2001)
90. S.J. Park, J.S. Kim, J. Colloid Interface Sci. **239**, 380 (2001)
91. G.J. Ehlert, H.A. Sodano. Int. SAMPE Tech. Conf. 2010 SAMPE Fall Technical Conference and Exhibition; Salt Lake City (2010)
92. T.C. Chang. J. Ind. Technol. **15**, 1 (1998–1999)
93. C.I. Su, R.S. Yeh, C.L. Wang, Text. Res. J. **74**, 776 (2004)
94. M.A. Montes-Morán, A. Martínez -Alonso, J.M.D. Tascón, R.J. Young. Compos. Part A **32**, 361 (2001)
95. S.J. Park, B.J. Kim, J. Colloid Interface Sci. **275**, 590 (2004)
96. B.J. Kim, S.J. Park, J. Colloid Interface Sci. **315**, 791 (2007)
97. Y.Z. Wan, Y.L. Wang, Y. Huang, H.L. Luo, J. Mater. Sci. **40**, 3355 (2005)
98. S.J. Park, K.S. Cho, C.G. Choi, J. Colloid Interface Sci. **258**, 424 (2004)
99. M.K. Seo, S.J. Park, S.K. Lee, J. Colloid Interface Sci. **285**, 306 (2005)

100. G.Y. Heo, S.J. Park, Carbohydr. Polym. **88**, 562 (2012)
101. S.J. Park, M.K. Seo, K.Y. Rhee, Carbon **41**, 592 (2003)
102. T. Ramanathan, A. Bismarck, E. Schulz, K. Subramanian, Compos. Sci. Technol. **61**, 599 (2001)
103. A. Bismarck, M.E. Kumru, J. Springer, J. Colloid Interface Sci. **210**, 60 (1999)
104. S.J. Park, M.K. Seo, J.R. Lee, J. Colloid Interface Sci. **268**, 127 (2003)
105. M.S. Dresselhaus, M. Endo, in *In Graphite Intercalation Compounds II*, ed. by H. Zabel, S. A. Solin (Springer, New York, 1992), p 347
106. T. Nakajima, J. Fluo. Chem. **105**, 229 (2000)
107. R.B. Mathur, V. Gupta, O.P. Bahl, A. Tressaud, S. Flandrois, Syn. Met **114**, 197 (2000)
108. N. Bartlett, T. Mallouk, F. Okino, G. Rosenthal, J. Verniolle, J. Fluo. Chem. **23**, 409 (1983)
109. S. Guruvenket, G.R.S. Iyer, L. Shestakova, P. Morgen, N.B. Larsen, G.M. Rao, Appl. Surf. Sci. **254**, 5722 (2008)
110. N. Dilsiz, J. Adhesion. Sci. Technol. **14**, 975 (2000)
111. Z. Xu, Y. Huang, C. Min, L. Chen, L. Chen, Raid. Phys. Chem. **79**, 839 (2010)
112. T.D. Burchell, W.P. Eatherly, J. Nucl. Mater. **179**, 205 (1991)
113. S. Blazewicz, J. Piekarczyk, J. Chlopek, J. Bloki, J. Michalowski, M. Stodulski, P. Zychowski. ibid. **40**, 721 (2002)
114. S. Tiwari, J. Bijwe, S. Panier, Wear **271**, 2184 (2011)
115. S.J. Park, M.K. Seo, Y.S. Lee, Carbon **41**, 723 (2003)
116. A. Tressaud, V. Gupta, C. Guimon, F. Moguet, Mater. Sci. Eng., B **30**, 61 (1995)
117. J. Yu, L. Meng, D. Fan, C. Zhang, F. Yu, Y. Huang, Comp. Part B Eng. **60**, 261 (2014)
118. A. Tressaud, B. Chevalier, L. Piraux, M. Cassart, T. Nakajima, *Fluorine–carbon and fluoride–carbon materials* (Dekker, New York, 1995)
119. H. Touhara, F. Okino, Carbon **38**, 241 (2000)
120. E.Z. Kurmaev, A. Moewes, D.L. Ederer, H. Ishii, K. Seki, M. Yanagihara, F. Okino, H. Touhara, Phys. Lett. A **288**, 340 (2001)
121. S.J. Park, M.S. Cho, Carbon **38**, 1053 (2000)
122. E. Jeong, J. Kim, S.H. Cho, Y.S. Bae, Y.S. Lee, J. Fluorine Chem. **132**, 291 (2011)
123. G. Wei, K. Fujiki, H. Saitoh, K. Shirai, N. Tsubokawa, Polym. J. **36**, 316 (2004)
124. F. Deng, L. Yu, G. Cheng, T. Lin, M. Sun, F. Ye, Y. Li, J. Power. Sourc. **251**, 202 (2014)
125. G.H. Waller, S.Y. Lai, B.H. Rainwater, M. Liu, J. Power Sourc. **251**, 411 (2014)
126. J. Du, R. Yue, F. Ren, Z. Yao, F. jiang, P. Yang, Y. Du. Biosens. Bioelectro. **53**, 220 (2014)
127. M.A. Shazed, A.R. Suraya, S. Rahmanian, M.A. Mohd, Salleh. Mater. Des. **54**, 660 (2014)
128. R.L. Zhang, Y.D. Huang, L. Liu, Y.R. Tang, D. Su, L.W. Xu, Appl. Surf. Sci. **257**, 3519 (2011)
129. Y.D. Huang, R.L. Zhang, L. Liu, D. Su, in *A method of preparation the waterborne epoxy resin sizing agent for the carbon fiber*, Patent, CN 201010300131.3, 2010
130. S.M. Goushegir, J.F. dos Santos, S.T. Amancio-Filho, Mater. Design **54**, 196 (2014)
131. S.J. Park, T.J. Kim, J. Appl. Polym. Sci. **80**, 1439 (2001)
132. J.F. Feller, Y. Grohens, Compos. Part A-Appl. S. **35**, 1 (2004)
133. A.B. Strong, in *The Magic of Composite Performance* (Brigham Young University)
134. A. Warrier, A. Godara, O. Rochez, L. Mezzo, F. Luizi, L. Gorbatikh, S.V. Lomov, A.W. VanVuure, I. Verpoest, Compos. Part A: Appl. Sci. Manuf. **41**, 532 (2010)
135. L.M. Gao, T.W. Chou, E.T. Thostenson, Z.G. Zhang, Carbon **48**, 3788 (2010)
136. L. Gao, T.-W. Chou, E.T. Thostenson, A. Godara, Z. Zhang, L. Mezzo, Carbon **48**, 2649 (2010)
137. L.M. Gao, T.W. Chou, E.T. Thostenson, Z.G. Zhang, M. Coulaud, Carbon **49**, 3382 (2011)
138. W. Liu, S. Zhang, L. Hao, F. Yang, W. Jiao, X. Li, R. Wang, Appl. Surf. Sci. **284**, 914 (2013)
139. J. Berg, F.R. Jones, Compos. Part A **29**, 1261 (1998)
140. H. Yuan, S. Zhang, C. Lu, S. He, F. An, Appl. Surf. Sci. **279**, 279 (2013)
141. M. Guigon, E. Klinklin, Composites **25**, 534 (1994)
142. N. Dilsiz, J.P. Wightman, Colloids Surf. A **164**, 325 (2000)

143. J. Ittipol, in *Scanning Tunneling Microscopy (STM) and Atomic Force Microscopy (AFM) Studies of CFs and Interfacial Bonding in Polyimide Composites*, A Dissertation Presented to the Graduate Faculty of the University of Akron. UMI Company, 1996
144. Z. Dai, F. Shi, B. Zhang, M. Li, Z. Zhang, Appl. Surf. Sci. **257**, 6980 (2011)
145. S.A. Meguid, Y. Sun, Mater. Design **25**, 289 (2004)
146. K.P. Gadkaree, J. Mater. Sci. **27**, 3827 (1992)
147. M. Hussian, A. Nakahira, K.M. Niihara, Mater. Lett. **26**, 185 (1996)
148. F.H. Su, Z.Z. Zhang, W.M. Liu, Mater. Sci. Eng., A **392**, 359 (2005)
149. Z.Z. Zhang, F.H. Su, K. Wang, W. Jiang, X.H. Men, W.M. Liu, Mater. Sci. Eng., A **404**, 251 (2005)
150. Y. Yang, C. Lu, X. Su, X. Wang, J. Mater. Sci. **42**, 6347 (2007)
151. http://www.torayca.com/en/techref/index.html
152. H. Guo, Y.D. Huang, L. Liu, X.H. Shi, Mater. Design **31**, 1186 (2010)
153. X. Cao, Y.F. Wen, S.H. Zhang, Y.G. Yang, New Carbon Mater. **4**, 337 (2006)
154. S. Wu, *Polymer Interface and Adhesion* (Marcel Dekker Inc, New York, 1982)
155. T.H. Cheng, J. Zhang, S. Yumitori, F.R. Jones, C.W. Anderson, Composites **25**, 661 (1994)
156. S. Yumitori, D. Wang, F.R. Jones, Composites **25**, 698 (1994)
157. S.L. Gao, E. Mäder, S.F. Zhandarov, Carbon **42**, 515 (2004)
158. A. Paipetis, C. Galiotis, Compos. Part A **27**, 755 (1996)
159. R.L. Zhang, Y.D. Huang, N. Li, L. Liu, D. Su, J. Appl. Polym. Sci. **125**, 425 (2012)

Chapter 5
Testing of Carbon Fibers and Their Composites

Abstract In this chapter, we will introduce an evaluation of carbon fibers and their composites with various analyses. The major types of methods of characterization techniques that work best for carbon fibers and their composite materials to understand their properties such as chemical, physical, and mechanical properties and advanced characterization methods for surface or morphology analyses such as transmission electron microscopy, scanning probe microscopy, and atomic force microscopy will be described. In addition, a concise summary of all the major characterization methodologies used on composite materials will be presented.

5.1 Introduction

Composite materials have made their way into every aspect of the modern technological society, particularly in applications requiring high strength and low weight, specifically in the aerospace industry. Because composites are hybrid heterogeneous materials, they can be difficult to characterize by a single methodology. In this chapter, (1) the major types of methods of characterization techniques that work best for composite materials to understand their properties such as chemical, physical, and mechanical properties and (2) advanced characterization methods for surface or morphology analyses such as transmission electron microscopy, scanning probe microscopy, and atomic force microscopy will be described. Finally, a concise summary of all the major characterization methodologies used on composite materials will be presented.

5.2 Evaluation of Carbon Fibers

5.2.1 Introduction

In this section, we describe the techniques and methods that are generally used to characterize the properties of carbon fibers for application in composite materials. Sophisticated experimental techniques are required for fiber characterization, and

S.-J. Park, *Carbon Fibers*, Springer Series in Materials Science 210,
https://doi.org/10.1007/978-981-13-0538-2_5

test laboratories must be well equipped for measuring the fiber properties. It is also recognized that in many cases, the measurement of a fiber property that manifests itself in the reinforced composite can best be accomplished by measuring the properties of the composites.

5.2.2 Elemental Analysis

In elemental analysis, a sample of some material (e.g., soil, waste or drinking water, bodily fluids, minerals, and chemical compounds) is analyzed for its elemental and occasionally, the isotopic composition. A variety of quantitative wet gravimetric and spectroscopic chemical analysis techniques may be applied to analyze the compositions of and the presence of trace elements in fibers [1–4].

A suitable standardized method for carbon and hydrogen analysis, modified to handle carbon and polymeric fibers, is provided by the ASTM test method D 3178 [5]. Carbon and hydrogen concentrations are determined by burning a preweighed quantity of the sample in a closed system and fixing the products of combustion in an absorption train after complete oxidation and purification from interfering substances.

Carbon and hydrogen concentrations are expressed as percentages of the total dry weight of the fiber. The ASTM test method D3174 [6] describes a related test in which metallic impurities may be determined by the analysis of the ash residue.

Li et al. [2] investigated the feasibility of activating vapor grown carbon fibers (VGCFs) with supercritical fluids (SCFs). The quantitative analysis of the VGCFs by elemental analysis is shown in Table 5.1. The original VGCF (VGCF0) contains highly pure carbon-containing material, which consists of 97.3% carbon, 0.08% hydrogen, and 0.22% nitrogen by weight. All the treated VGCFs showed the absence of any significant changes in terms of hydrogen and nitrogen; however, in VGCF3, 4, and 5, other elements such as oxygen or contaminants from the corrosion of the reactor were present in significant amounts.

5.2.3 X-Ray Photoelectron Spectroscopy

X-ray photoelectron spectroscopy (XPS), also known as electron spectroscopy for chemical analysis (ESCA), is the most widely contemporary surface

Table 5.1 Elemental analysis of VGCFs [2]

Element	VGCF0	VGCF1	VGCF2	VGCF3	VGCF4	VGCF5
C (%)	97.32	97.61	97.53	64.78	71.46	92.59
H (%)	0.08	0.01	0.01	0.26	0.70	0.01
N (%)	0.22	0.18	0.15	0.23	0.15	0.09
Others (%)	2.38	2.20	2.31	34.73	27.69	7.31

characterization method [7, 8]. A large amount of information can be acquired from each spectrum (Table 5.2), and the technique is flexible enough to be used on a large variety of sample types.

Each atom on the surface of a material (except for hydrogen) consists of valence electrons, which are involved in chemical bonding, in addition to core electrons. These core electrons possess a unique binding energy, which is characteristic of the type of atom to which it is bound. By analyzing the binding energies of the electrons and the peak areas, quantitative elemental surface analysis is possible [2].

Because the electrons can only travel a short distance through the sample without undergoing inelastic collisions resulting in a drastic loss of energy, XPS is considered to be highly surface sensitive. Usually, only 50–100 Å of the sample surface is analyzed using this technique [2, 8].

Surface analysis by XPS begins by placing the sample in an ultrahigh vacuum environment ($\sim 10^{-10}$ Torr) and then irradiating the material with a source of low-energy X-rays. If the frequency of the excitation X-rays is greater than the

Table 5.2 Information obtained by XPS from Ratner and Castner [7]

In the outermost 10 nm of a surface, XPS can provide:

- ✓ Identification of all elements (except H and He) present at concentrations greater than 0.1 at%
- ✓ Semiquantitative determination of the approximate elemental surface composition (with an error $< \pm 10\%$)
- ✓ Information about the molecular environment (such as oxidation state and bonding atoms)
- ✓ Information about aromatic or unsaturated structures from shake-up (p*® p) transitions
- ✓ Identification of organic groups using derivatization reactions
- ✓ Nondestructive elemental depth profiles 10 nm into the sample and surface heterogeneity assessment using (1) angular-dependent XPS studies and (2) photoelectrons with differing escape depths
- ✓ Destructive elemental depth profiles several hundred nanometers into the sample using ion etching (for organics)

binding energy of each element, photoemission will occur. Schematics of the processes which occur during XPS are shown in Figs. 5.1 and 5.2 [7, 8].

The resulting photoelectrons are emitted from the surface with kinetic energy (E_k) measured by a hemispheric analyzer. Using the incident X-ray energy (hv), the binding energy (E_b) is calculated using Einstein's relationship given in 5.1 [3, 7, 8], where ϕ is the work function of the spectrometer.

$$E_b = hv - E_k - \phi \tag{5.1}$$

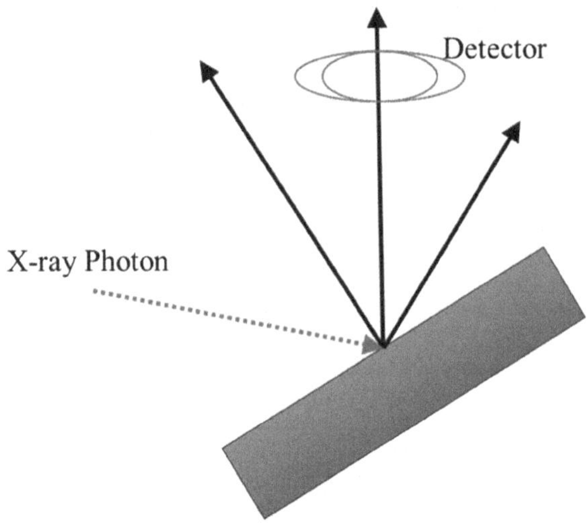

Fig. 5.1 Schematic representing processes occurring during XPS

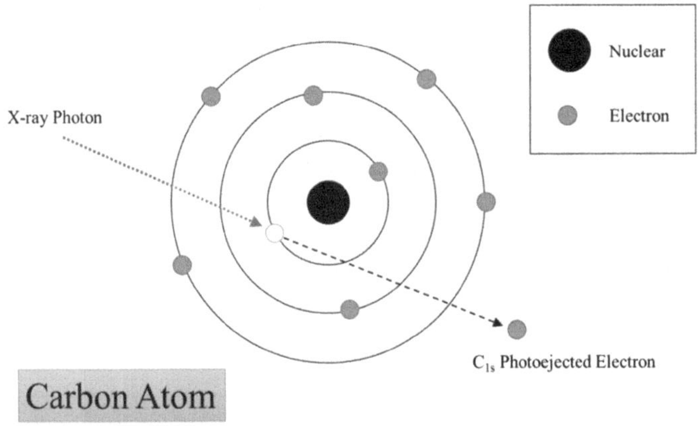

Fig. 5.2 Schematic of a carbon atom undergoing photoelectron emission [3]

The photoelectrons are then separated according to their energy, counted, and related to the atomic and molecular environment from which they were ejected. A spectrum of the number of emitted electrons versus binding energy is obtained.

Specific electrons of an atom have certain binding energies; however, the chemical environment of the atom can create variations in the values. These changes in the energy values, or chemical shifts, represent the presence of covalent or ionic bonds between atoms and help deduce the chemical structure of the material surface. Some common binding energy peak assignments for the carbon 1 s peak and the oxygen 1 s peak are seen in Tables 5.3 and 5.4 [3, 7, 8].

Various researchers have explored curve-fitting approaches with a vast majority of the studies concentrating on core spectrum studies. Over 250 papers have addressed the curve-fitting area, and hence, only a brief description of the type of information that can be obtained by XPS will be presented here. A complete review of the curve-fitting approach can be found in papers referenced by Sherwood [1].

Wang and Sherwood [9] showed that XPS can be used to reveal the interfacial chemistry in the carbon fiber matrix system. The use of propionaldehyde under acidic conditions as a coupling agent is seen to yield an interfacial chemical reaction

Table 5.3 Common binding energy assignments for carbon 1 s peak from Ratner and Castner [3]

Functional group		Binding energy (eV)
Hydrocarbon	C–H, C–C	285.0
Amine	C–N	286.0
Alcohol, Ether	C–O–H, C–O–C	286.5
Carbonyl	C=O	288.0
Amide	N–C=O	288.2
Acid, Ester	O–C=O	289.0
Urea	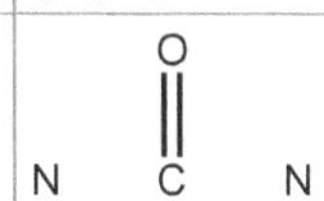	289.2
Carbamate	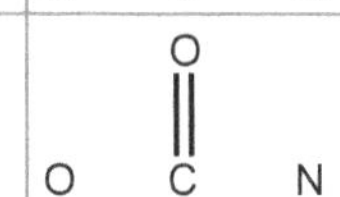	289.9
Carbonate	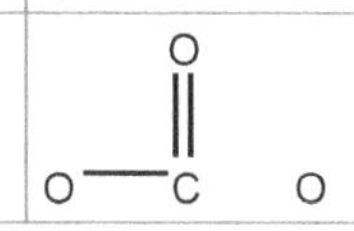	290.3

Table 5.4 Common binding energy assignments for oxygen 1 s peak from Ratner and Castner [3]

Functional group		Binding energy (eV)
Carbonyl	C=O, O–C=O	532.2
Alcohol, Ether	C–O–H, C–O–C	532.8
Ester	C–O–C=O	533.7

that can be interpreted to correspond to acetal cross-linking. The core and valence band XPS data can be interpreted by ab initio Hartree–Fock calculations. Air oxidation monitored by thermogravimetric analysis shows that the cross-linked samples exhibit enhanced oxidation resistance.

In Fig. 5.3, the XPS profile of the C_{1s} region of a carbon fiber electrochemically oxidized in phosphoric acid and subsequently treated with acidified PAL is presented. The separations between the component peaks are compared with the separations calculated from an ab initio calculation [9].

Previous work by researchers on the highly resolved core carbon 1 s spectrum of untreated carbon fibers showed the different bonds of carbon, where the main component peak (the lowest binding energy peak) corresponded to the carbon fiber and C=C bonds. Other carbon bonds were identified by higher binding energy features corresponding to oxidation, i.e., alcohol (C–OH), carbonyl (C=O), and carboxyl (COO) and to amino C–N (Fig. 5.4).

5.2.4 X-Ray Diffraction

X-ray diffraction (XRD) is a versatile nondestructive technique, which reveals detailed information about the chemical composition and crystallographic structure of natural and manufactured materials. A crystal lattice is a regular three-dimensional distribution (e.g., cubic and rhombic) of atoms in space. The lattices are arranged such that they form a series of parallel planes separated from one another by distance d, which varies with the nature of the material. For any crystal, planes exist in a number of different orientations, each with its own specific d-spacing.

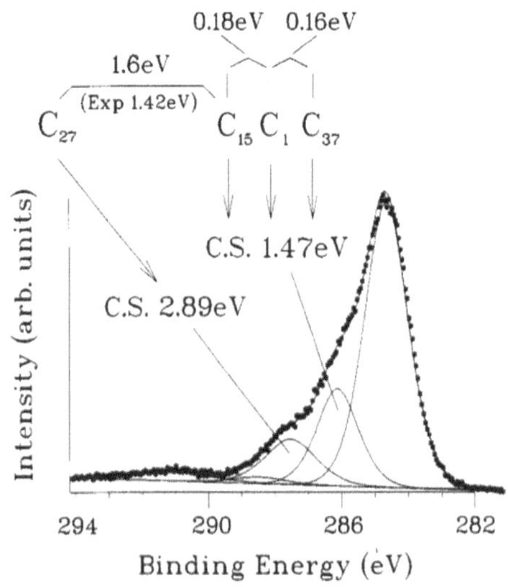

Fig. 5.3 XPS profile of C_{1s} region of a carbon fiber electrochemically oxidized in phosphoric acid [9]

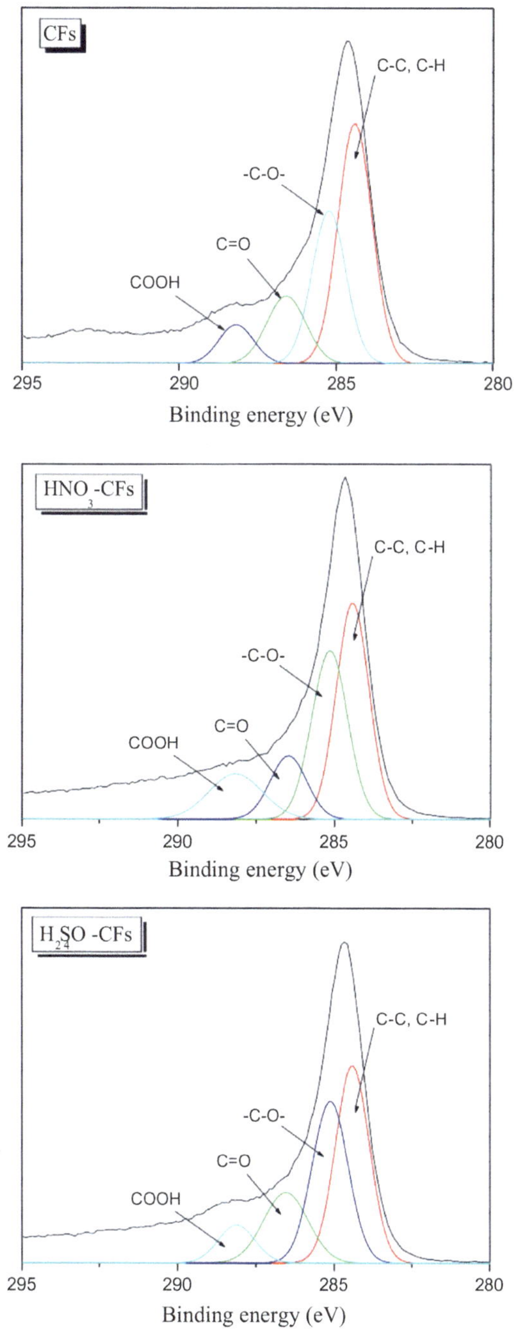

Fig. 5.4 High-resolution carbon 1 s peak of untreated carbon fibers

By varying the angle theta, the Bragg's Law conditions are satisfied by different d-spacings in polycrystalline materials. Plotting the angular positions and intensities of the resultant diffracted peaks of radiation produces a pattern, which is characteristic of the sample. When a mixture of different phases is present, the resultant diffractogram is formed by the addition of the individual patterns.

Using the principles of XRD, a wealth of structural, physical, and chemical information about the material investigated can be obtained. A host of XRD techniques is available for various classes of materials, each revealing its own specific details of the sample studied.

Iwashita et al. [10] investigated the application of a standard procedure of XRD measurements on carbon materials. The various peaks include the one near 29° corresponding to the 002 reflection, near 57° corresponding to the 004 reflection, the one near 75° corresponding to the 110 reflection, and the one near 89° corresponding to the 112 reflection, as shown in Fig. 5.5, respectively.

5.2.5 *Raman Spectroscopy*

Raman spectroscopy is a laser-based method using inelastically scattered light to investigate the vibrational, rotational, and other low-frequency modes of interaction between molecules. Raman bands are sharp peaks in the spectra that can be associated with vibration modes at the molecular level. Early studies on graphitic

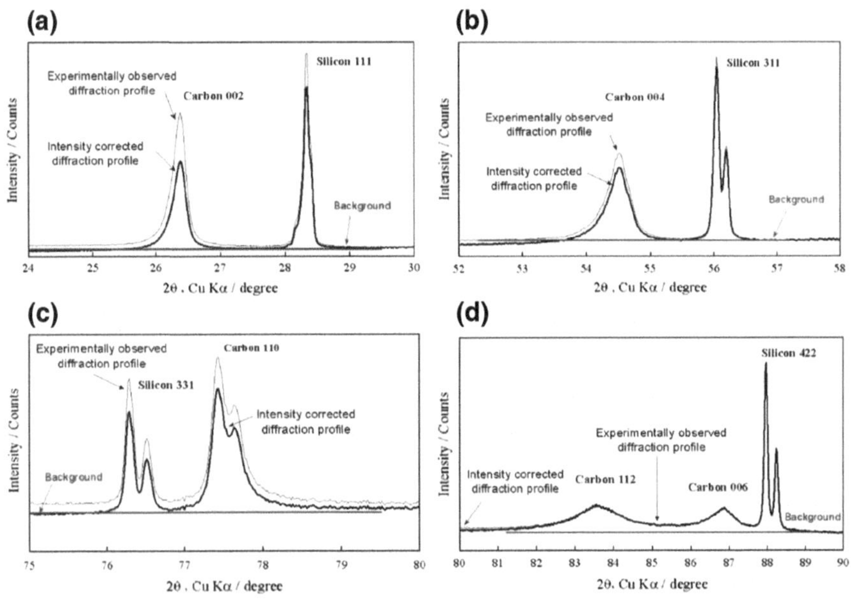

Fig. 5.5 XRD patterns of carbon materials [10]

materials revealed that a Raman band at ~1585 cm^{-1} can be related to C–C vibrations in graphite and is present in all carbon fibers. Another band in polycrystalline graphite was found at ~1330 cm^{-1} and can be related to the boundaries of the graphite crystals. Thus, this peak could be related to the particle size and structural disorder effects (Figs. 5.6 and 5.7), [11, 12].

5.2.6 Auger Electron Spectroscopy

An alternative method for chemical surface analysis is Auger electron spectroscopy (AES). During bombardment of the surface under investigation with a beam of electrons in the energy range 1–5 keV, vacancies are created in the core level. These vacancies represent excited ions in the surface region, which may undergo de-excitation, resulting in the creation of Auger electrons. Each electron has energy characteristic to a specific element. By analyzing all back-scattered Auger electrons

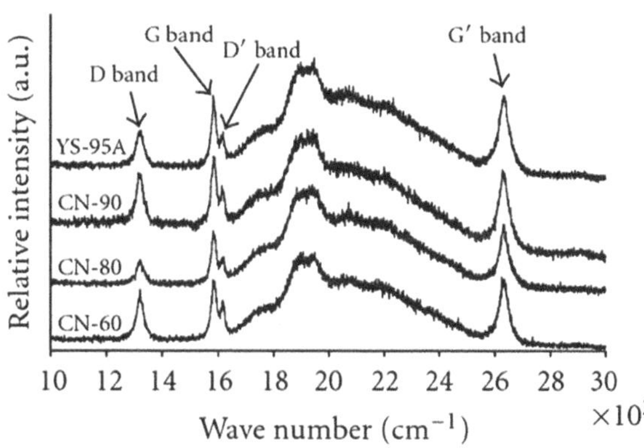

Fig. 5.6 Raw spectra with D, G, D′, and G′ Raman bands identified in pitch-based Nippon Graphite Fibers using an excitation wavelength of 752 nm [11]

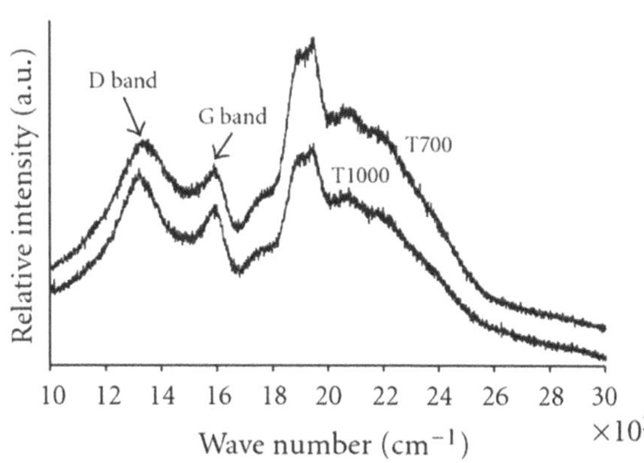

Fig. 5.7 Raw spectra with D and G Raman bands identified in polyacrylonitrile (PAN)-based fibers using an excitation wavelength of 752 nm [11]

in the energy range 0–1 keV, a complete picture of the elemental composition of the outermost atomic layers is obtained. The surface sensitivity originates from the fact that the escape depth of electrons in the energy range 50–1000 eV is between 0.5 and 1.5 nm. In recent years, AES is widely used in the fields of corrosion, catalysis, metallurgy, thin-film devices, and tribology [13].

5.2.7 Scanning Tunneling Microscopy (STM)

STM is based on quantum tunneling. When a conducting tip is brought in close proximity to the surface to be examined, a bias (voltage difference) applied between the two can allow electrons to tunnel through the vacuum between them. The resulting tunneling current is a function of the tip position, applied voltage, and the local density of states (LDOS) of the sample. Information is acquired by monitoring the current as the tip scans across the surface, and the data are usually displayed in the form of an image. STM can be a challenging technique because it requires extremely clean and stable surfaces, sharp tips, excellent vibration control, and sophisticated electronics; nonetheless, many scientists have built their own STMs [14, 15].

Paredes et al. [15] investigated the surface modification of ultrahigh-modulus carbon fibers by STM. Representative atomic-scale STM images of the fresh untreated material are presented in Fig. 5.8. In Fig. 5.8a, for example, the triangular atomic-scale pattern, typical to the STM images of the pristine graphite basal plane, can be identified. Undulations in the topography can be observed in Fig. 5.8b. Figure 5.8c shows another type of structure rather commonly encountered on the untreated fiber surface. Atomic-sized spots can be distinctly perceived in this image; however, their arrangement differs from that of perfect graphite both in the long range (as observed in Fig. 5.8a) as well as locally (as in Fig. 5.8b).

5.2.8 Atomic Force Microscopy

Atomic force microscopy (AFM) or scanning force microscopy (SFM) is a very high-resolution scanning probe microscopy, with allows for characterization at resolutions on the order of fractions of a nanometer, which is more than 1000 times better than the optical diffraction limit. The precursor to the AFM, the scanning tunneling microscope, was developed by Gerd Binnig and Heinrich Rohrer in the early 1980s at IBM Research laboratory, Zurich, which earned them the Nobel Prize for Physics in 1986. Binnig, Quate, and Gerber invented the first AFM in 1986. The first commercially available AFM was introduced in 1989. The AFM is one of the foremost tools for imaging, measuring, and manipulating matter at the nanoscale (Figs. 5.9 and 5.10) [16, 17].

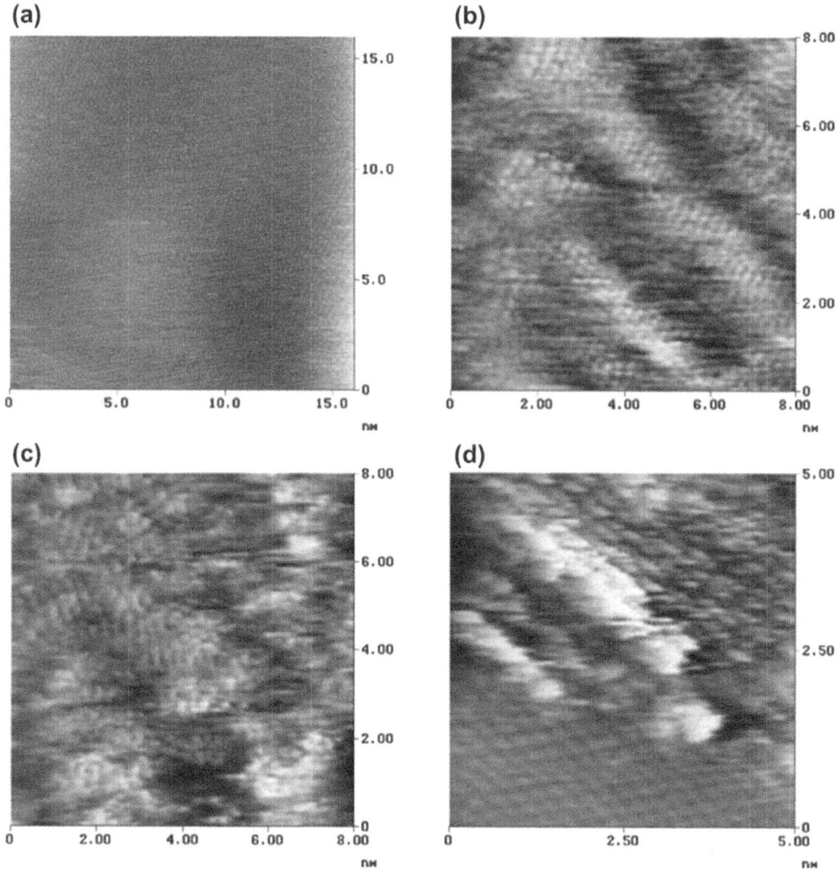

Fig. 5.8 Atomic-scale STM images of untreated ultrahigh-modulus carbon fiber surface. *Z* scale is 3 nm [15]

Fig. 5.9 Surface contour plot of a carbon fiber [16]

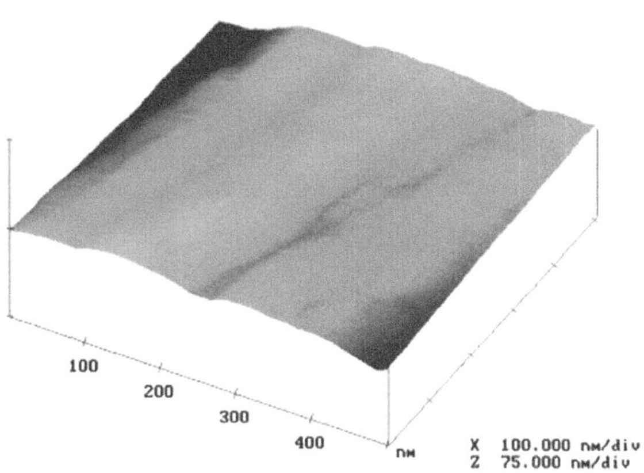

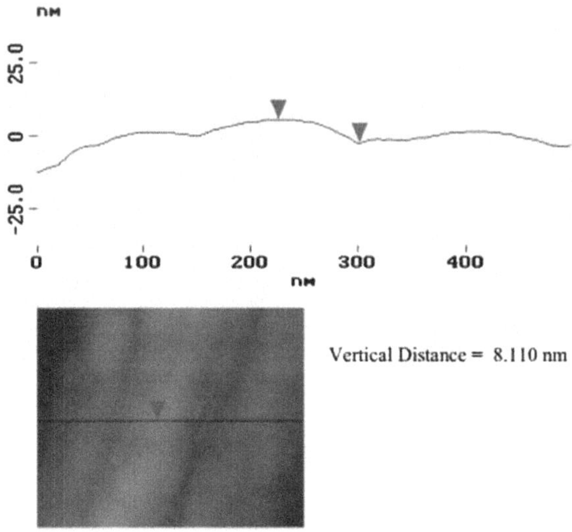

Fig. 5.10 Horizontal section plot of a carbon fiber [16]

Smiley and Delgass used AFM to investigate the topographical changes of carbon fibers exposed to low-temperature low-power oxygen plasma treatments (Figs. 5.11 and 5.12) [17]. Through their research, it was found that the grooves present in the AFM images of untreated fibers were the same as those seen by the SEM. The axial grooves were unevenly spaced by approximately 40–120 nm apart and showed a depth distribution ranging from 1 to 7 nm. The fibers treated with oxygen plasma for 2 and 15 min were also imaged by AFM. It was found that the fiber surfaces go through an initial roughening after the short treatment, followed by an overall smoothening with increase in the treatment time.

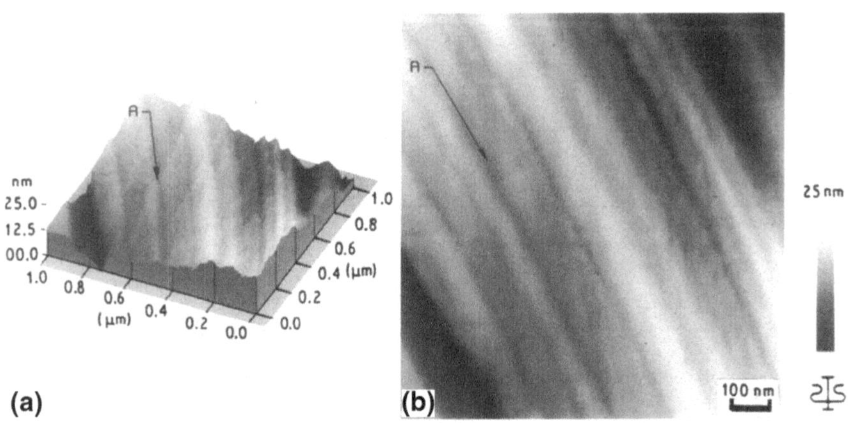

Fig. 5.11 AFM image of untreated Cellion 6000 carbon fiber. **a** Tilted view and **b** top view [17]

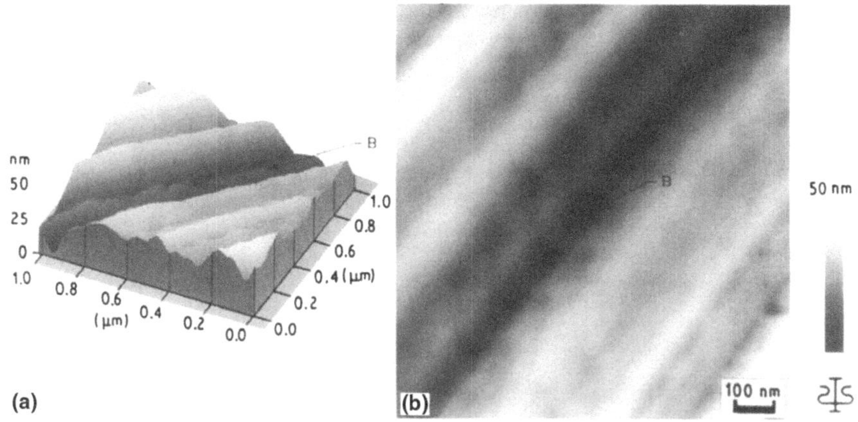

Fig. 5.12 AFM image of a Cellion 6000 carbon fiber treated in oxygen plasma for 2 min: **a** tilted view and **b** top view [17]

5.2.9 *Titration*

The potential chemical activity of surface groups on fibers may be determined by titration techniques. For example, the relative concentration of hydrolyzable groups introduced during the manufacturing or posttreatment of carbon fibers may be determined by measuring the pH. However, titration techniques are typically not used on commercial carbon fibers because of the low levels of surface functional groups present.

5.2.10 *Moisture Content*

The moisture content or moisture regains of fibers or textiles may be determined using common methods. Care must be exercised when applying the procedure because volatile materials in addition to moisture may also be removed. If possible, tests should be performed on fibers that have not been sized. Moisture content is expressed as weight percentage moisture based on the dry weight of the specimen (Fig. 5.13).

5.2.11 *Thermal Stability and Oxidative Resistance*

The susceptibility of fibers and fiber surfaces to undergo oxidation is measured as the weight loss under specific conditions of time, temperature, and atmosphere. Thermogravimetric analysis (TGA) is a process that uses heat and stoichiometry

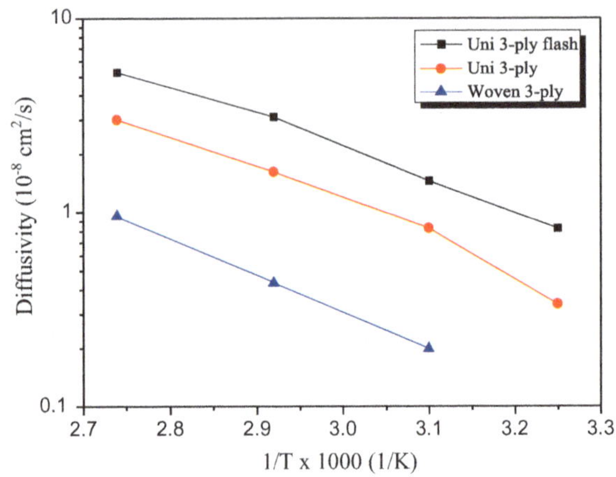

Fig. 5.13 Moisture diffusivity versus 1/*T* for various neat resins and composites [18]

ratios to determine the percent by mass of the various components. Analysis is carried out by raising the temperature of the sample gradually and plotting the weight (percentage) against temperature. The temperature in many testing methods routinely reaches 1000 °C or greater. After the data are obtained, curve smoothing and other operations may be carried out to find the exact points of inflection.

High-resolution TGA is often employed to obtain greater accuracy in areas at which the derivative curve peaks. In this method, the temperature increase slows as weight loss increases. This is done to more accurately identify the exact temperature at which a peak occurs. Several modern TGA devices can vent burn-off to an infrared spectrophotometer to analyze the gas composition (Table 5.5).

5.2.12 *Filament Diameter*

Perfect graphite has three-dimensional periodicity and belongs to the hexagonal crystal system. However, Johnson [31] reported that face-centered cubic sequence is also observed. In carbon fibers, within each basal plane of aromatic rings, the carbon atoms are placed on a hexagonal lattice, but the corresponding atoms on an adjacent plane may be translated by an arbitrary displacement. Hence, carbon fibers generally have only two-dimensional order referred to as "turbostratic" structure [32]. Interplanar spacing in carbon fibers processed even at higher temperatures is significantly higher than that of perfect graphite. However, many researchers [33–36] have reported the evidence of three-dimensional order in certain carbon fibers. Kumar et al. [33] have reported that most PAN-based carbon fibers exhibit a particulate morphology, whereas pitch-based carbon fibers exhibit a sheet-like morphology. They also reported that a sheet-like morphology does not necessarily imply a three-dimensional order; on the other hand, a fiber with particulate morphology can display three-dimensional order. Rayon-based fibers show particulate morphology

Table 5.5 General condition during PAN fiber oxidative stabilization

Reference	Oxidative stabilization temperature (°C)	Types of stabilization process	Stretching ratio/load	Heating rate (°C/min)	Air flow rate (L/min)	Prestabilization treatment
Chen and Harrison [19]	230	Batch process, total stabilization time: 5 h	Underrestrained at constant length (tow was tied at each end)	1	4	DMF solution was used as plasticizer
Wangxi et al. [20]	200–400	Batch process	Fixed-length method	2	–	–
Hou et al. [21]	300	Batch process	Free shrinkage, tow precursor was tied to Kevlar thread at both ends	5	3	–
Yu et al. [22, 23]	195–280	Continuous process: ten separated zones, 6 min for each zone	Stretching was applied on PAN fibers by controlling speed differences between feed rollers and take-up rollers	–	–	–
Gupta and Harrison [24]	200–500, optimum temperature: 380	Batch process	Free shrink/stretch with 5-Ib load cell	1	4	–
Hou et al. [25]	200–280	Batch process, isothermal for 22 h	Free shrinkage, tow precursor was tied to Kevlar thread at both ends	2	3	–
Mathur et al. [26]	230	Batch process, isothermal for 100 min	Underrestrained at a constant load (1 mg) per filament tow	5	1	CuCl solution was used as plasticizer
Wu et al. [27]	160–230	Continuous process: three separated zones	Underrestrained at a constant load of 2 N	1	–	–

(continued)

Table 5.5 (continued)

Reference	Oxidative stabilization temperature (°C)	Types of stabilization process	Stretching ratio/load	Heating rate (°C/min)	Air flow rate (L/min)	Prestabilization treatment
Ge et al. [28]	190–275	Continuous process: ten separated zones, 6 min for each zone, total stabilization time: approximately 60 min	5% stretching ratio	–	–	–
He et al. [29]	190–270	Continuous process: ten separated zones, total stabilization time: approximately 40 min	10% stretching ratio	–	–	–
Fazlitdinova et al. [30]	245–290	Batch process, isothermal for various residence times [0.5, 1, 2, 4, 6, and 8 h]	Underrestrained at a constant load of 0.6 g/tex	–	–	–

and may show three-dimensional order. SEM images of PAN-based, pitch-based, and rayon-based carbon fibers are shown in Figs. 5.14, 5.15, and 5.16 [33], respectively.

Extensive graphitized small areas have been observed to exist even in fibers, which have never been exposed to temperatures beyond 1100 °C [34]. Preferential orientation of graphitic sheets along the fiber axis increases with heat-treatment temperature. The orientation and size of these sheets are greater on the surface than in the interior of the carbon fibers [33, 34]. The modulus of the fibers increases with increase in orientation [34]. Carbon fibers are reported to consist of up to 20% voids

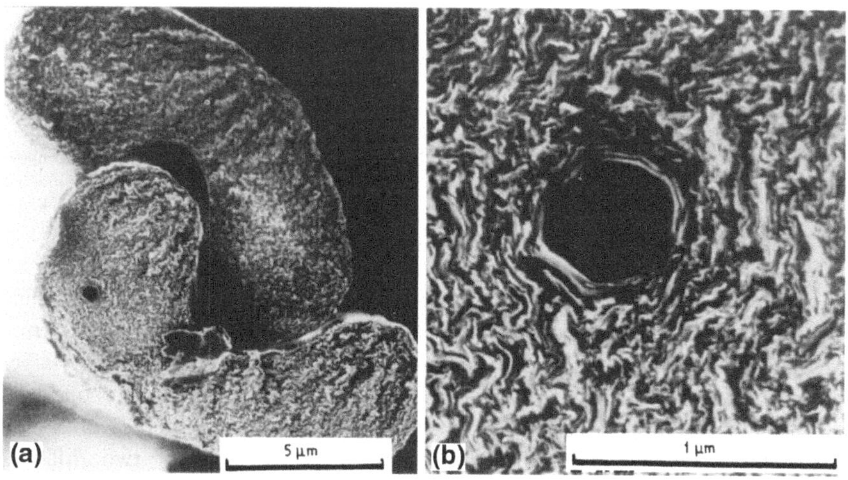

Fig. 5.14 SEM images of PAN-based carbon fibers [33] at **a** low and **b** high magnifications

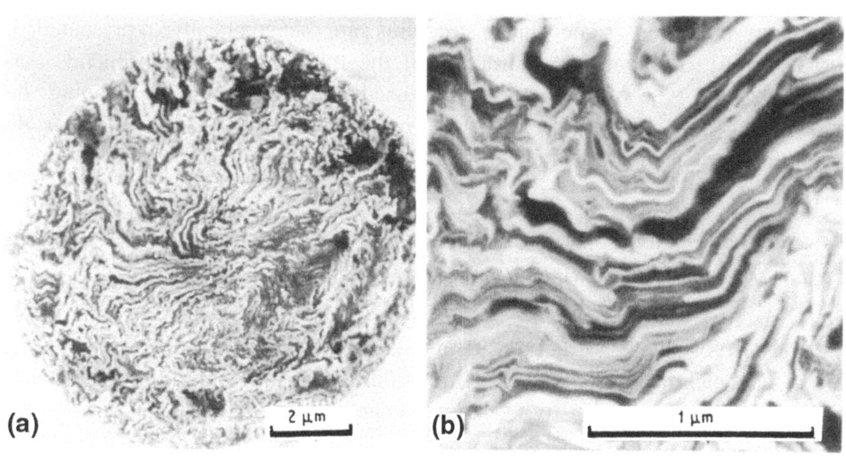

Fig. 5.15 SEM images of pitch-based carbon fibers [33] at **a** low and **b** high magnifications

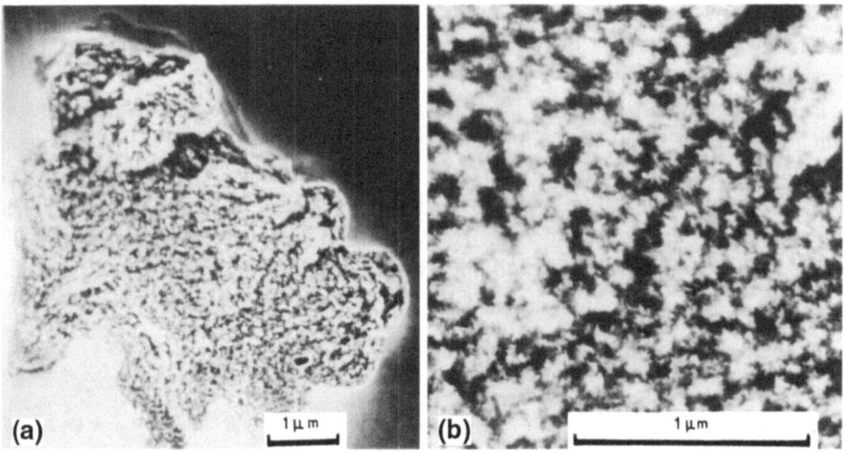

Fig. 5.16 SEM images of rayon-based carbon fibers [33] at **a** low and **b** high magnifications

elongated in along the fiber axis [32]. There is a transition from many small pores to fewer larger pores with an increase in the heat-treatment temperature [35].

PAN-based carbon fibers show higher tensile and compressive strengths than pitch-based carbon fibers because PAN-based fibers consist of particle-like structures and smaller crystals in comparison to the sheet-like structure and larger crystals in pitch-based fibers [35–40]. The cross-sectional structure of carbon fibers also plays an important role in determining the compressive properties of carbon fibers. Hayes et al. [36] studied the compressive behavior of two different pitch-based carbon fibers and proposed that fibers with a folded-radial texture show higher compressive strengths than fibers with a flat-layer structure. The two structures have been reproduced in Fig. 5.17 [36]. Knibbs [41] identified three different types of structures for PAN-based carbon fibers prepared under different processing conditions. The three structures are schematically represented in Fig. 5.18 [34]. The transverse structure of the final carbon fibers depends very much upon the type of spinning process used, the temperature of spinning, the shape of the spinneret, and the use of a stirrer and its shape [34, 42]. However, the structure is independent of the heat-treatment temperature [34].

5.2.13 Electrical Resistivity

The presence of an electric field inside a material will cause the flow of an electric current. The electrical resistivity (ρ, Ω/m) is defined as the ratio of the electric field (E, V/m) to the current density (J, A/m^2) generated.

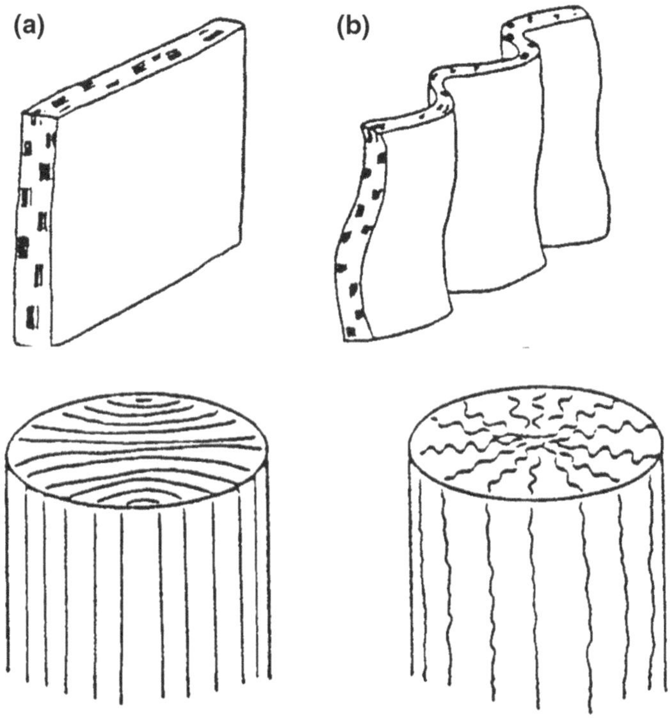

Fig. 5.17 Structure of two different pitch-based carbon fibers: **a** flat and **b** folded, proposed by Endo [36]

$$\rho = \frac{E}{J} \tag{5.2}$$

Park et al. [43] investigated the effects of the thickness of a nickel coating on the electric properties of nickel/carbon hybrid fibers. To confirm the effects of the thickness of the nickel layer on the electric resistance of the nickel/carbon hybrid fibers, the four-point probe method was employed and the results are shown in Fig. 5.19. The specific resistance was observed to decrease with increase in plating time.

Also, while the specific resistivity of the as-received sample was higher than $0.80 \times 10^{-3}\ \Omega$ cm, the value decreased dramatically with increase in plating time to up to 60 s. From the good linearity of results in the range 5–30 s, there seems to be a strong correlation between the electric conductivity and the plating time.

5.2.14 Coefficient of Thermal Expansion

The coefficient of thermal expansion describes how the size of an object changes with change in temperature, and the value represents the fractional change in size

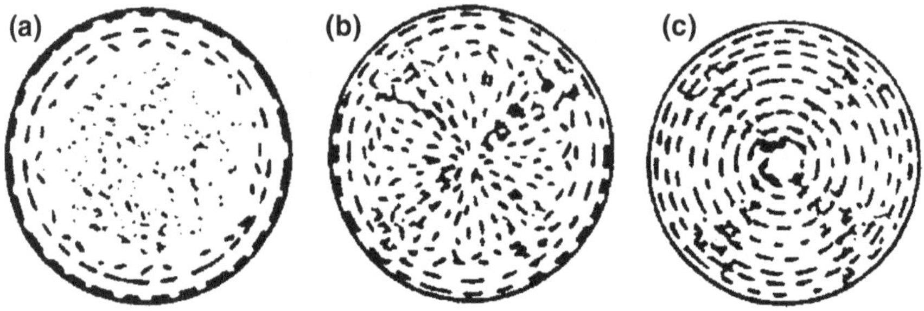

Fig. 5.18 Three different structures for PAN-based carbon fibers prepared under different processing conditions, as identified by Knibbs [41]

Fig. 5.19 Specific electric resistivity of Ni-plated carbon fibers as a function of plating time [43]

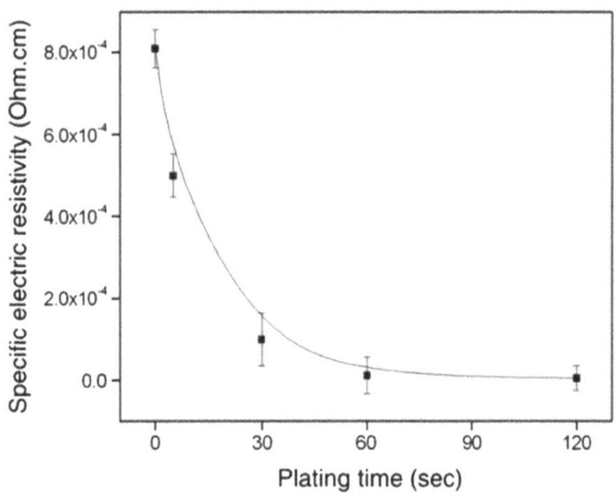

per degree change in temperature at a constant pressure. Several types of thermal expansion coefficients, including volumetric, area, and linear, have been developed. The type of coefficient used depends on the particular application and the dimensions, which are considered important. For solids, one might only be concerned with the change along a length or over some area.

Korab et al. [44] investigated the thermal expansion of cross-ply and woven carbon fiber-copper matrix composites. The in-plane thermal expansion behaviors of the cross-ply samples with and without copper foils are compared in Fig. 5.20. Samples were oriented in such a way that in three layers, fibers were parallel, and in two layers, the fibers were perpendicular to the direction of measurement. The in-plane response of the cross-ply composites is qualitatively similar to the axial response of the unidirectional specimens. Curves showed hysteresis between the heating and cooling parts of the loop, which could be ascribed to the plastic deformation of the copper matrix at higher temperatures. Further, in the case of the sample with copper foil (50% C_f), the first loop was not closed, while in the case of the sample without copper foils (57% C_f), all three loops were closed.

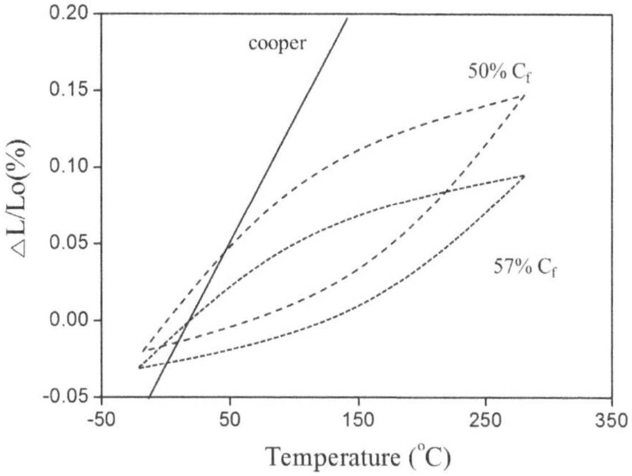

Fig. 5.20 In-plane thermal expansion of a cross-ply composite made of five layers [44]

5.2.15 Thermal Conductivity

Heat transfer across materials with high thermal conductivity occurs at a higher rate than across materials with low thermal conductivity. Correspondingly, materials with high thermal conductivity are widely used in heat sink applications, and materials with low thermal conductivity are used as thermal insulation. Thermal conductivity of materials is temperature dependent. The reciprocal of thermal conductivity is thermal resistivity.

Karaipekli et al. [45] investigated the improvement in the thermal conductivity of stearic acid using expanded graphite and carbon fiber additives. Figure 5.21 shows the variation of thermal conductivity of stearic acid/expanded graphite and the steric acid/carbon fiber composite phase change materials with different mass fractions of expanded graphite and carbon fibers. As clearly seen from these results, the thermal conductivity of stearic acid increased with increase in the mass fraction of expanded graphite and carbon fibers.

5.2.16 Specific Heat

Specific heat is measured in a calorimeter such as that described in ASTM test method D2766 [46]. The measurement of specific heat is complex and is best left to experienced laboratories.

5.2.17 Thermal Transition Temperature

Differential scanning calorimetry, differential thermal analysis, or thermal mechanical analysis may be applied to measure the glass transition temperature

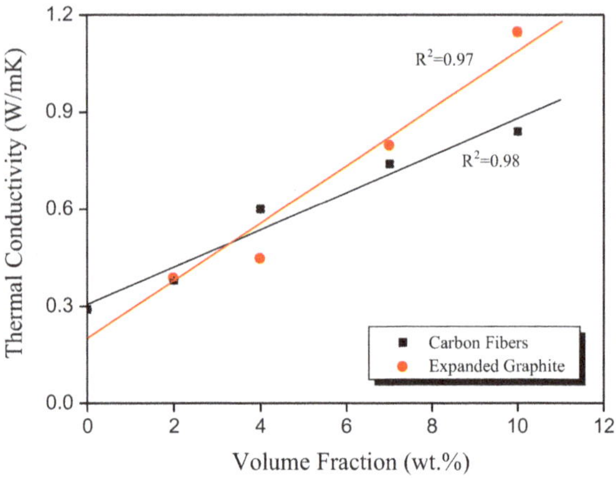

Fig. 5.21 Variation of effective thermal conductivity of stearic acid containing various volume fractions of expanded graphite and carbon fiber additives [45]

(T_g), and if the fiber is semicrystalline, its crystalline melting temperature (T_m) can be measured. The general procedures for measuring T_g and T_m of organic fibers are provided in ASTM test methods D3417 and D3418 [47, 48].

5.2.18 Tensile Properties

Tensile properties are a measure of the maximum stress that a material can withstand while being stretched or pulled before necking, which is when the specimen's cross section begins to significantly contract. Tensile strength is the opposite of compressive strength, and the values can be quite different. Tensile strength is defined as stress, which is measured as force per unit area. For some nonhomogeneous materials (or for assembled components), it can be reported simply as a force or as a force per unit width. In the SI system, the unit is pascal (Pa) (or a multiple thereof, often megapascals (MPa)) or, equivalently, Newton per square meter (N/m^2). The customary unit is pound-force per square inch (lbf/in^2 or psi) or kilo-pounds per square inch (ksi, or sometimes kpsi), which is equal to 1000 psi. When measuring tensile strengths, ksi is commonly used for convenience.

5.3 Evaluation of Composites

5.3.1 Introduction

The function of the matrix in a composite is to hold the fibers in the desired position and to provide a path for introducing external loads into the fibers. Because the strengths of the matrix materials are generally lower than the fiber strengths by an order of magnitude or more, it is desirable to orient the fibers within a composite structure so that the fibers will carry the major external loads. Although the success of a composite largely relies on this, the strength and other properties of matrix materials cannot be ignored. Matrix material properties can significantly affect how a composite will perform, particularly with respect to in-plane compression, in-plane shear, resistance to impact damage, and other interlaminar behaviors, and especially when exposed to moisture and elevated temperatures.

This section focuses on methods of testing and evaluating matrix materials and their constituents. Chemical, physical, thermal, and mechanical properties are considered, in addition to methods for test specimen preparation and environmental conditioning of the test specimens.

5.3.2 Coefficient of Thermal Expansion

The thermal expansion of composite materials is an important design parameter in cryogenic technology. A large range of expansion can be achieved by varying the fiber type, content, and arrangement and the matrix. Generally, the thermal expansion of composite materials is anisotropic. For carbon fiber composites, the anisotropy results not only from the fiber arrangement but also from the fiber anisotropy. Carbon fibers show a small negative coefficient of thermal expansion in the fiber direction and a large positive coefficient perpendicular to it. For angle-plies, the stresses, thermally induced between layers, strongly influence the thermal expansion behavior. In many cases, it is possible to match the thermal expansion of fiber composites to their application at least in one direction, and it should be noted that often the same thermal expansion can be achieved by two or more different fiber arrangements. The solution with the lowest internal stresses is favored. At a fiber angle of $\pm 30°$, thermally induced stresses induce a negative expansion coefficient for carbon fiber composites. This might be useful for compensating the thermal expansion of different materials [49].

5.3.3 Thermal Conductivity

The thermal conductivity of metals, alloys, or composites with = $0, 2, \ldots, 100$ W/mK can be measured by comparative method with steady-state longitudinal heat flow from room temperature to up to approximately 1000 °C. The comparative instrument measures heat flow based on the known thermal properties of standard reference materials. The test specimen is sandwiched between two identical reference samples. This stack is placed between two heating elements controlled at different temperatures. A guard heater is placed around the test stack to ensure a constant heat flux through the stack and to eliminate lateral heat flow losses. As heat flows from the hot element to the cold element, the temperature gradient (T/L) across the stack is measured with thermocouples. Once the specimen reaches a state of thermal equilibrium, its thermal conductivity is calculated from the following expression:

$$\lambda = \frac{LQ}{A\Delta T} \tag{5.3}$$

In the above equation, Q is the heating power of the heater. The experimental error is in the range of approximately ±10%. The specimen geometry is cylindrical (with a diameter of 25 or 50 mm and a height of approximately 10–40 mm).

Shim et al. [50] investigated the thermal conductivity of carbon fiber-reinforced composites with various types of cross sections. Figure 5.22a, b show the thermal conductivity of the carbon fiber-reinforced composites in the directions parallel and perpendicular to the reinforcement direction, respectively, as a function of the fiber mass fraction. From the figures, we can see that the thermal conductivity of the composites depends on three parameters, i.e., measurement temperature, fiber content, and type of fiber cross section. C-CF/EP (C-type carbon fiber-reinforced epoxy composites), in particular, shows the highest value of thermal conductivity in a direction parallel direction to the fiber, *k//*, while H-CF/EP (hollow-type carbon fiber-reinforced epoxy composites) shows the lowest value of thermal conductivity in the transverse direction to the fiber, $k\perp$. The difference between *k*(*//*, 120 °C) and *k*(*//*, 40 °C) of C-CF/EP is relatively larger than that of R-CF/EP and C-CF/EP, while the difference between $k(\perp$, 120 °C) and $k(\perp$, 40 °C) of R-CF/EP is relatively larger than that of C-CF/EP and H-CF/EP.

5.3.4 Poisson's Ratio

There is very limited information available on the Poisson's ratio of fiber-reinforced concrete. In most analytical studies, the Poisson's ratio is generally assumed to be the same as that of concrete. This may be a reasonable assumption provided the composite remains in the elastic range of behavior. As soon as cracking occurs, the confining effects of the fibers bridging the cracks will have a significant effect on the lateral deformation, and thus, the value of the measured Poisson's ratio will be affected. To the best of our knowledge, no investigation is known to have addressed this issue.

5.3.5 Rheological Analysis

Newtonian fluids can be characterized by a single coefficient of viscosity for a specific temperature. However, this viscosity will change with temperature, although it does not change with the flow rate or strain rate. Only a small group of fluids exhibits such constant viscosity and is known as Newtonian fluids. However, for a large class of fluids, the viscosity changes with the strain rate (or relative velocity of flow) and the fluids are termed as non-Newtonian fluids. Rheology generally accounts for the behavior of non-Newtonian fluids by characterizing the

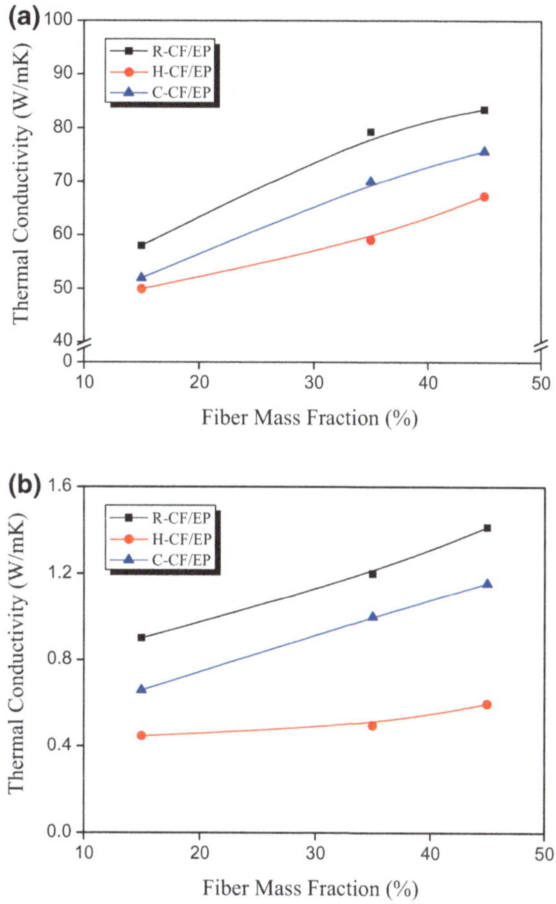

Fig. 5.22 Thermal conductivity of carbon fiber-reinforced composites in a direction **a** parallel and **b** perpendicular to that of reinforcement [50]

minimum number of functions that are needed to relate the stresses with the rate of change of strains or strain rates. For example, the viscosity of ketchup is reduced by shaking (or by other forms of mechanical agitation, where the relative movement of the different layers in the material actually causes a reduction in the viscosity), while that of water is not. Ketchup is a shear-thinning material because an increase in relative velocity causes a reduction in the viscosity, while some other non-Newtonian materials show the opposite behavior. That is, the viscosity increases with relative deformation, and these materials are called shear thickening or dilatant materials. Since Sir Isaac Newton drew the origins of the concept of viscosity, the study of liquids with strain-rate-dependent viscosity is also often termed as non-Newtonian fluid mechanics (Fig. 5.23).

The viscoelastic properties of polymers are determined by the effects of many variables, including temperature, pressure, and time. Other important variables include chemical composition, molecular weight and weight distribution, degree of branching and crystallinity, types of functional groups, component concentration,

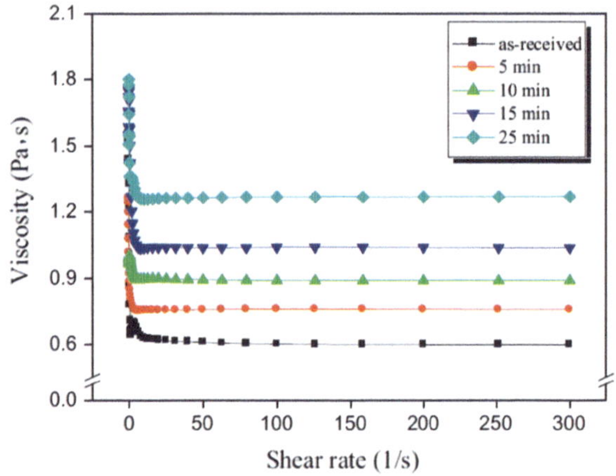

Fig. 5.23 Behavior of viscosity of Ni-coated carbon nanofibers-reinforced epoxy matrix nanocomposites [51]

dilution with solvents or plasticizers, and mixture with other materials to form composite systems. Using molecular theory, the dependence of viscoelastic properties on these variables can be simplified by introducing additional concepts such as free volume, monomeric friction coefficient, and spacing between the entanglement loci to provide a qualitative understanding, and in many cases, a quantitative prediction of how to achieve the desired physical and chemical properties and microstructure.

Viscoelastic behavior reflects the combination of viscous and elastic responses of materials under mechanical stress, which are intermediate between liquids and solids. Fundamentally, viscoelasticity can be related to the motions of the flexible polymer molecules and their entanglements and network junctions, which forms the molecular basis of viscoelasticity. Thus, rearrangements on a local scale (kinks) are relatively rapid, while those on a long-range scale (convolutions) are very slow. In addition, a new assortment of configurations is obtained under stress. The response to the local aspects of the new distribution is rapid, while the response to the long-range aspects is slow. Thus, the range of time scales covering the response of such a system to externally applied stress is wide and continuous. By measuring the viscoelastic properties of polymers, information can be obtained about the nature and the rates of change of the configurational rearrangements and the nature of the (macro) molecular interactions over a range of time scales (Fig. 5.24).

Examples may be given to illustrate the potential applications of these principles to practical problems in the processing and use of rubbers, plastics, and fibers. Polymers are the basic materials of rubber and plastic industries and are of vital importance to textile, petroleum, automobile, paper, and pharmaceutical industries. Their viscoelastic properties determine the mechanical performance of the final products of these industries and the success of processing methods at intermediate stages of production.

In viscoelastic materials such as most polymers and plastics, the presence of liquid-like behavior depends on the properties of the applied load and hence varies

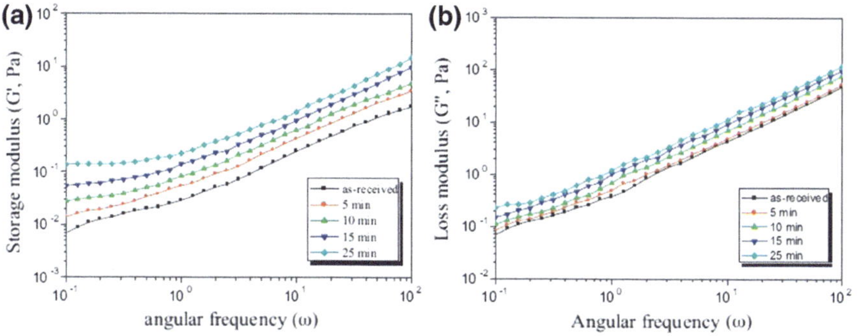

Fig. 5.24 Storage **a** and loss **b** moduli of Ni-coated carbon nanofibers-reinforced epoxy matrix nanocomposites measured with respect to angular frequency [51]

with the rate of applied load, i.e., how quickly a force is applied. The silicone toy "Silly Putty" behaves quite differently depending on the rate of application of a force. On pulling slowly, silicone exhibits continuous flow similar to that evidenced in a highly viscous liquid. Alternatively, when hit hard and directly, the material shatters like a silicate glass.

In addition, conventional rubber undergoes glass transition (often called rubber-glass transition). The Space Shuttle Challenger disaster was caused by rubber O-rings that were being used well below their glass transition temperature on an unusually cold Florida morning. The O-rings could not flex adequately to form proper seals between the sections of two solid-fuel rocket boosters.

5.3.6 *Tensile Behavior*

Tensile strength is the maximum stress that a material can withstand while being stretched or pulled before necking, which is when the cross section of the specimen begins to significantly contract. Tensile strength is the opposite of compressive strength, and the values can be quite different (Fig. 5.25).

Tensile strengths are rarely used in the design of ductile members; however, the tensile strength is important for brittle materials. The values exhibited by common materials such as alloys, composite materials, ceramics, plastics, and wood are tabulated.

Fig. 5.25 Specimens used for tensile tests

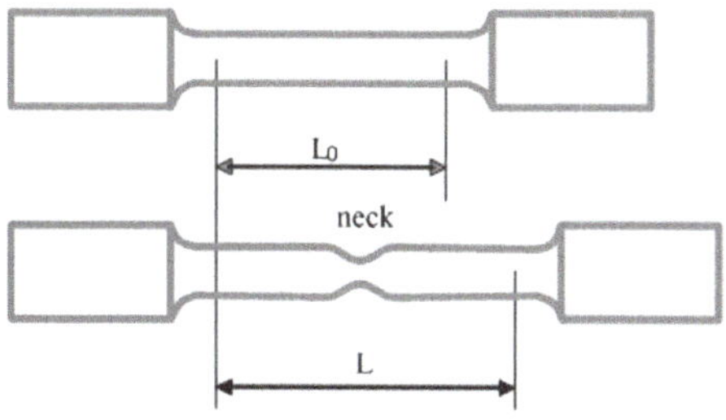

Tensile strength is defined as tress, which is measured as force per unit area. For some nonhomogeneous materials (or for assembled components), the tensile strength can be reported just as a force or as a force per unit width. In the SI system, the unit of tensile strength is Pascal (Pa) (or a multiple thereof, often megapascals (MPa)) or equivalently Newton per square meter (N/m^2). The customary unit is pound-force per square inch (lbf/in^2 or psi) or kilo-pounds per square inch (ksi, or sometimes kpsi), which is equal to 1000 psi. Kilo-pounds per square inch is commonly used for convenience when measuring tensile strengths

Fu et al. [52] investigated the tensile properties of polypropylene composites reinforced with short-glass fibers and short-carbon fibers (i.e., SGF/PP and SCF/PP, respectively). The stress–strain curves of the SGF/PP composites and SCF/PP composites are shown in Fig. 5.26. It can be seen that the SGF/PP composites and SCF/PP composites exhibit brittle fracture and show linear deformation at lower stresses and nonlinear deformation at higher stresses. The strains at failure shown by SCF/PP composites are lower than those shown by SGF/PP composites

5.3.7 *Shear Strength*

Shear strength is the strength of a material or component against the type of yield or structural failure where the material or component fails in shear. A shear load is a force that tends to produce a sliding failure on a material along a plane parallel to the direction of the force.

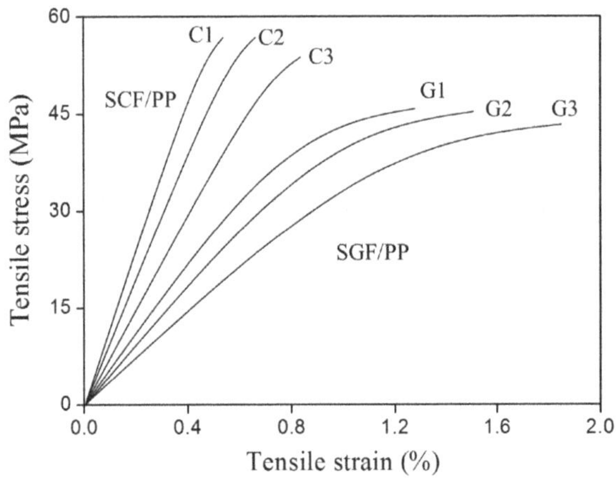

Fig. 5.26 Typical tensile stress–strain curves of SGF/PP and SCF/PP composites. SGF/PP composites: curve G1, behavior of a composite containing 25 vol.% glass fibers; curve G2, behavior of a composite containing 16 vol.% glass fibers; and curve G3, behavior of a composite containing 8 vol.% glass fibers. SCF/PP composites: curve C1, behavior of a composite containing 25 vol.% carbon fibers; curve C2, behavior of a composite containing 16 vol.% carbon fibers; and curve C3, behavior of a composite containing 8 vol.% carbon fibers [52]

5.3.7.1 Interlaminar Shear Strength

The interlaminar shear strength (ILSS) is the interfacial shear stress or shear strength of the matrix materials and is measured with the three-point bend test (ASTM D 2344), which is ideal for outline testing. Figure 5.27 shows a schematic of the test and shows how shear, tensile, and compressive forces are involved [53].

Park et al. [54] investigated the effect of anodization of carbon fibers on the interfacial mechanical properties of epoxy matrix composites. Figure 5.28 shows the ILSS of the carbon fiber-reinforced composites. A good relationship can be observed between the characteristics of the anodized carbon fiber surfaces and the resulting fiber matrix adhesions on the mechanical interfacial properties of the composite under the employed experimental conditions. The ILSS values increase with enhanced wettability of the fibers for adhesion at interfaces caused by the anodization, which can be attributed to the increase of the polarity of the fiber surfaces.

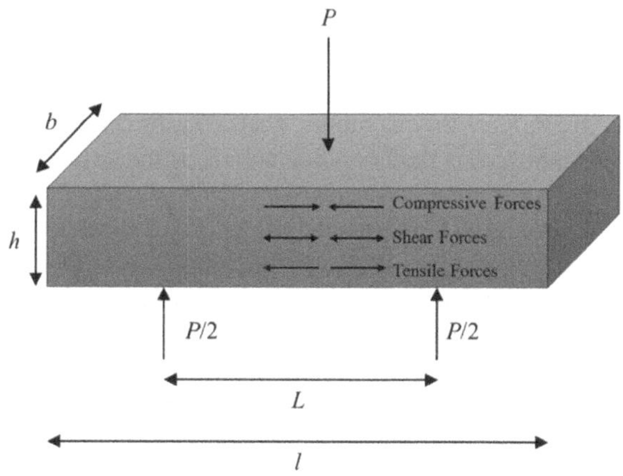

Fig. 5.27 Test configuration of a short-beam shear test [53]

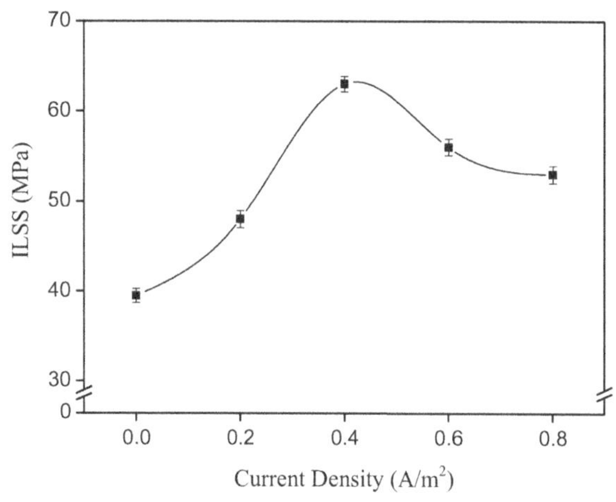

Fig. 5.28 ILSS values of carbon fiber-reinforced composites [54]

5.3.7.2 Interfacial Shear Strength

The interfacial shear strength (IFSS) is a reliable quantitative index of the adhesion between the fiber and matrix. Despite the wide application of transverse compression, double cantilever beam (DCB), and other flexural tests to measure the ILSS of the composites, single-filament composite (SFC) and microbond pull-out tests are more effective because these techniques directly measure the IFSS values of the fibers and the matrix (Fig. 5.29). Until now, most studies, which use the microbond pull-out technique, are devoted to carbon fiber/epoxy composites.

IFSS can be calculated according to the following equation, which is based on the assumption of a constant IFSS between the fiber and its surrounding matrix [55]:

$$\mathrm{IFSS} = \frac{F_d}{\pi d_f L_e} \tag{5.4}$$

$$= \frac{\sigma_d F_d}{4L_e} \tag{5.5}$$

Here, F_d is the maximum fiber axial force recorded at the onset of microdroplet debonding, d_f is the fiber diameter, L_e is the embedded fiber length and $\pi d_f L_e$ is the embedment area, and σ_d is the tensile stress at debonding.

Li et al. [55] investigated the effective IFSS of CNT-coated carbon fibers in an epoxy matrix. The IFSS of the carbon fiber/epoxy composites with and without deposited CNTs was tested by the single-fiber composite fragmentation test (SFFT), and the measured data are provided in Fig. 5.30. As shown in Fig. 5.30, the deposition of the COOH–CNTs leads to a significant increase of the IFSS (approximately 43%) for the T700SC composite. Moreover, the IFSS of the T300B composite increases by approximately 12% and 11% after the deposition of COOH–CNTs and OH–CNTs, respectively. However, the T700SC composite shows a decrease in the IFSS by 17% after the deposition of OH–CNTs.

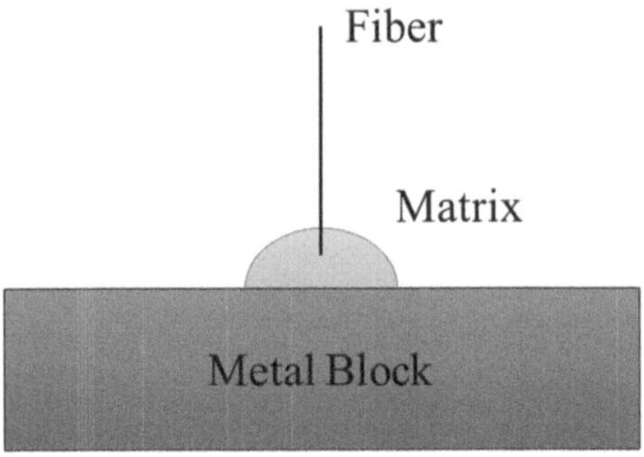

Fig. 5.29 Schematic of single-fiber pull-out test specimen

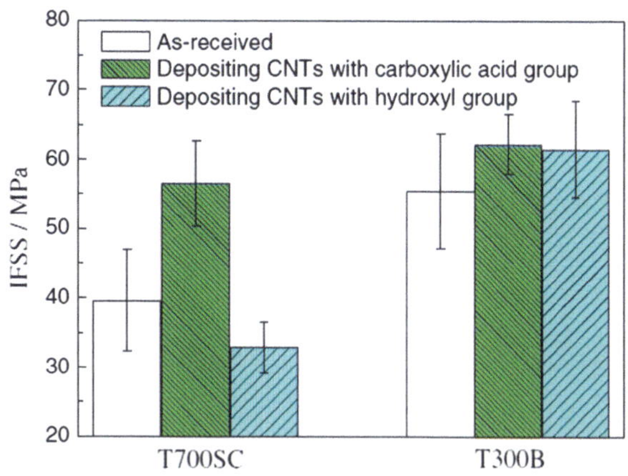

Fig. 5.30 Effects of deposited CNTs on IFSS of single-fiber composites. Error bars indicate standard deviation [55]

5.3.8 *Flexural Behavior*

The flexure test method measures the behavior of materials subjected to simple beam loading [56]. It is also called the transverse beam test with some materials. The maximum fiber stress and maximum strain are calculated for increments of load, and the results are plotted in a stress–strain diagram. Flexural strength is defined as the maximum stress in the outermost fiber. This is calculated at the surface of the specimen on the convex or tension side. Flexural modulus is calculated from the slope of the stress versus deflection curve. If the curve has no linear region, a secant line is fitted to the curve to determine the slope. There are three relevant standards, i.e., ASTM D 790 M, BSI 2782 Method 1005, and CRAG, and all of them cover a four-point procedure.

A flexure test produces tensile stress in the convex side of the specimen and compression stress in the concave side. This creates an area of shear stress along the midline. To ensure that the primary failure originates from the tensile or compression stress, the shear stress must be minimized. This is ensured by controlling the span to depth ratio (S/d), which is the length of the outer span divided by the height (depth) of the specimen. For most materials, an S/d of 16 is acceptable. Some materials require an S/d of 32–64 to keep the shear stress sufficiently low.

Flexure testing is often done on relatively flexible materials such as polymers, wood, and composites. There are two test types: three-point flex (Fig. 5.31) and four-point flex (Fig. 5.32). In a three-point test, the area of uniform stress is quite small and concentrated under the center loading point. In a four-point test, the area of uniform stress exists between the inner span loading points (typically half the outer span length).

For a rectangular sample under a load in a three-point bending setup,

$$\sigma = \frac{3FL}{2bd^2} \tag{5.6}$$

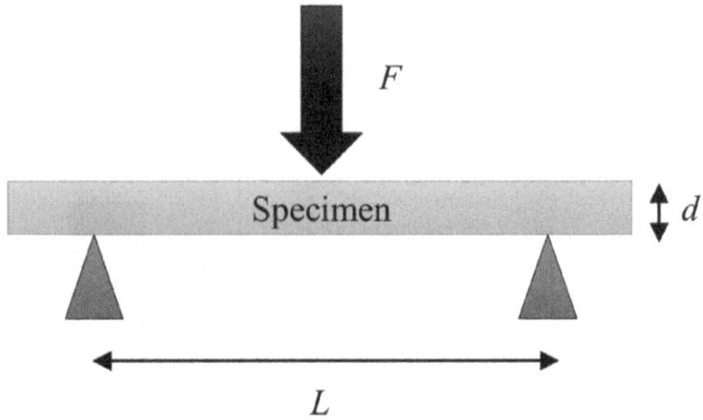

Fig. 5.31 Schematic of beam under three-point bending

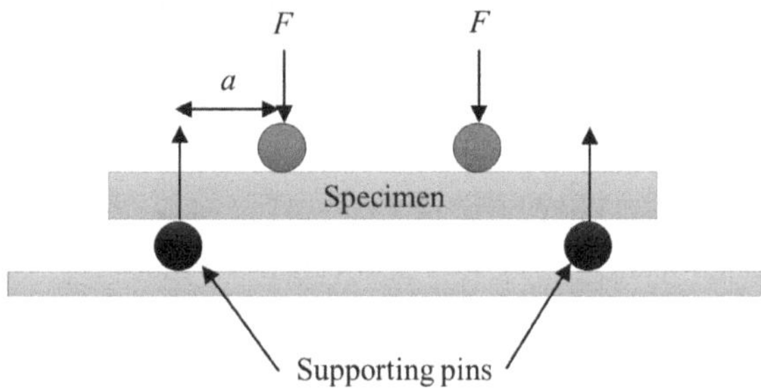

Fig. 5.32 Schematic of beam under four-point bending

In the above equation, F is the load (force) at the fracture point (N), L is the length of the support span (mm), b is the width (mm), and d is the thickness (mm).

For a rectangular sample under a load in a four-point bending setup,

$$\sigma = \frac{3Fa}{bd^2} \tag{5.7}$$

In the above equation, F is the load (force) at the fracture point (N), a is the distance between the supporting and loading pins (mm), b is the width (mm), and d is the thickness (mm).

Mannocci et al. [57] investigated the structure of the posts by confocal microscopy, and to evaluate the flexural strength of fiber posts stored under different conditions, a three-point bending test was used simultaneously to observe the failure mode of the posts by confocal microscopy (Fig. 5.33).

5.3.9 *Uniaxial Compressive Behavior*

Compressive strength is the capacity of a material or structure to withstand axially directed pushing forces. It provides data (or a plot) of force versus deformation for the conditions used in a test. When the limit of compressive strength is reached, brittle materials are crushed.

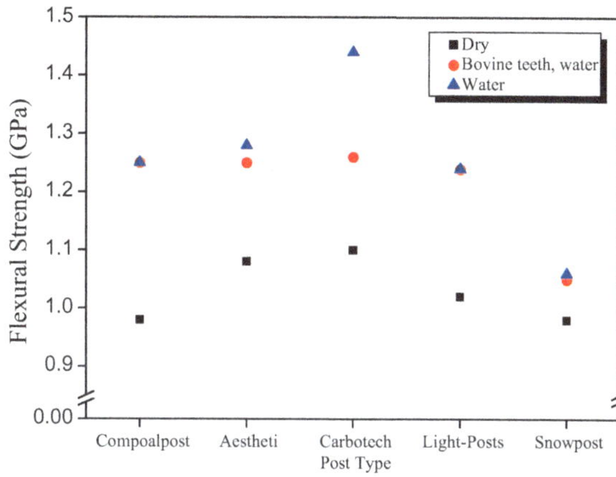

Fig. 5.33 Graphical representation showing results of flexural strength test [57]

Kumar et al. [58] investigated that the uniaxial compression behavior of adhesively bonded composite scarf-joints. Figure 5.34 shows the typical geometry of our compression test specimens. These were 12.5 mm wide and had a gauge length of 10 mm in order to ensure that there was a compromise between the need to overcome Euler macrobuckling and end-tab effects. If the gauge length was too short, the end-tabs would become too close to each other and will not allow uniform distribution of stresses in the central region.

Figure 5.35 shows the unidirectional compressive scarf-joint strength as a function of scarf-angle, θ. As shown in Fig. 5.35, smaller (i.e. steeper) scarf-angles yielded higher compressive joint strengths compared to those with larger scarf-angles. The knockdown in compressive strength is most prominent for scarf-joint angles less than about 1°. As the scarf-angle of the joint increases, the additional reduction in compressive strength becomes more gradual.

5.4 Fatigue

Fatigue is determined under cyclic loading when a crack is initiated and then grows to a critical size to result in failures occurring at stresses below the values obtained for normal static tests. In general, any static test adopted for fatigue testing is much more demanding. The shape of the test specimen is chosen to fail within the gage length in a manner akin to the failure of a structural component. Fatigue tests can be undertaken in the tensile, flexural, compression, and shear modes and are generally carried out in a servo-hydraulic test machine. Optical microscopy, ultrasonics, X-ray, and infrared thermography are used for detecting the onset and study of fatigue failure [53].

Khan et al. [59] investigated the fatigue damage behaviors of carbon fiber-reinforced epoxy composites containing nanoclay. Figure 5.36 shows the damage index, D, plotted as a function of the fatigue cycles for hybrid composites

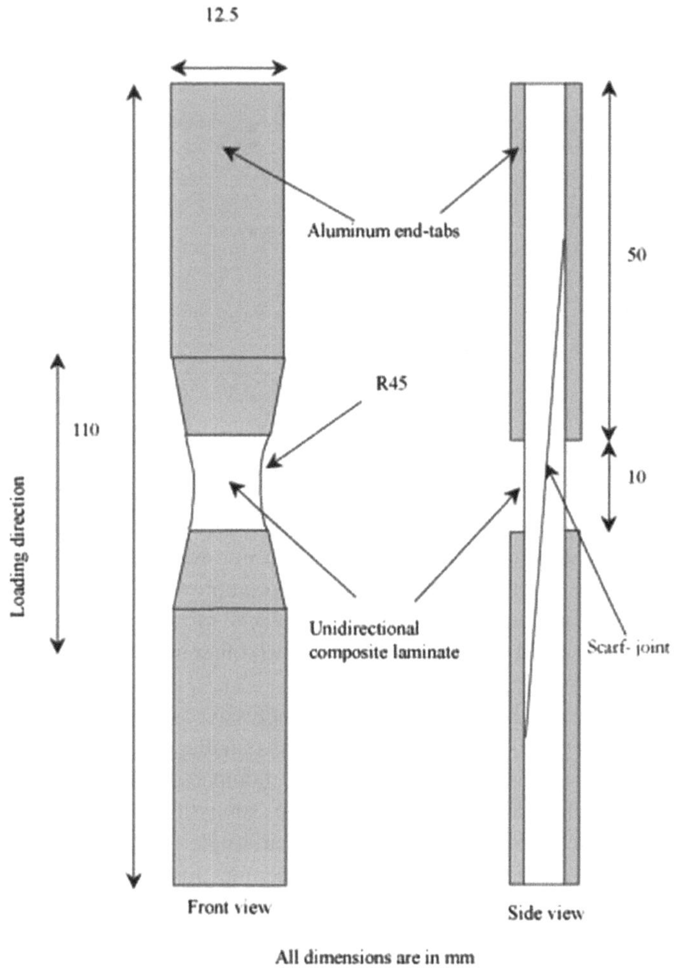

Fig. 5.34 A typical adhesively bonded scarf joint specimen geometry used for compression testing [58]

containing different clay contents. It can be seen that at the early stage of fatigue (say 0–12,500 cycles), the hybrid composites in general exhibited marginally more damage than the neat composites. After the initial damage period, the hybrid composite specimens sustained a relatively longer stable period with low damage indices for the rest of their fatigue life. The final failure took place much earlier in the neat composite than in the clay-CFRP hybrid composites. Further, the higher the clay content, the longer was the fatigue life, with the exception of hybrid composites containing 5 wt% clay. The diminishing improvement in fatigue life in the case of composites with clay contents higher than 3 wt% is attributed to the higher possibility of the formation of unwanted agglomerates of a relatively large size.

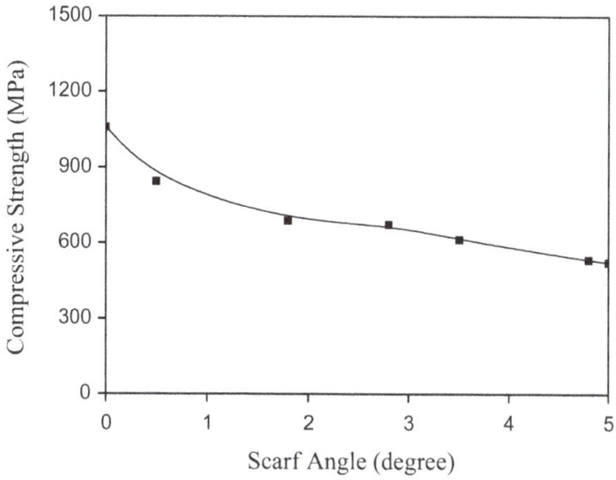

Fig. 5.35 Experimental data showing variation of composite scarf-joint strength in compression [58]

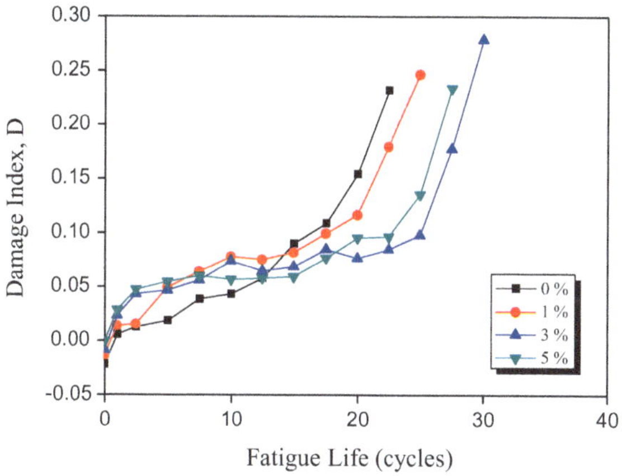

Fig. 5.36 D plotted as a function of fatigue life [59]

5.4.1 *Creep*

The application of a force to a material over a sustained period of time induces creep, and if this force eventually leads to cracking, fracture, or rupture, the process is termed as stress cracking, static fatigue, or creep rupture. Further, if the environment speeds up the process, the rupture is termed as environmental stress cracking. The creep of a composite material will depend on the reinforcement and the matrix. Carbon fibers are not significantly affected by creep, although some loss of strength can be expected to occur under sustained loading. However, the matrix will exhibit creep, and a thermoplastic will exhibit creep to a higher extent than a thermosetting resin. However, if the fiber is well aligned in a unidirectional composite, it will have good resistance to creep along the fiber axis, although it would display creep during torsion and flexure. Plastics do tend to recover when the stress is removed, as long as no damage has occurred [53].

5.4.2 Impact Behavior

The Izod impact test is shown schematically in Fig. 5.37. The test setup and procedure are similar to those outlined above. In the Izod test, the specimen is clamped in the vertical plane as a cantilever beam and impacted by a swinging pendulum at the unsupported end. The test suffers similar problems to those reported above and again is best suited as a tool for ranking the impact resistance of composite materials.

The Charpy impact test is a standardized high strain-rate test, which determines the amount of energy absorbed by a material during fracture (Fig. 5.38). The specimen is supported in a horizontal plane and impacted by a swinging pendulum directly opposite the notch. The energy dissipated during impact is usually recorded by a dial on the test apparatus. Further information can be obtained by instrumenting the impactor with a strain gauge, thereby enabling the determination of the variation of the impact force with time. The energy absorbed during impact can also be determined by integrating the force/time signal.

Therefore, the low-velocity impact test is performed with a drop-weight impact tester instead of Izod and Charpy tests [60] . Park et al. [61] investigated the impact strength of C/C composites as a function of the $MoSi_2$ content in the prepreg resin using a drop weighting impact tester. The impact properties of C/C composites are improved on increasing the $MoSi_2$ content, which coincides with the results obtained for fracture toughness. Hence, the C/C composites with $MoSi_2$ affect the impact damage conditions in comparison to composites without $MoSi_2$. During impact damage in C/C composites with low-velocity impact (1.8 m/s), the damage

Fig. 5.37 Izod impact test

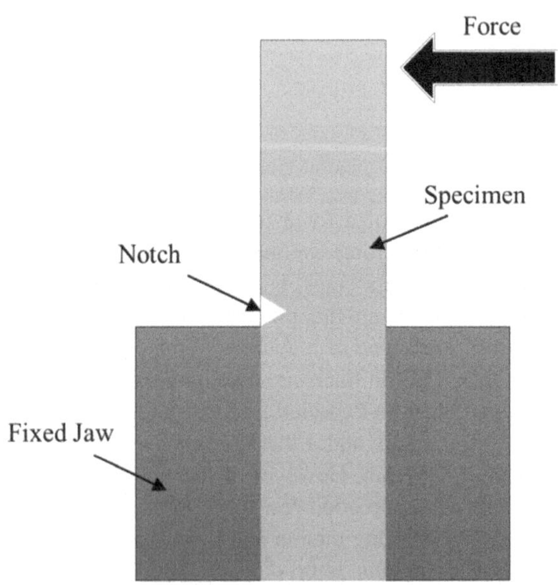

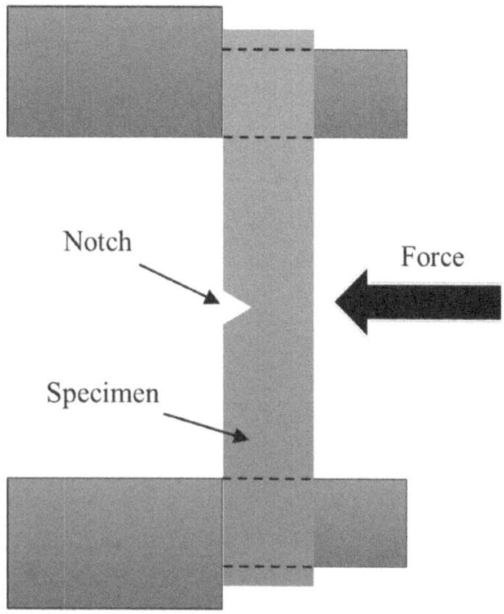

Fig. 5.38 Charpy impact test

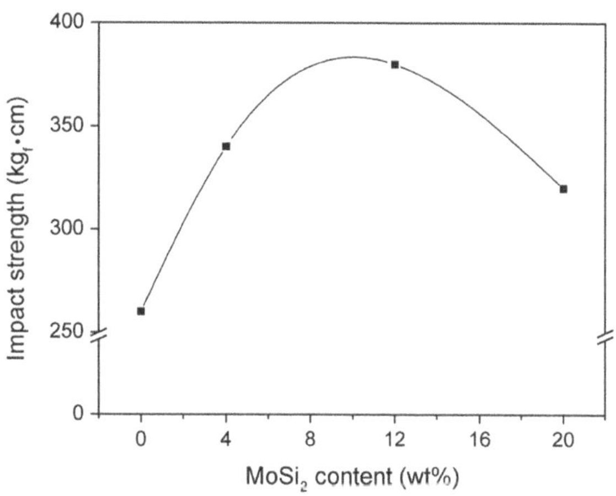

Fig. 5.39 Impact strength of C/C composites as a function of $MoSi_2$ content in prepreg resin [61]

zone acts as a soft region within a stiff laminate and magnifies the stress locally because of matrix cracking, fiber debonding delamination, and fiber fracture (Fig. 5.39).

5.4.3 Fracture Toughness

The fundamental fracture events occurring in the fragmentation test are illustrated in Fig. 5.40. In the initial state (0), there is no damage, and the specimen compliance is C_0. The energy released by the first fiber fracture, G, is determined from the difference

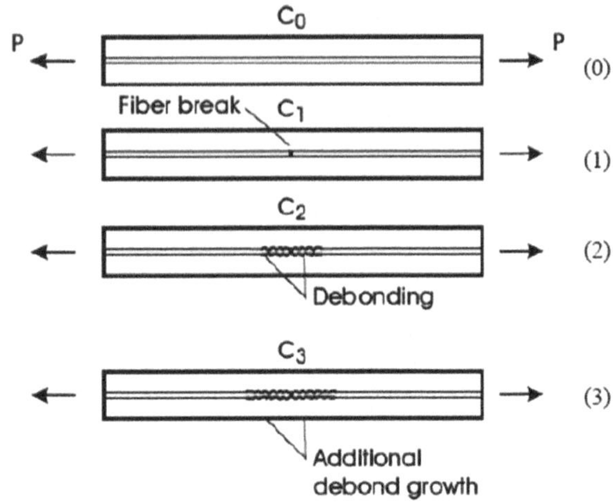

Fig. 5.40 Schematic of fracture events and associated compliances of SFFT specimen [62]

in the specimen compliance after (1) and before the first fiber fracture (i.e., C_1–C_0). The second fracture state (2) considers the debonding, which follows the first fiber fracture. An expression for G for debonding which occurs subsequent to a fiber break can be derived in terms of the compliance difference between C_2 and C_1.

Finally, in the third state (3), the specimen is supposed to be unloaded and then loaded again until the existing debonding begins to grow. The energy release rate, G_d, corresponding to growth of the debonding after unloading and reloading the specimen is determined from the compliance change (i.e., C_3–C_2). One important condition for this analysis is that the debonding grows before further fiber breakage occurs in the single-fiber fragmentation test specimen [62].

The expression for the calculation of mode II critical strain energy release rate, G_{IIC}, was derived from the Irwin and Kies expression for fracture energy [63, 64] as follows:

$$G_c = \frac{P^2}{2W} \cdot \frac{dC}{da} \tag{5.8}$$

In the above equation, P is the load, C is the compliance (displacement/load = δ/P), and a is the crack length. In this study, G_{IIC} was calculated using the direct beam theory method [63, 64], which used the end notched flexure (ENF) compliance expression given by [63, 65] the following equation:

$$C = \frac{2L^3 + 3a^2}{8E_f W h^3} \tag{5.9}$$

In the above equation, L is the half-span length, a is the crack length, E_f is the flexural modulus, and h is the half the specimen thickness. From (5.8) and (5.9), with a as the sole variable for C, the following equation can be obtained:

$$G_{IIC} = \frac{9a^2P^2}{16EW^2h^3} \tag{5.10}$$

For beams under small deflection, an expression for E can be obtained from (5.9), which may be substituted into (5.10), yielding (5.11).

$$G_{IIC} = \frac{9a^2P\delta}{2W(2L^3 + 3a^3)} \tag{5.11}$$

The maximum load was used as the critical load PC to determine the critical energy release rate and fracture toughness.

Figure 5.41 shows the typical relationship between the load and displacement at the central loading point of the ENF specimens. Elastic behavior is observed in the ENF specimens. Therefore, the maximum load is used as the critical load to determine the critical energy release rate and fracture toughness.

5.5 Relationship Between Surface and Interfacial Properties in Composites

5.5.1 *Surface Free Energy and Work of Adhesion*

It is well known that the knowledge of surface energetics at a given temperature of a solid has recently allowed significant progress in many academic and scientific fields involving two nonidentical molecular interactions at a certain intermolecular

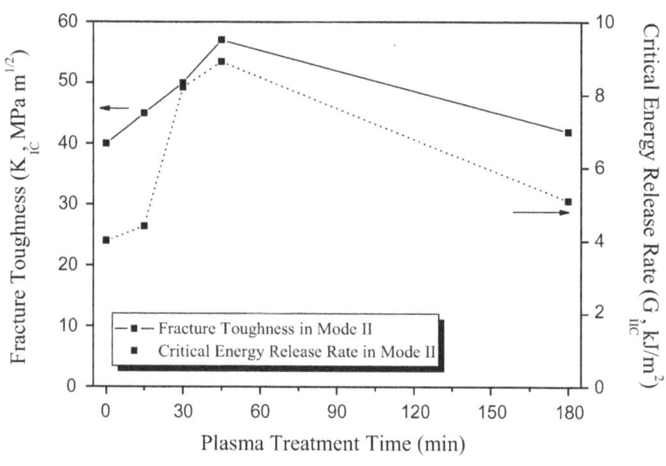

Fig. 5.41 **a** Typical load–displacement ENF curves and **b** fracture toughness and critical energy release rate in mode II interlaminar fracture tests for carbon fiber-reinforced epoxy matrix composites along the fiber direction [66]

distance, such as adsorption (gas–solid), wettability (liquid–solid), and adhesion (solid–solid). The surface free energy or surface tension (r_s) of a solid (subscript S) can be expressed by the sum of the London dispersive (r^L) and specific (or polar, r_S^{SP}) components.

$$r_s = r_S^L + r_S^{SP} \tag{5.12}$$

The superscripts L and SP are the components of the surface free energy resulting from intermolecular interactions of the London force of van der Waals attraction and the specific force (Debye, Keesom, hydrogen bonding, and other small polar effects), respectively. When the surface free energy of a solid can be determined on the basis of the contact angle measurements using the geometric mean, the work of adhesion, W_A, according to the Fowkes' proposition [67] based on the solid (S)–liquid (L) droplet–air system can be described by the following equation:

$$W_A = \gamma_L(1 + \cos\theta) \tag{5.13}$$

$$= 2\sqrt{\gamma_L^L \cdot \gamma_S^L} + 2\sqrt{\gamma_L^{SP} \cdot \gamma_S^{SP}} \tag{5.14}$$

θ is the contact angle of a liquid droplet in the solid state. For use on a large variety of polar solids, at least two nonidentical liquids with polar surface energies such as water, diiodomethane, ethylene glycol, and glycerol, are needed if the values of r^L and r^{SP} are known.

Park et al. [61] investigated the effects of the MoSi content on the work of adhesion of the C/C composites. In Fig. 5.42, the MoSi filler increases the work of adhesion, which is strictly influenced by the London dispersive component, W^L. In each case, it can also be seen that $W^{\ L}$ is more predominant than the specific component (W^{SP}) with respect to the work of adhesion of the composites. This is due to the carbonaceous nature of the composites caused by the decrease of oxygen functional groups, since carbonization leads to a very low content of oxygen.

5.5.2 Surface Free Energy Analysis Using Linear Fit Method

The wettability of carbon fibers was determined by measuring the wicking rates either by the mass pickup technique or by the surface velocity method. The contact angle used in this investigation was then calculated using Washburn's equation [68], which defines the flow of a liquid through a capillary as follows:

$$\frac{m}{t} = \frac{c \cdot \rho_2 \cdot \gamma_L \cdot \cos\theta}{2\eta} \tag{5.15}$$

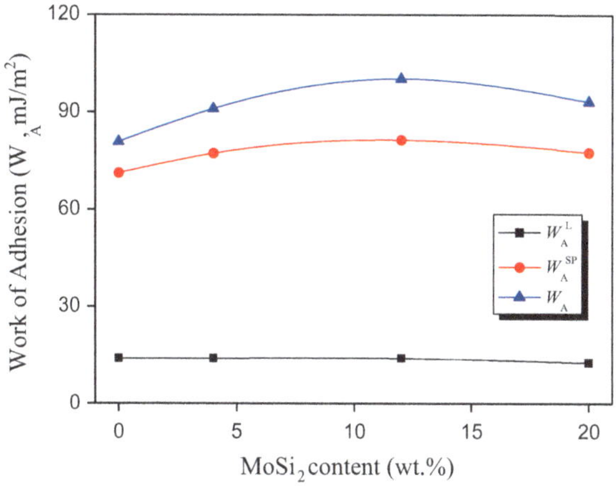

Fig. 5.42 Work of adhesion (W_A) of C/C composites as a function of MoSi content of the prepreg resin [61]

In (5.15), m is the weight of the penetrating liquid, t is the flow time, *c* is the packing factor, γ_L is the surface tension of liquid, and η and ρ_2 are the liquid density and viscosity, respectively. The London dispersive and specific components of surface free energy of carbon fibers are determined by measuring the contact angles of a variety of testing liquids with known London dispersive and specific components and by analyzing the results in accordance with the method proposed by Owens and Wendt [69] and Kaelble [70], using the geometric mean.

$$\frac{\gamma_L(1-\cos\theta)}{2(\gamma_L^L)^{1/2}} = (\gamma_S^{SP})^{1/2}\cdot\left(\frac{\gamma_L^{SP}}{\gamma_L^L}\right)^{1/2} + (\gamma_S^L)^{1/2} \tag{5.16}$$

In the above equation, the subscripts *L* and *S*, respectively, represent the liquid and solid states, γ_L^L represents the London dispersive component, and γ_L^{SP} corresponds to the specific component.

Park et al. [71] investigated the surface free energies and their oxyfluorinated carbon fiber components are listed in Table 5.6. A decrease in the surface free energies is observed. The higher the fluorine content on the carbon fiber surfaces,

Table 5.6 Surface free energy (γ_S), London dispersive (γ_S^d), and specific (γ_S^p) component of oxyfluorinated carbon fibers as functions of oxyfluorination condition (unit: mJm^{-2}) [71]

Specimens	γ_S^d	γ_S^P	γ_S	X_p
As-received	40.35	3.06	43.41	0.07
CFO-RT	30.80	8.61	39.41	0.22
CFO-100	29.70	11.68	41.38	0.28
CFO-300	28.62	9.51	38.13	0.25
CFO-400	28.45	9.10	37.55	0.24

the weaker the surface free energies. In other words, seemingly the polar component of the surface free energy increases at higher degrees of fluorination and then decreases slightly again for lower degrees of fluorination and with increasing fluorination temperatures. All of the experimental results listed in Table 5.6 are those usually observed for fluorinated carbon materials, considering that C–F bonding varies from chemical to physical bonding with increasing fluorination temperatures (for CFx prepared at 100 °C or higher).

5.5.3 Weibull Distribution

Both the tensile and pull-out strength of carbon fibers are usually assessed by single-filament tensile tests [72]. The experimental data generated by these tests have high scatter, mainly because of the presence of flaws along the fibers. Thus, the interpretation of the data must be done statistically. Several statistical distributions have been used to describe the tensile data, the most flexible so far being the Weibull distribution [73, 74].

$$F(\sigma, \sigma_0, m) = 1 - \exp\left(-\left(\frac{\sigma}{o_0}\right)^m\right) \tag{5.17}$$

In (5.17), σ is the load factor, σ_0 is the Weibull scale, and m represents the shape parameters. The flaw-induced nature of fiber failure leads to a length dependence of its tensile strength; i.e., the longer the fiber length, the larger the number of flaws that are present and the higher is the probability of occurrence of a severe flaw. The Weibull distribution has to be adapted to account for this dependence.

This also assumes that the strength distribution of each independent link is described by a simple Weibull distribution characterized by identical parameters. This leads to fiber survival probabilities equal to the product of the survival probabilities of each link. On the basis of these considerations, the Weibull cumulative distribution function F (σ; σ_0, m) and the corresponding mean Weibull strength ($\bar{\sigma}$), adapted to account for the gauge length dependence of the fibers (L), may be described by the following equations:

$$F(\sigma, \sigma_0, m) = 1 - \exp\left(-L\left(\frac{\sigma}{o_0}\right)^m\right) \tag{5.18}$$

$$\bar{o} = \sigma_0 L^{-1/m} \Gamma\left(1 + \frac{1}{m}\right) \tag{5.19}$$

In the above equation, Γ is the gamma function.

The parameter estimate for the two-parameter Weibull distribution is made for the strength data obtained at all gauge lengths simultaneously. In this work, the maximum likelihood theory is used to determine the Weibull parameters.

The method used is described in Stoner's thesis [75] and in the calculation programs explained in the thesis. In this way, for each type of fiber, a single set of parameters is obtained that fits all the gauge lengths tested [76]. The estimate of the tensile strength at any gauge length needed for the interfacial shear strength calculations is performed simply by substituting the calculated parameters into (5.19), for the specified fiber length.

Also, the load factor (σ) and L taken as part of the σ_0 parameter are redefined with the unit of stress (MPa) such that the Weibull distribution function can be expressed as follows:

$$P = 1 - \exp[-(\sigma/\sigma_0)^m] \tag{5.20}$$

By taking logarithm of the survival probability (1− P) twice, we also obtained the following equation:

$$\ln[1/(1-P)] = m \ln(\sigma) + \ln[(1+\sigma_0)^m] \tag{5.21}$$

The parameters m and σ_0 defining the Weibull distribution are usually determined either graphically or numerically by the least-squares method. Therefore, we can assign a failure probability P_i to each value of σ_i after ranking all the measured values in ascending order (i taking values from 1 to n, which corresponds to the number of measurements of the sample tested), according to one of the principal probabilities used.

$$P = i/(n+1) \tag{5.22}$$

After the tensile and pull-out tests on the carbon fibers, Park et al. [77] ordered the fracture strength values for each sample and assigned a probability P_i to each value σ_i using 5.22. The different cumulative probability distributions are shown in Fig. 5.43. For an initial failure probability at 6%, the fracture strength is increased

Fig. 5.43 Relationship between the failure probability and fracture strength in Ar^+ ion-irradiated carbon fibers [77]

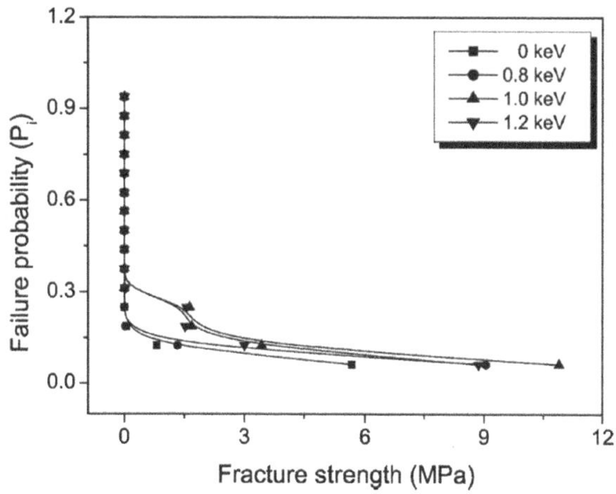

Table 5.7 Maximum IFSS (τ_{max}), interfacial toughness (G_i), and Weibull distribution parameters of a single carbon fiber with different ion beam irradiations [77]

Ion Beam Energy (keV)	τ_{max} (MPa)	G_i (kJ)	Parameter (m)	Scale Parameter (σ_0)
0	5.65	9.25	2.66	1.43×10^{-2}
0.8	9.04	12.77	2.41	1.22×10^{-2}
1.0	10.90	14.33	2.25	1.16×10^{-2}
1.2	8.86	11.76	2.17	1.24×10^{-2}

from 5.6 MPa for the as-received state to 10.9 MPa after a 1.0 keV Ar^+ ion treatment. The distributions for entire Ar^+ ion-irradiated carbon fibers are comparably closer to each other, with the exception of a few high extreme values. Also, the Weibull parameter m and the scale parameter σ_0 of all the ion-irradiated carbon fibers after pull-out tests exhibit similar behavior, compared with as-received carbon fibers, as listed in Table 5.7. This means that the interfacial shear strength (τ) or failure strength of Ar^+ ion-irradiated single-filament carbon fiber is dependent on the failure mechanism of carbon fibers in an epoxy matrix system.

References

1. P.M.A. Sherwood, J. Electron Spectrosc. Relat. Phenom. **81**, 319 (1996)
2. Y.Y. Li, K. Mochidzuki, A. Sakoda, M. Suzuki, Carbon **39**, 2143 (2001)
3. J. C. Vickerman, I. S. Gilmore, B. D. Ratner, and D. G. Castner, *Surface analysis—the principal techniques,* 2nd ed. (Wiley, 2009)
4. J.M. Walls, *Methods of surface analysis* (Cambridge University Press, Australia, 1989)
5. ASTM D 3178, ASTM Book of Standards. ASTM, Philadelphia, 05.05 (1984)
6. ASTM D 3174, ASTM Book of Standards. ASTM, Philadelphia, 05.05 (1984)
7. D. Briggs, M.P. Seah, *Practical surface analysis by Auger and X-ray photoelectron spectroscopy* (Wiley, Chichester, 1983)
8. J.P. Eberhart, *Structural and chemical analysis of materials* (Wiley, New York, 1991)
9. T. Wang, P.M.A. Sherwood, Chem. Mater. **6**, 788 (1994)
10. N. Iwashita, C.R. Park, H. Fujimoto, M. Shiraishi, M. Inagaki, Carbon **42**, 701 (2004)
11. G. Washer, F. Blum Jr., Adv. Mater. Sci. Eng. **2008**, 693207 (2008)
12. N. Melanitis, P.L. Tetlow, C. Galiotis, J. Mater. Sci. **31**, 851 (1996)
13. F. Hopfgarten, Fibre. Sci. Technol. **11**, 67 (1978)
14. H.J. Jacobasch, K. Grundke, P. Uhlmann, F. Simon, E. Mader, Compos. Interfaces **3**, 293 (1996)
15. J.I. Paredes, A. Martínez-Alonso, J.M.D. Tascón, J. Colloid Interface Sci. **258**, 276 (2003)
16. Z. Dai, B. Zhang, F. Shi, M. Li, Z. Zhang, Y. Gu, Appl. Surf. Sci. **257**, 8457 (2011)
17. R.J. Smiley, W.N. Delgass, J. Mater. Sci. **28**, 3601 (1993)
18. L.R. Bao, A.F. Yee, Polymer **43**, 3987 (2002)
19. J.C. Chen, I.R. Harion, Carbon **40**, 25 (2002)
20. W. Zhang, L. Jie, J. Liang, J. Mater. Sci. Technol. **20**, 369 (2004)
21. Y. Hou, T. Sun, H. Wang, D. Wu, J. Text. Res. **78**, 806 (2008)
22. M.J. Yu, C.G. Wang, Y.J. Bai, Y.X. Wang, Q.F. Wang, H.Z. Liu, Polym. Bull. **57**, 525 (2006)
23. M.J. Yu, Y.J. Bai, C.G. Wang, Y. Xu, P.Z. Guo, Mater. Lett. **61**, 2292 (2007)

24. A. Gupta, I.R. Harrion, Carbon **35**, 809 (1997)
25. Y. Hou, T. Sun, H. Wang, D. Wu, J. Mater. Sci. **43**, 4910 (2008)
26. R.B. Mathur, T.L. Dhami, O.P. Bahl, Polym. Degrad. Stab. **14**, 179 (1986)
27. G. Wu, C. Lu, L. Ling, A. Hao, F. He. J. Appl. Polym. Sci. **96**, 1029 (2005)
28. H. Ge, H. Liu, J. Chen, C. Wang, J. Appl. Polym. Sci. **113**, 2413 (2009)
29. D.X. He, C.G. Wang, Y.J. Bai, N. Lun, B. Zhu, Y.X. Wang, J. Mater. Sci. **42**, 7402 (2007)
30. A.G. Fazlitdinova, V.A. Tyumentsev, S.A. Podkopayev, G.P. Shveikin, J. Mater. Sci. **45**, 3998 (2010)
31. W. Johnson, *Strong fibers* (Elsevier, Amsterdam, 1985)
32. S. Damodaran, P. Desai, A.S. Abhiraman, J. Text. Inst. **81**, 384 (1990)
33. S. Kumar, D.P. Anderson, A.S. Crasto, J. Mater. Sci. **28**, 423 (1993)
34. J.B. Donnet, R.C. Bansal, *Carbon fibers* (Marcel Dekker, New York, 1990)
35. H. Shinohara, T. Sato, F. Saito, J. Mater. Sci. **28**, 6611 (1993)
36. G.J. Hayes, D.D. Edie, J.M. Kennedy, J. Mater. Sci. **28**, 3247 (1993)
37. M. Furuyama, M. Higuchi, K. Kubomura, H. Sunago, ibid. 28 (1993) 1611
38. M. G. Northolt, L. H. Veldhuizen, H. Jansen, in *Proceedings of 20th Biennial Conference on Carbon* (American Carbon Society, 1991) p. 222
39. V.R. Mehta, S. Kumar, J. Mater. Sci. **29**, 3658 (1994)
40. H.J. Jacobsch, K. Grundke, P. Uhlmann, F. Simon, E. Maeder, Comp. Interfaces **3**, 293 (1996)
41. R.H. Knibbs, J. Microsc. **94**, 273 (1971)
42. M.S. Dresselhaus, G. Dresselhaus, K. Sugihara, I.L. Spain, H.A. Goldberg, *Graphite fibers and filaments* (Springer, Berlin, 1988)
43. B.J. Kim, W.K. Choi, M.K. Um, S.J. Park, Surf. Coat. Tech. **205**, 3416 (2011)
44. J. Korab, P. Stefanik, S. Kavecky, P. Sebo, G. Korb, Comp. A **33**, 133 (2002)
45. A. Karaipekli, A. Sari, K. Kaygusuz, Renew. Energy **32**, 2201 (2007)
46. ASTM D 2766, ASTM Book of Standards. ASTM, Philadelphia, 5.02 (1984)
47. ASTM D 3417-83, ASTM Book of Standards. ASTM, Philadelphia, 8.03 (1984)
48. ASTM D 3418-82, ASTM Book of Standards. ASTM, Philadelphia, 8.03 (1984)
49. G. Schwarz, F. Krahn, G. Hartwig, Cryogenics **31**, 244 (1991)
50. H.B. Shim, M.K. Seo, S.J. Park, J. Mater. Sci. **37**, 1881 (2002)
51. B.J. Kim, K.M. Bae, M.K. Seo, K.H. An, S.J. Park, Mater. Sci. Eng., A **528**, 4953 (2011)
52. S.Y. Fu, B. Lauke, E. Mader, C.Y. Yue, X. Hu, Compos. Pt. A **31**, 1117 (2000)
53. S.J. Park, M.K. Seo, *Interface science and technology* (Elsevier, Nederland, 2011)
54. S.J. Park, Y.H. Chang, Y.C. Kim, K.Y. Rhee, J. Nanosci. Nanotechnol. **10**, 117 (2010)
55. M. Li, Y. Gu, Y. Liu, Y. Li, Z. Zhang, Carbon **52**, 1091 (2013)
56. L.Y. Yuan, S.S. Shyu, J.Y. Lai, J. Appl. Poly. Sci. **42**, 2525 (1991)
57. F. Mannocci, M. Sherriff, T.F. Watson, J. Endod. **27**, 758 (2001)
58. S.B. Kumar, S. Sivashanker, A. Bag, I. Sridhar, Mater. Sci. Eng., A **412**, 117 (2005)
59. S.U. Khan, A. Munir, R. Hussain, J.K. Kim, Compos. Sci. Technol. **70**, 2077 (2010)
60. I.S. Kim, Mater. Res. Bull. **33**, 1069 (1998)
61. S.J. Park, M.K. Seo, D.R. Lee, Carbon **41**, 2991 (2003)
62. F.A. Ramirez. L.A. Carlsson, B.A. Acha, Compos. Pt. A, 40 (2009) 679
63. P. Compstona, P.Y.B. Jar, P.J. Burchill, K. Takahashi, Compos. Sci. Technol. **61**, 321 (2001)
64. G.R. Irwin, J.A. Kies, J. Welding **19**, 193 (1954)
65. L.A. Carlsson, J.W. Gillespie, *Application of fracture mechanics to composites* (Elsevier, Amsterdam, 1989)
66. S.J. Park, M.K. Seo, J.R. Lee, Compos. Interf. **13**, 249 (2006)
67. F.M. Fowkes, J. Phys. Chem. **67**, 2538 (1963)
68. E.W. Washburn, Phys. Rev. **17**, 273 (1921)
69. D.K. Owens, R.C. Wendt, J. Appl. Polym. Sci. **13**, 1741 (1969)
70. D.H. Kaelble, J. Adhes. **2**, 66 (1970)
71. M.K. Seo, S.J. Park, J. Colloid Interface Sci. **330**, 237 (2009)
72. A. Hampe, G. Klinka, S. Meretz, E. Schulz, Composites **26**, 40 (1995)

73. E. Mader, K.H. Freitag, Composites **21**, 397 (1990)
74. W. Weibull, J. Appl. Mech. **18**, 293 (1951)
75. E.G. Stoner, *The effect of shape on the tensile strength of pitch-based carbon fibers* (Clemson University, Clemson, SC, 1991)
76. S.J. Park, Y.S. Jang, J. Colloid Interface Sci. **237**, 91 (2001)
77. S.J. Park, M.K. Seo, H.Y. Kim, D.R. Lee, J. Colloid Interface Sci. **261**, 393 (2003)

Chapter 6
Manufacture of Carbon Fiber Composites

Abstract Currently, the use of carbon fibers and their composites in space and aerospace applications continues to grow. However, the vast majority of composite molding is used in automotive, marine, industrial, and recreational applications. In this chapter, a brief explanation of various processing methods will be provided and the manufacturing processes typically used to make products found in various applications will be covered. Further, it is classified into two types of open molding and closed molding process will be described.

6.1 Selection of Manufacturing Process for Carbon Fiber Composites

In many cases, composite processing uses the same technique as matrix processing, which includes injection molding, compression molding, and extrusion. There are other techniques unique only to composite processing. These include filament winding, pultrusion, and hand lay-up. Although some techniques are used commonly in polymer processing, the operational conditions can be very different; thus, it is important not to directly transfer knowledge without careful consideration [1, 2]. In this section, a brief explanation of various processing methods will be provided and the manufacturing processes typically used to make products found in construction/civil infrastructure market will be covered.

The creation of a product from many different manufacturing processes is unique to the composites industry. There are a wide variety of processes available to the composites manufacturer to produce cost-efficient products. Each of the fabrication processes consists of characteristics that define the type of products to be produced. This is advantageous because these characteristics allow the manufacturer to provide the best solution for the customer. In order to select the most efficient manufacturing process, the manufacturing team considers several factors such as user needs, performance requirements, size of the product, surface complexity, appearance, production rate, total production volume, economic targets/limitations, labor, materials, tooling/assembly, and equipment.

S.-J. Park, *Carbon Fibers*, Springer Series in Materials Science 210,
https://doi.org/10.1007/978-981-13-0538-2_6

The identification of the process route is essential during the early stages of design because the process selection, material selection, and design features are intimately linked and for practical purposes, are inseparable.

The benefits of identifying the process route include the following:

- Maximized process efficiency and quality
- Total process development and solutions achieve concept to prototype to product transition
- Ensuring that process robustness is factored into new product developments
- Requirement of shorter durations to market new developments
- Absence of exposure to the risk associated with the capital investment of equipment
- Informed, impartial advice and support as and when required.

The development of a robust process technology is fundamental to the success of any new composite product venture. Drawing on a wealth of experience, covering virtually every sector of the polymer industry, it is possible to receive guidance on all aspects of the process from tool design and scale-up through to a complete line layout.

There is a variety of processing routes for obtaining composites. Selecting the right process for the product is paramount to achieve success. It is vital that throughout the design process, there is a fundamental understanding of the range of processes available along with their advantages, disadvantages, and limitations. For instance, a poor design may exclude a more cost-effective production route. Key factors influencing the process selection are

- Number of parts required
- Matrix type
- Reinforcement type
- Shape and complexity of the product
- Size of the product
- Cost per unit part
- Dimensional accuracy and tolerances
- Appearance/finishing of the product
- Strength and stiffness
- Postprocessing/assembly techniques.

6.2 Type of Reinforcement

Regardless of the materials, reinforcements are available in forms to serve a wide range of processes and end-product requirements. Materials supplied as reinforcements include roving, milled fiber, chopped strands, and continuous, chopped, or thermoformable mats. Reinforcement materials can be designed with unique fiber architectures and can be preformed (shaped) depending on the product requirements and manufacturing process.

6.2.1 Multiend and Single-End Rovings

Rovings are used primarily in thermoset compounds, but can be used in thermoplastics. Multiend rovings consist of many individual strands or bundles of filaments, which are then chopped and randomly deposited into the matrix resin. Processes such as sheet molding compound, preform, and spray-up use multiend roving. Multiend rovings can also be used in some filament winding and pultrusion applications. The single-end roving consists of many individual filaments wound into a single strand. The product is generally used in processes that use a unidirectional reinforcement, such as filament winding and pultrusion.

6.2.2 Mats

Reinforcing mats are usually described by weight per unit area. For instance, a 2-oz chopped strand mat will weigh 2 oz per square yard. The type and amount of binder used to hold the mat together dictate the differences between mat products. In some processes such as hand lay-up, it is necessary for the binder to dissolve. In other processes, particularly in compression molding, the binder is required to withstand the hydraulic forces and the dissolving action of the matrix resin during molding. Therefore, two categories of mats, i.e., soluble and insoluble, are produced.

6.2.3 Woven, Stitched, and Braided Fabrics

Many types of fabrics can be used to reinforce resins in a composite. Multidirectional reinforcements are produced by weaving, knitting, stitching, or braiding continuous fibers into a fabric from twisted and plied yarns. Fabrics refer to all flat-sheet, roll goods, irrespective of whether they are strictly fabrics. Fabrics can be manufactured by using almost any reinforcing fiber. The most common fabrics are constructed with fiberglass, carbon, or aramid. Fabrics are available in several weave constructions and thicknesses (from 0.0010 to 0.40 inches). Fabrics offer oriented strengths and high reinforcement loadings often found in high-performance applications (Fig. 6.1).

Fabrics are typically supplied on rolls of 25–300 yards in length and 1–120 inches in width. The fabric must be inherently stable enough to be handled, cut, and transported to the mold, but pliable enough to conform to the mold shape and contours. When properly designed, the fabric will allow for quick wet-out and wet through of the resin and will stay in place once the resin is applied. Fabrics such as rovings and chopped strands come with specific sizings or binder systems, which promote adhesion to the resin system.

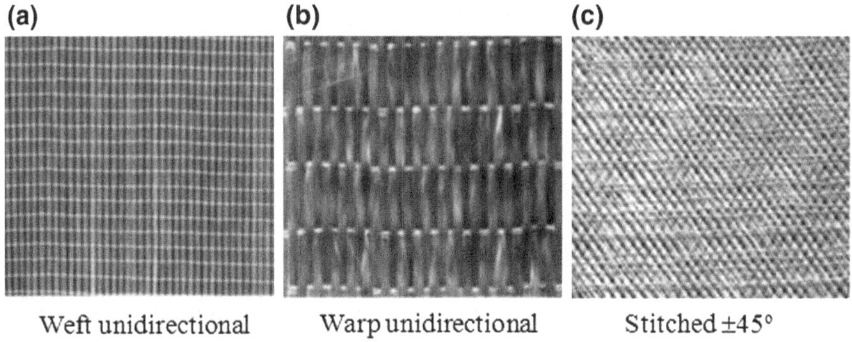

Fig. 6.1 Common fabrics used for open and closed molds

Fabrics allow for the precise placement of the reinforcement. This cannot be done with milled fibers or chopped strands and is only possible with continuous strands using relatively expensive fiber placement equipment. Because of the continuous nature of the fibers in most fabrics, the strength-to-weight ratio is much higher than that encountered in the case of cut or chopped fiber versions. Stitched fabrics allow for customized fiber orientations within the fabric structure. This can be of great advantage when designing for shear or torsional stability.

Woven fabrics are fabricated on looms in a variety of weights, weaves, and widths. In a plain weave, each fill yarn or roving is alternately crossed over and under each warp fiber, allowing the fabric to be more drapeable and to conform to curve surfaces. Woven fabrics are manufactured, wherein half of the strands of fiber are laid at right angles to the other half (0°–90°) (Fig. 6.2). Woven fabrics are commonly used in boat manufacturing.

Stitched fabrics, also known as nonwoven, noncrimped, stitched, or knitted fabrics, have optimized strength properties because of the fiber architecture. In the woven fabric, two sets of interlaced continuous fibers are oriented in a 0° and 90° pattern, where the fibers are crimped and are not straight. Stitched fabrics are produced by assembling successive layers of aligned fibers. Typically, the available fiber orientations include the 0° direction (warp), 90° direction (weft or fill), and

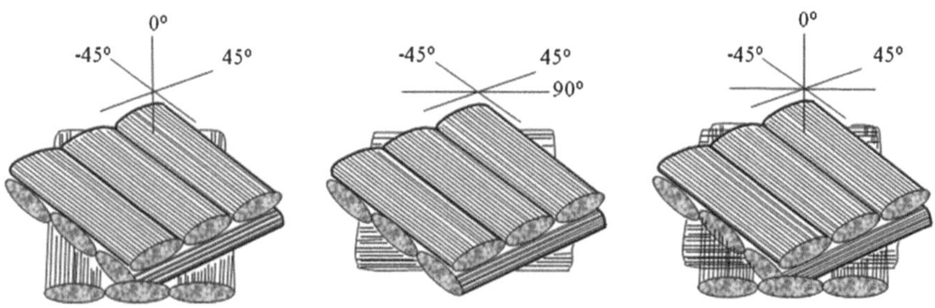

Fig. 6.2 Diagram of stitched triaxial and quadraxial fabrics

±45° direction (bias). The assembly of each layer is then sewn together. This type of construction allows for load sharing between the fibers so that a higher modulus, both tensile and flexural, is typically observed. The fiber architecture construction allows for optimum resin flow when composites are manufactured. These fabrics have been traditionally used in boat hulls for 50 y. Other applications include light poles, wind turbine blades, trucks, busses, and underground tanks. These fabrics are currently used in bridge decks and column repair systems. Multiple orientations provide a quasi-isotropic reinforcement.

Braided fabrics are engineered with a system of two or more yarns intertwined in such a way that all of the yarns are interlocked for optimum load distribution. Biaxial braids provide reinforcement in the bias direction only with fiber angles ranging from ±15° to ±95° (Fig. 6.3). Triaxial braids provide reinforcement in the bias direction with fiber angles ranging from ±10° to ±80° and axial (0°) direction.

6.2.4 Unidirectional

Unidirectional reinforcements include tapes, tows, unidirectional tow sheets, and rovings (which are collections of fibers or strands). Fibers in this form are all aligned parallel in one direction and uncrimped providing the highest mechanical properties (Fig. 6.4).

Composites using unidirectional tapes or sheets have high strength in the direction of the fiber. Unidirectional sheets are thin, and multiple layers are required for most structural applications.

In some composite designs, it may be necessary to provide a corrosion or weather barrier to the surface of the product. A surface veil is a fabric made from nylon or polyester that acts as a very thin sponge, which is capable of absorbing

Fig. 6.3 Biaxial and triaxial braided fabrics

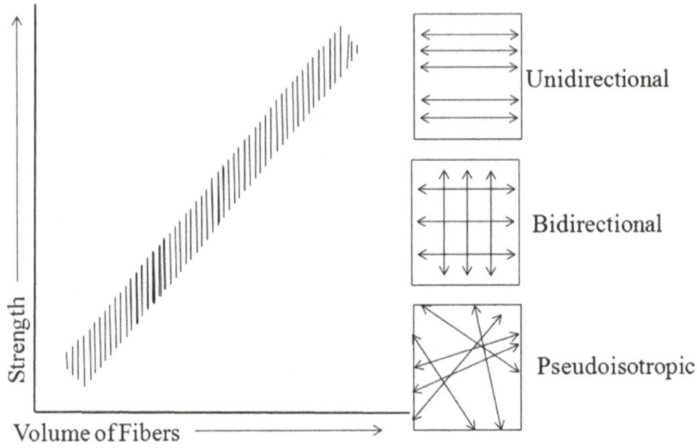

Fig. 6.4 Relationship between strength and fiber orientation

resin to 90% of its volume. This helps to provide an extra layer of protective resin on the surface of the product. Surface veils are used to improve the surface appearance and ensure the presence of a corrosion-resistance barrier for typical composite products such as pipes, tanks, and other chemical process equipment. Other benefits include increased resistance to abrasion, ultraviolet (UV), and other weathering forces. Veils may be used in conjunction with gel coats to provide reinforcement to the resin.

6.2.5 *Sandwich Construction*

A typical sandwich construction consists of two strong and relatively thin outer sheets or faces separated by but adhered to a layer of a less-dense and low-cost core material, which has lower strength and lower stiffness (see Fig. 6.5). The faces bear

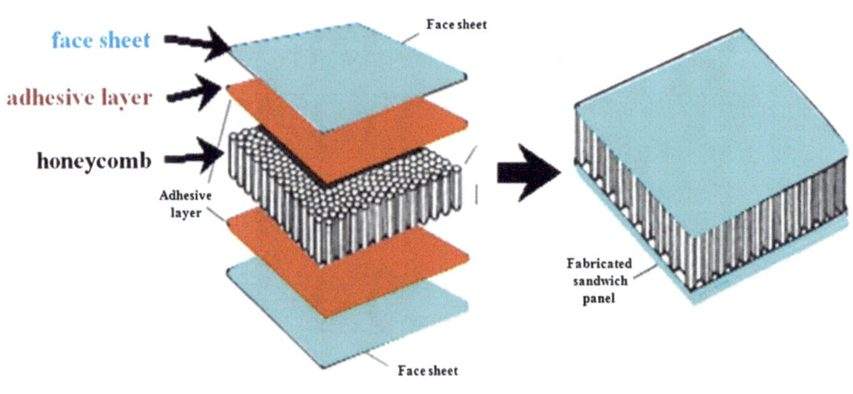

Fig. 6.5 Typical sandwich construction for carbon composites

most of the in-plane loading and transverse bending stresses. The core separates the faces and resists deformation perpendicular to the face, while providing a certain degree of shear rigidity along planes perpendicular to the faces. The face materials could be metals (e.g., titanium, steel, and aluminum alloys), plywood, or a fiber-reinforced composite. The core could be made of materials such as foamed plastics, synthetic rubbers, balsa wood, honeycomb, and truss. The sandwich architecture provides exceptional flexural stiffness compared to monocoque structures, while reducing weight and cost.

6.2.6 Prepregs

Prepregs are a ready-made material made of a reinforcement form and polymer matrix. Passing reinforcing fibers or fabrics through a resin bath is used to make a prepreg. The resin is saturated (impregnated) into the fiber and then heated to advance the curing reaction to different curing stages. Thermoset or thermoplastic prepregs are available and can be either stored in a refrigerator or at room temperature depending on the constituent materials. Prepregs can be manually or mechanically applied at various directions based on the design requirements.

6.3 Type of Matrix

6.3.1 Polymer Matrix Selection

The matrix of composites works as a binder transferring the loads through the fiber network. It maintains the fiber orientation and protects the fibers from environmental effects, redistributing the load to the surrounding fibers when an individual fiber breaks. Important criteria to be considered when selecting a resin candidate are the stiffness (elastic modulus), the yield and ultimate strengths, and toughness properties. Other factors such as thermal properties, processability, cost, availability, and health concerns are also of great importance [3].

The resin must be compatible with the processing method. Resin transfer molding (RTM) is the main process required during polymer matrix selection. This process involves a two-part mold in which a fiber preform is placed and the mold is closed. The resin is then pumped under low pressure through injection ports into the mold, filling the mold, and completely wetting out the reinforcement. Both the mold and resin can be heated depending on the type of resin. Currently, the aerospace industry is a major user of RTM components, and the automotive industry has made limited use of RTM for decades [4].

RTM is also gaining popularity in infrastructure, sports, and military industries. The advantages of RTM relative to hand lay-up are improved quality, higher production rates, reduced labor, and lower volatiles emissions, and while the main disadvantages are higher equipment costs and the need for low-viscosity resins.

The application of composite materials to primary structure to reduce structural weight is forcing structural designers and materials engineers to look for new and tough resin systems. Thermosets, elastomers, and thermoplastics are the three main categories of polymers. Thermoset polymers dominate as matrices in structural composite applications because of their good mechanical and thermal properties, good bonding to reinforcement, low cost, low viscosity, and ease of processing. Interest in thermoplastics is increasing because of their advantages such as toughness, potential processing advantages, recyclability and low-volatile emissions, while their high viscosity and poor bonding to reinforcement are disadvantageous [5]. Tough resins are generally formulated by adding elastomeric or thermoplastic compounds to the more brittle thermoset resin base. Elastomers generally have too low of an elastic modulus to serve as a matrix for rigid structural composites.

The selection of a resin involves several factors. Chemical characteristics such as resin viscosity, glass transition temperature (T_g), gel time, cure cycle, injection pressure, thermal stability, shelf-life, environmental resistance, and volatile emissions during processing are some of the parameters that need to be considered to determine operating and processing conditions for a specific resin. Mechanical properties such as strength and elastic modulus in certain directions, interlaminar fracture toughness, and environmental resistance are major composite properties to which the matrix must contribute [5, 6].

The most common thermoset resins used as composite matrices are unsaturated polyesters, epoxies, and vinyl esters. These resins offer good processability for liquid processing techniques such as RTM. The nature of the RTM process and the requirements of the wind turbine blade applications demand that the resin system should meet the target requirements. Of these, the resin modulus is important in maintaining composite compressive strength, particularly under hot and wet conditions.

Preferred resin characteristics include

- Low cost
- Resin elastic modulus of 2.75 GPa or higher
- Resin viscosity ranging from 100 to 500 cps
- T_g of 70 °C or higher
- Low moisture absorption
- Gel time of at least 20 min
- Preferably curable at room temperature
- High toughness.

Currently, unsaturated polyester resins are the most common systems used in composites by the wind turbine industry for the manufacture of blades. They are the most affordable, are easily processed, and possess adequate mechanical properties. However, most polyesters are brittle resins and have low-temperature resistance and significant moisture sensitivity. Vinyl esters are a chemical mixture of unsaturated polyesters and epoxy resins. The result is a resin with mechanical, thermal, and chemical properties similar to epoxies, with the ease of processing and high rate of cross-linking of unsaturated polyesters [5]. Vinyl ester resins are also stiff and brittle, but are tougher than polyesters because of the presence of the epoxy backbone[5]. Epoxy resins are widely used for high-performance composites, especially in aerospace, military, and sports industries [5, 7, 8–11]. Epoxy resins generally offer an increase in mechanical properties compared to polyesters and vinyl esters, but at a higher cost [3]. Another disadvantage of epoxies is their relatively high water absorption rate in comparison to vinyl esters [5]. The nature of curing for thermosets is explained in the following section. Details of each major thermoset resin material are described later.

6.3.2 Overview of Polymers

A polymer is a long molecule containing atoms held together primarily by covalent bonds along the molecule, while secondary bonds act between molecules [12]. The secondary bonds are an order of magnitude weaker than the covalent bonds. In general, thermoplastic polymers consist of separate molecules held together by secondary bonds. Thermoset polymers, when cured, form a three-dimensional network of covalent bonded segments, with secondary bonds forming between adjacent segments of the cross-links [13–19].

Thermoplastics can be separated into two subgroups, semicrystalline and non-crystalline (amorphous). Thermoplastics are linear or branched polymers, which melt upon heating when the thermal energy is adequate to overcome the secondary bonds. When melted, thermoplastics have a relatively high viscosity, which restricts available processing methods. Thermosets are cross-linked network polymers, which are amorphous and cannot be melted once the network is formed during curing. Thermosets have a relatively low viscosity prior to curing, which allows for convenient processing with adhesives and composites. They are also very reactive prior to curing, which allows for good bonding to the reinforcement [20]. Curing occurs after the product is in its final form.

In amorphous polymers, molecules can slip relative to each other without breaking covalent bonds. Chain slippage provides high strain to failure and toughness and damage tolerance. Semicrystalline polymers possess an increased strength and temperature and environmental resistance compared to amorphous thermoplastics. In thermosets, cross-linking is the process by which covalent bonds are formed between molecules through a chemical reaction creating a giant three-dimensional network. The polymer chains between the cross-links are now not as free to slip relative to each other, and thermosets have improved elastic modulus, creep resistance, and thermal/environmental resistance relative to thermoplastics, but at the expense of relatively brittle behavior [21].

When cross-links are formed in thermosets, the liquid polymer starts losing its ability to flow because the molecules can no longer slip past one another. Curing is the process of extending the polymer chain length and cross-linking chains together into a network. The molecular weight increases with the growth of the chain and then chains are linked together into a network of nearly infinite molecular weight. Curing is evident when there is a sudden change of the resin from a liquid to viscoelastic mass called a gel [22]. From the processing point of view, gelation is a critical factor because the polymer does not flow and is no longer processable beyond this point. The mechanism of curing differs for each polymer group, as will be discussed later.

The reason for coating the fibers is to improve the fiber/matrix interfacial strength and moisture resistance through both physical and chemical bonds and to protect the fiber surface from abrasion during handling conditions. Curing parametersand chemical agents for cross-linking a resin are different for each specific type of resin. A system requiring only a catalyst to begin the curing process is said to be promoted. A system requiring chemical compounds for the catalyst to initiate the cross-linking reaction of a resin is termed as an unpromoted system.

Epoxy resins are usually obtained in a two- or three-part system, which reacts when mixed together at the proper temperature. The reason suppliers often provide unpromoted resins to users is because the amount of promoter added to a resin will directly affect the processing time and shelf-life.

6.3.3 Properties of Polymers

6.3.3.1 Thermal Properties

Properties at elevated temperatures and the maximum use temperature are major concerns in the application of composite materials. These properties are dictated by the polymer matrix. T_g is defined as the temperature at which mobility between the molecules and segments in the amorphous regions becomes possible.

Above this temperature, the polymer is rubbery; below this temperature, the polymer is rigid. A partially crystalline polymer retains some rigidity up to the melting temperature, T_m, which is higher than T_g, even though the amorphous part of the material is soft and rubbery. T_g is the point at which there is adequate thermal energy to overcome secondary bonds; thus, segments of chains are then free to move, while being restrained at points of cross-linking (thermosets), chain entanglement (amorphous thermoplastics), or crystallites (semicrystalline thermoplastics). The polymer softens significantly with the temperature increasing to the proximity of T_g. The maximum use temperature for an amorphous polymer used as a composite matrix is usually below T_g [19].

The specific heat capacity of a polymer is higher when the molecules are free to move, and hence, the value of specific heat capacity decreases with decrease in cross-linking and increases with increase in temperature to the proximity of T_g. A differential scanning calorimeter (DSC) can be used to measure T_g by measuring the change in heat capacity. The DSC measures the difference in enthalpy and weight between a sample and a reference material, both subjected to a controlled temperature program [19]. Measurement of T_g of the thermosets used in this study was difficult, particularly when the samples were wet.

Another method to estimate the temperature at which a polymer softens is by the heat deflection temperature (HDT) measurement [22]. This technique determines the temperature when bending deflection at a constant stress increases rapidly. Details of this technique are described later.

6.3.3.2 Tension and Compression

Tension and compression tests are used to determine the yield and ultimate strengths and the ductility of a material. For a composite material, the stress–strain response is a function of the matrix and fiber properties. For a unidirectional composite, the slope of the stress–strain curve (Fig. 6.6), the longitudinal elastic modulus, E_{11}, can be accurately predicted by the rule of mixtures [23].

$$E_{11} = V_f \times E_f + V_m \times E_m \tag{6.1}$$

where E_f is the fiber modulus, E_m is the matrix modulus, and V_f is the fiber volume fraction, $V_m = (1 - V_f)$, if no porosity is present.

In the transverse direction perpendicular to the fiber axis, the modulus E_{22} is approximated by the Halpin–Tsai relationship [24] :

$$E_{22} = E_m \times \left(1 + z \times h \times V_f\right) / \left(1 - h \times V_f\right) \tag{6.2}$$

where $h = (E_{12f}/E_m - 1)/(E_{12f}/E_m - z)$, and $E_{12f} = E_{f/2} \times (1 + u)$; n is the Poisson ratio, E_{12f} is the shear modulus of the fiber, z is the curve-fitting factor given as 2 for E_{22} [24], and h is the curve-fitting factor.

In polymer matrix composites, the transverse modulus is dominated by the matrix modulus, while the longitudinal modulus is dominated by the fiber modulus. The stress–strain curve for unidirectional materials is usually approximately linear to failure. The tensile strength in the longitudinal direction occurs approximately

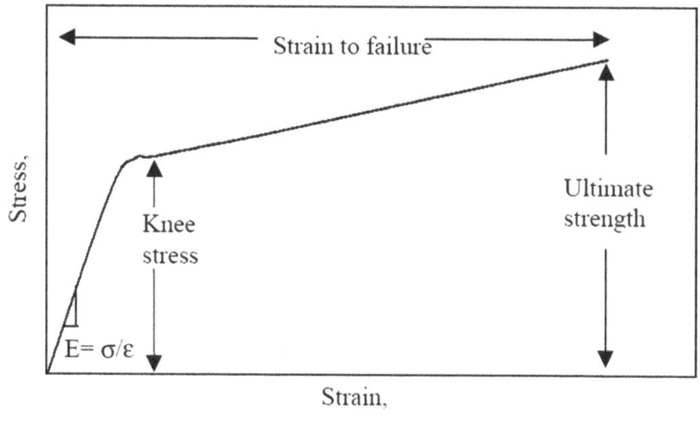

Fig. 6.6 Stress–strain curve of laminate

when the strain in the fiber reaches a value close to the fiber ultimate strain. The transverse strength (and shear strength) is dominated by the matrix, with the mode of failure being a crack growing parallel to the fibers in the matrix and fiber/matrix interface. The limiting value for the transverse tensile strength is the ultimate strength of the matrix. For brittle resins and/or poorly bonded fibers, the transverse strength will be lower than the matrix strength [25]. The compressive strength of unidirectional composites in the longitudinal direction is also a matrix-dominated property Polymers for most glass fiber composites [25]. Failure occurs when the fibers locally buckle or kink in the matrix; the matrix provides lateral resistance against buckling. The compressive strength can be approximated by the following equation:

$$s = G_m/2\left(1 + v_f\right) \tag{6.3}$$

where $G_m = E_m/2(1 + v)$; s is the predicted compressive strength; G_m is the shear modulus of the resin; n is the Poisson ratio of the resin, and E_m is the tensile modulus of the resin.

This formula assumes perfect fiber alignment and tends to significantly overestimate the compressive strength [24]. Most composites are used with layers in various directions. The ply lay-up used for multidirectional laminates in this thesis was mostly [0/+45/0]*s*, where *s* indicates symmetry about the mid-thickness; thus, this was an eight-ply laminate. This laminate was tested at both the 0° and 90° directions. The stress–strain curve for a multidirectional laminate is a function of the stress–strain behavior of each ply, transformed to the overall laminate coordinates. The stress–strain response is usually predicted by a laminated plate theory-based software program [22]. A typical stress–strain curve for a multidirectional laminate in tension would then include nonlinear responses where off-axis plies cracked, with the ultimate strength dominated by 0° layers if any (Fig. 6.6).

In the absence of 0° fibers in the direction considered, the knee strength is the most important criterion for design. It affords the designer an estimate of how much elastic elongation the material can tolerate prior to significant matrix cracking. The stress at which knee strain occurs is called the 0.2% offset knee stress, calculated by drawing a parallel line (that has an origin at 0.2% strain) to the linear portion of the stress–strain curve until it intersects the curve (Fig. 6.7) [26]. This is similar to the usual method used to define the yield stress in metals and polymers. The neat matrix yield stress was calculated in this manner in this study.

6.3.4 Polymer Chemistry

The primary functions of the resin are to transfer stress between the reinforcing fibers, act as a glue to hold the fibers together, and protect the fibers from the mechanical and environmental damage. To enable the fibers to carry maximum load, the matrix must have a lower modulus against elongation than the

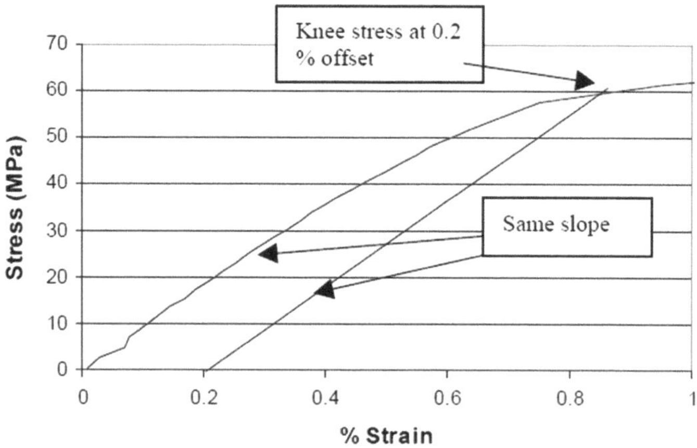

Fig. 6.7 Knee stress at 0.2% strain

reinforcement. The matrix determines the service operating temperature of the composites as well as the processing parameters for part manufacturing.

Polymer matrix resins are divided into two major groups known as thermoset and thermoplastic. Thermoset resins are usually liquids or low-melting solids in their initial form. When used to produce finished goods, these thermosetting resins are "cured" by the use of a catalyst, heat, or a combination of the two. Once cured, the solid thermoset resins cannot be converted back to their original liquid form. Unlike thermoplastic resins, cured thermosets will not melt and flow but will soften when heated (and lose hardness) and once formed they cannot be reshaped. Thermoplastic resins, on the other hand, become soft when heated and may be shaped or molded while in a heated semifluid state and become rigid when cooled. Maximum continuous-use temperatures of the various types of thermoset and thermoplastic resins are shown in Table 6.1.

6.3.4.1 Thermoset Resins

Thermosets are cross-linked polymers cured or set using heat or heat and pressure, which once cured cannot be remelted or reformed. During curing, they form three-dimensional molecular chains, called cross-links, as shown in Fig. 6.8. The higher the number of cross-linkings, the more rigid and thermally stable the materials will be. Cured thermoset resins may soften when heated but do not melt or flow. They generally show a higher resistance to heat than thermoplastics. Because thermosets are brittle, they are used with filler materials such as powders and fibers to provide improved strength and/or stiffness [27–30]. Fibers can be either chopped or wound, and commonly include glass, glass fibers, or cloth. Some products contain solid lubricant fillers such as graphite and molybdenum disulfide. Others contain aramid fibers, metal powders, or inorganic fillers with ceramics and silicates.

Table 6.1 Maximum continuous-use temperature for various thermosets and thermoplastics

Materials	Maximum continuous-use temperatures (°C)
Thermosets	60–150
Vinyl ester	60–150
Polyester	70–150
Phenolics	80–215
Epoxy	150–250
Cyanate esters	230–320
Bismaleimide	
Thermoplastics	50–80
Polyethylene	50–75
Polypropylene	70–95
Acetal	75–100
Nylon	70–120
Polyester	120–220
Poly(phenylene sulfide)	120–250
Poly(ether ether ketone)	200–260
Teflon	

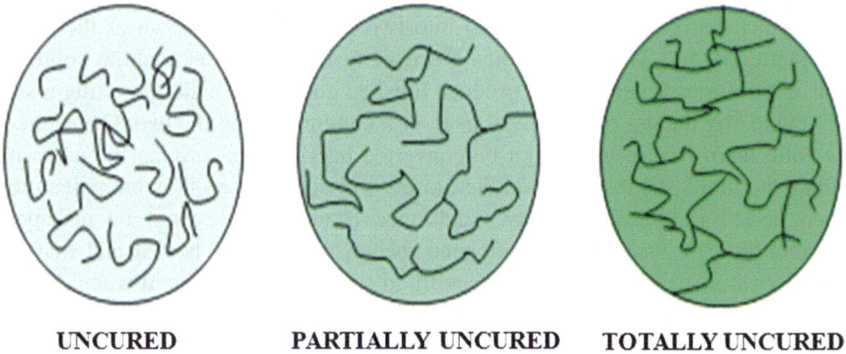

Fig. 6.8 Cross-linking of thermoset molecules during curing

Thermoset resins provide easy processability and better fiber impregnation because the liquid resin is used at room temperature for various processes such as filament winding, pultrusion, resin transfer molding, and reaction injection molding. Thermosets offer greater thermal and dimensional stability, better rigidity, and higher electrical, chemical, and solvent resistance. Some examples of the most common resins used in thermoset composites are epoxy, polyester, vinyl ester, phenolics, cyanate esters, bismaleimides, and polyurethane. Some of the basic properties of thermoset resins are listed in Table 6.2.

1. Polyester resins

Polyester resins are formed by reacting a diacid and a dialcohol by condensation polymerization to form an ester. Orthophthalic polyesters are prepared by combining

Table 6.2 Typical properties of thermoset resins

Resins	Density (g/cm^3)	Tensile modulus (GPa)	Tensile strength (MPa)
Epoxy	1.2–1.4	2.5–5.0	50–110
Phenolics	1.2–1.4	2.7–4.1	35–60
Polyesters	1.1–1.4	1.6–4.1	35–90
Vinyl esters	1.04–1.1	3.2–3.6	68–80

phthalic anhydride with either maleic anhydride or fumaric acid. A combination using isophthalic acid or terephthalic acid results in an isophthalic polyester, which has better thermal stability, chemical resistance, and mechanical properties than an orthophthalic polyester, although the cost is higher. The number of repeating units in a typical polyester is in the range 10–100. Because double-bonded carbon–carbon bonds are called unsaturated bonds, the thermoset polyesters containing these bonds are called unsaturated polyesters [31–34]. After the polymerization and depending on the number of units, a highly viscous liquid may result. For further processing, the polyesters are dissolved in low-molecular-weight monomers such as styrene (the most widely used), also known as solvents. Unsaturated polyesters usually contain 35–50% monomer by weight.

Polyesters are cured by using organic peroxides as initiators such as methyl ethyl ketone peroxide (MEKP) and benzoyl peroxide (BPO). The initiator reacts with the carbon–carbon double bond forming a new bond and another free radical on the carbon (Fig. 6.9). This new radical reacts with another carbon–carbon double bond to form a new bond and another free radical. A typical concentration of initiators is 1–2%. Higher or lower concentrations of the initiator will result in incomplete cross-linking Polyester resins with inferior properties. Cross-linking takes place when the carbon–carbon double bonds from separate molecules are linked together, creating a giant three-dimensional molecule, which increases the molecular weight of the polymer. Monomers also take part in the cross-linking reaction because they contain active carbon–carbon double bonds and serve as bridges between polyester molecule chains. One disadvantage of the solvents is that they are volatile and their vapors are deposited in the environment when processing. One advantage of polyester is that cross-linking does not generate by-products, which renders molding easy (this is true for epoxies and vinyl esters as well) [5].

The mobility of molecules decreases with increase in molecular weight and the viscosity increases; consequently, the reaction stops when free radicals are prevented from finding new double bonds. An increase in temperature during the curing process will allow for increased mobility and the creation of more free radicals. Postcuring is a process that increases the T_g of a resin because it allows the completion of cross-linking by eliminating reactive sites. Often the highest temperature reached by a room-temperature cross-linking polyester (with exothermic curing) will become its T_g [12].

Mechanical properties of cured polyester resins are affected by the monomer type and amount, acids, and curing temperatures. Orthophthalic polyesters are the least expensive form of unsaturated polyesters, but they have limited mechanical

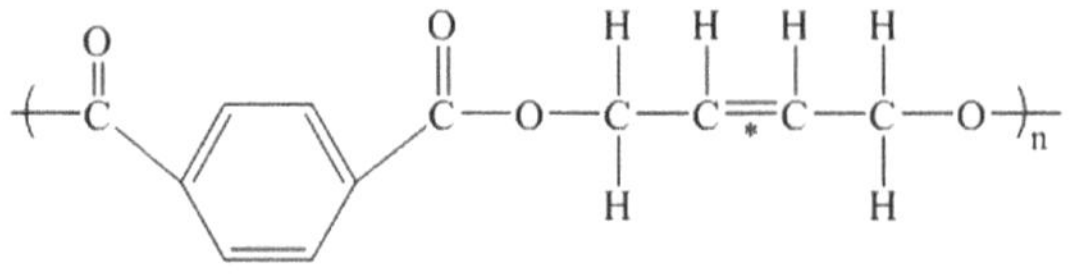

(a) Unsaturated polyester with active C=C site(*)

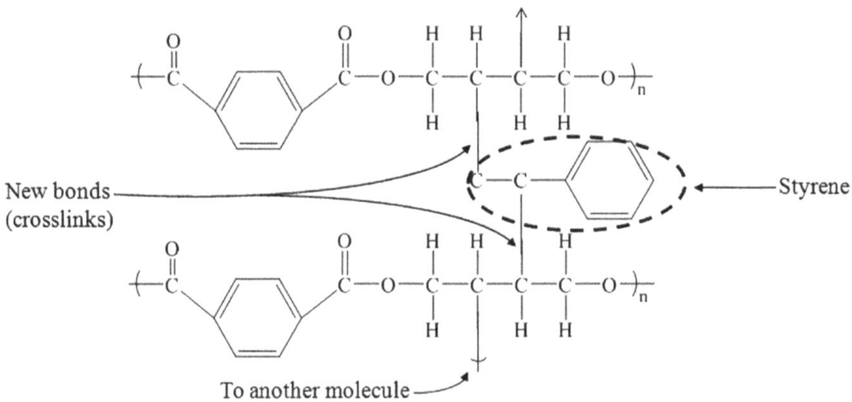

(b) Crosslinking of unsaturated polyester

Fig. 6.9 Unsaturated polyester showing **a** reactive carbon–carbon double bond and **b** cross-linking reaction

properties and sensitivity to environmental conditions. Isophthalic polyesters are more costly, but they show higher tensile and flexural properties because of the higher molecular weight and higher number of linear chains [3]. The reaction between a polyester resin and a free radical (provided by the catalyst) is shown in Fig. 6.9.

2. Vinyl ester resins

Vinyl ester resins are obtained by reacting an unsaturated acid with an epoxy. The reaction of methacrylic acid and bisphenol A (BPA) epoxy resin dissolved in styrene monomer is the most common version of vinyl esters [35]. An advantage of vinyl esters is that the cross-linking reaction is identical to the free radical cross-linking of unsaturated polyesters. The structure of the BPA vinyl ester is shown in Fig. 6.10. The cross-linking density of BPA vinyl esters decreases with increase in the molecular weight of the epoxy because the methacrylate sites of cross-linking are at the ends of the molecular chain.

Novolac epoxy vinyl ester resins offer an increased number of cross-linking sites along the backbone, which increases the final T_g of the resin and temperature resistance. The cross-linking reaction of vinyl esters is identical to the free radical cross-linking of unsaturated polyesters and the reaction also uses similar initiators and inhibitors. The doubly bonded carbon–carbon located at the end of the units

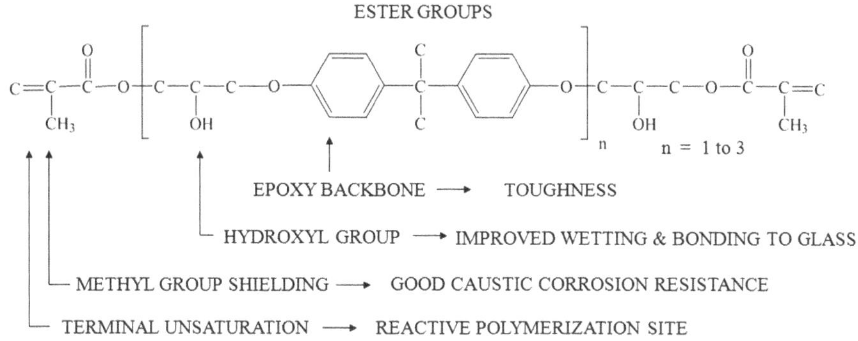

Fig. 6.10 Bisphenol A vinyl ester

only (Fig. 6.10). MEKP, BPO, and Trigonox are common catalysts for vinyl esters used from 1 to 2% volume. Trigonox catalyst is known for its nonfoaming character with vinyl esters. Cobalt naphthalene is a promoter and is usually added to the resin at 0.2–0.4% by weight.

Vinyl esters are well known for the resistance to environmental conditions because their high reactivity achieves complete curing more easily and rapidly than for polyesters. Vinyl esters show higher elongation at break than polyesters, which also makes them tougher. The chemical resistance of vinyl esters is generally greater than that of polyesters because of the influence of the methyl group [5].

3. Epoxy resins

Epoxy resins generally consist of the three-membered epoxy group ring. The most common type of epoxy used is the diglycidyl ether of bisphenol A (DGEBA) (Fig. 6.11). Epoxy groups could be located in different locations other than the ends [31]. At least two epoxy groups have to be on the polymer molecule for cross-linking. Epoxies usually show high viscosities at room temperature; therefore, diluents that also contain epoxide groups are used to lower the viscosity.

Hardeners are used to cross-link epoxies. Amine hardeners are the most common. A hardener should be added in amounts such that the number of epoxide groups is equivalent to the number of cross-linking sites provided by the hardener [5]. If the hardener is added in appropriate amounts, a well cross-linked structure with maximum properties will result. Some epoxies are formulated to cross-link at room temperature, but most epoxies used in composite applications require high temperatures to initiate the cross-linking [3]. Physical and mechanical properties are also improved by increasing the molecular weight when curing. In the case of polyester resins, no condensation by-products are formed during epoxy curing reactions.

The toughness of the epoxies depends on the length of the polymer chain between the epoxy groups. Longer chains (higher molecular weight) result in tougher polymers. One disadvantage of long chains is that there are a small number of cross-links per unit length (low cross-link density), which results in less stiff and

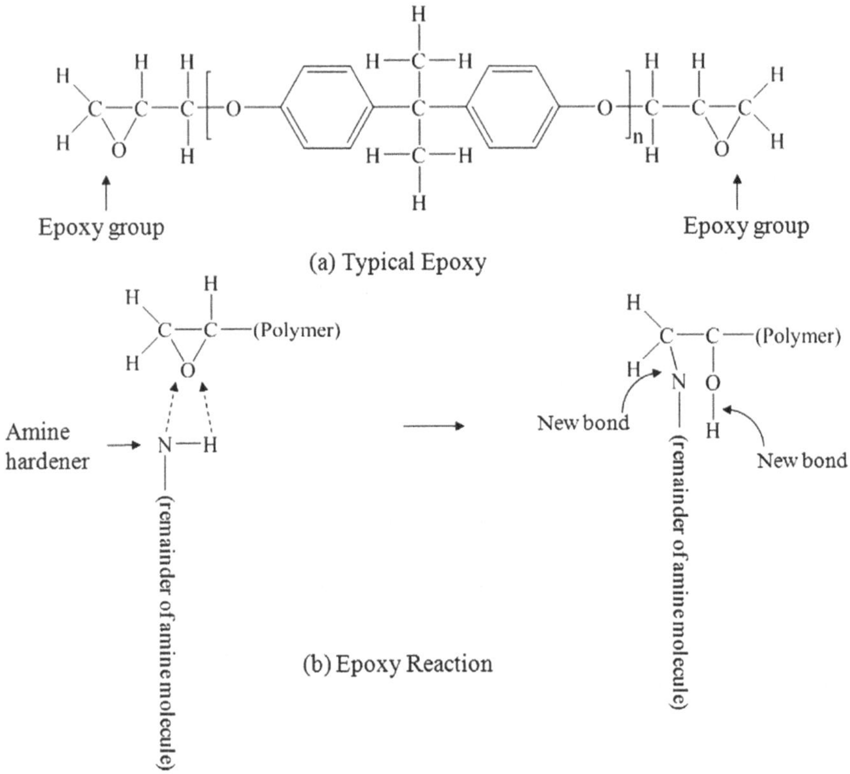

Fig. 6.11 **a** Typical epoxy polymer-formation and **b** epoxy-formation reactions

less strong materials, with lower modulus and heat resistance. Rubber polymers are added to epoxy resins to increase the toughness.

Epoxies are usually more expensive than unsaturated polyesters, but show important advantages. Epoxies are stronger, stiffer, tougher, more durable, and more solvent resistant and have a higher maximum operating temperature than polyester thermosets [5].

4. Polyurethane resins

Polyurethane resins can be either thermoset or thermoplastic. Polyurethanes are formed by reacting two monomers, each with at least two reactive groups. Polyol and isocyanate monomers are generally liquids that are combined to form the polyurethane. A typical polyurethane molecule can be seen in Fig. 6.12. Polyurethanes are very versatile polymers. The role of the polyol in polyurethane chemistry is like the role of the epoxy molecule in epoxy chemistry. The role of the isocyanate in polyurethane chemistry is similar to that of the hardener in epoxy chemistry. Polyols have OH groups at the ends of the branches. Polyurethanes show superior toughness and elongation to failure; therefore, they are used in the automotive industry, for example, to manufacture car bumpers [36].

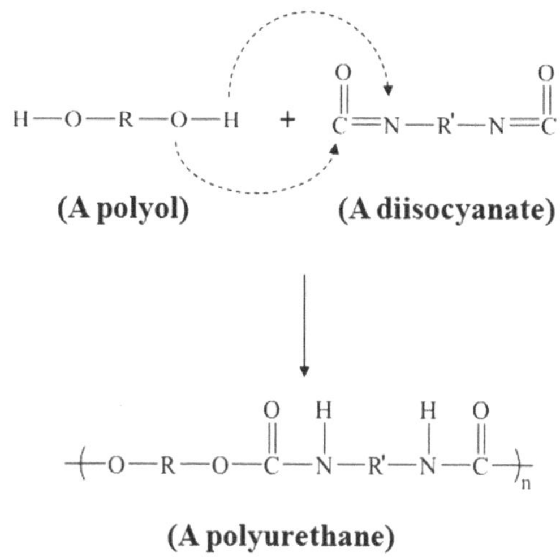

Note : R is usually a multifunctional polyether or polyester but can also be a small organic group.

R' is usually a large aromatic group.

Fig. 6.12 Polyurethane-formation reaction

Mechanical properties of polyurethanes will depend on the type of the monomer used. Ether-based polyurethanes show the highest mechanical properties and are also known for their short and fast solidification times, which makes them suitable for processing methods with faster injection times, such as reaction injection molding (RIM) in comparison to RTM [31]. There are semirigid and rigid polyurethanes. A low T_g caused by the flexible polyol chains is a characteristic of semirigid polyurethanes, which results in good flexibility. Rigid polyurethanes can be used at temperatures up to 150 °C because of the cross-link structure of the matrix material [3].

5. Phenolic resins

Phenolics are a class of resins commonly formed by the reaction of phenol (carbolic acid) and formaldehyde, and catalyzed by an acid or a base. Urea, resorcinol, or melamine can be used instead of phenol to obtain different properties. Their curing characteristics are different from other thermosetting resins such as epoxies because water is generated during the curing reaction. Phenolics are used for aircraft interiors, stowbins, and galley walls, as well as other commercial markets that require low-cost, flame-resistant, and low-smoke products. Pigmented applications are limited to red, brown, or black colors.

Phenolic composites have many desirable performance qualities, including high temperature resistance, creep resistance, excellent thermal insulation and sound damping properties, corrosion resistance, and excellent fire/smoke/smoke toxicity properties. Phenolics are used for various composite manufacturing processes such as filament winding, resin transfer molding, injection molding, and compression molding. Phenolics provide easy processability, tight tolerances, reduced machining, and high strength. Because of their high temperature resistance, phenolics are used in exhaust components, missile parts, manifold spacers, commutators, and disk brakes.

6. Cyanate ester resins

Cyanate esters offer excellent strength and toughness, better electrical properties, and lower moisture absorption compared to other resins. If they are formulated correctly, their high-temperature properties are similar to bismaleimide and polyimide resins. They are used for a variety of applications, including in space crafts, aircrafts, missiles, antennae, radomes, microelectronics, and microwave products. Cyanate esters are formed by the reaction of bisphenol esters and cyanic acid that cyclotrimerize to produce triazine rings during the second curing. Cyanate esters are more easily cured than epoxies. The toughness of cyanate esters can be increased by adding thermoplastics or spherical rubber particles.

7. Bismaleimide resins

Bismaleimide is produced by a vinyl-type polymerization of a prepolymer terminated with two maleimide groups. Bismaleimide has an intermediate temperature capability (between epoxy and polyimide). Monomers are usually synthesized from maleic anhydride and an aromatic diamine; the bismaleamic acid formed is cyclodehydrated to a bismaleimide resin. The double bond of the maleimide is very reactive and can undergo chain extension reactions. These addition-type polyimides or maleimide-based polyimides are used in high-performance structural composites requiring higher temperature use and increased toughness. Bismaleimides are hardly used in their pure solid form. In most applications, they come in conjunction with reactive comonomers (e.g., vinyl and allyl compounds, allyl phenols, and aromatic amines). A great advantage of these mixtures over the pure powder is better processability. With the help of comonomers, bismaleimide powders can be turned into paste-like liquids and cast to the desired shape. However, the viscosity of such systems is often very high so that solvents are added in order to make processing easier or even to make processing possible at all. A typical field of application of such bismaleimide systems is their use as matrix resins for high-performance fiber-reinforced composite materials in the aviation and space industries. Very often these matrix resins require high curing temperatures (more than 200 °C) and long curing times.

6.3.4.2 Thermoplastic Resins

Thermoplastic resins are materials that soften when heated and harden again when cooled. These materials are, in general, ductile and tougher than thermoset resins and are used for a wide variety of nonstructural applications without fillers and reinforcements [42–45]. Thermoplastics can be melted by heating and solidified by cooling, which render them capable of repeated reshaping and reforming. Thermoplastic molecules do not cross-link and therefore, they are flexible and reformable. Thermoplastics can be either amorphous or semicrystalline, as shown in Fig. 6.13.

After the rapid development of new types of fibers with extremely good properties during the final decades of the last century, the commercial availability of thermoplastic matrix composites represents an important recent innovation in the field of composites. The success of these composites can be attributed to different reasons and is interesting from the viewpoint of fabrication technologies, eco-sustainability of the entire process, and final performances.

In particular, the use of thermoplastic matrices avoids the presence of dangerous vapors in the working spaces in comparison to those encountered in the case of thermosetting matrices. This represents a strong environmental improvement and an economical challenge for the fabrication, the latter being further strongly improved if one considers the necessity of expensive hardware and procedures needed in the use of prepreg thermosetting composites. Thermoplastic composites imply the reversibility of thermal actions on materials during fabrication of the final element. In other words, it is possible to prefabricate semifinished items and later take them to final shape and dimensions without any time limit. Several thermoplastic polymers can be considered as possible matrices for composites, and many innovations are expected in the near future.

Presently, the most interesting polymers for actual applications are polyethylene, polypropylene, polyester, nylon, polycarbonate, polyetherimide, poly(phenylene sulfide), and poly(ether ether ketone), which are produced in different places in the world. There are also producers of various semifinished products using these matrices, which can be used to produce the final element [46–58]. Some typical properties of common thermoplastic matrices are listed in Table 6.3.

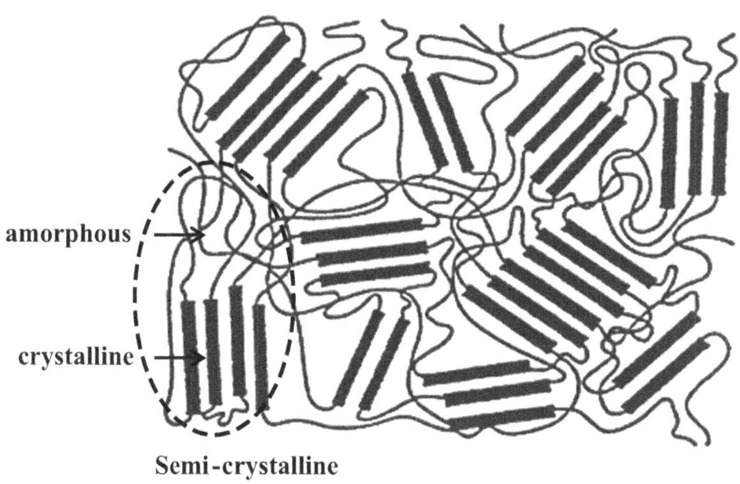

Fig. 6.13 Molecular arrangements in thermoplastic polymers

Table 6.3 Typical properties of thermoplastic resins

Resins	Density (g/cm^3)	Tensile modulus (GPa)	Tensile strength (MPa)
Polypropylene	0.90	1–1.4	25–38
Polyethylene	0.9–1.0	0.7–1.4	20–35
Polycarbonate	1.06–1.2	2.2–2.4	45–70
Nylon	1.1	1.3–3.5	55–90
Polyester	1.3–1.4	2.1–2.8	55–60
Poly(ether ether ketone)	1.3–1.35	3.5–4.4	100
Poly(phenylene sulfide)	1.3–1.4	3.4	80
Polyetherimide	1.27	3	105

1. Polyethylene

Polyethylene, which is a semicrystalline polymer with a linear molecular structure of repeating $-CH_2-CH_2-$ units, is the most commonly used plastic in the world. In general, polyethylene is considered as a strong lightweight thermoplastic with very good chemical resistance. Further, the polymer elongates before breaking, which enhances the toughness of the porous parts made from polyethylene. Polyethylene is classified into several categories based mostly on its density and branching. The mechanical properties of polyethylene significantly depend on variables such as the extent and type of branching, crystal structure, and molecular weight. Products made from standard polyethylene grades typically have pore size diameters ranging from 7 to 150 μm. Polyethylene is resistant to concentrated acids, alkalis, and many organic solvents. Depending on the crystallinity and molecular weight, T_m and T_g may or may not be observable. The temperatures at which these occur vary strongly with the type of polyethylene. For common commercial grades of medium- and high-density polyethylene, T_m is typically in the range 120–130 °C. The T_m for average commercial low-density polyethylene is typically 105–115 °C.

2. Polypropylene

Polypropylene is in many respects similar to high-density polyethylene. However, instead of forming a long polymer chain made up of repeating $-CH_2-$ components, the polypropylene molecule has a series of CH_3 groups that hang off the main carbon backbone. In isotactic polypropylenes, the CH_3 groups are oriented on one side of the carbon backbone. This orientation creates a greater degree of crystallinity, making polypropylene stiffer and more resistant to creep (the tendency to flow under stress) than polyethylene. Generally, polypropylene has demonstrated certain advantages in terms of improved strength, stiffness, and higher temperature capability over polyethylene. Because of its high specific strength, polypropylene has been very successfully applied to the formation of fibers, which is the single largest use of polypropylene. Available in molded shapes, sheets, and tubes, the nominal pore size in polypropylene ranges from 80 to 150 μm. Polypropylene is highly resistant to most acids and alkalis and is resistant to most organic solvents below 80 °C.

When not under stress, polypropylene may be used at temperatures ranging from 93 to 149 °C. Further, polypropylene can normally be subjected to steam sterilization.

Polypropylene is an inexpensive, ductile, low-strength material with reasonable outdoor performance. The material surface is soft wax-like and scratches easily. Stiffness and strength are often improved by reinforcement with glass, chalk, or talc. Polypropylene is opaque and white, but the material can be dyed in many colors.

3. Nylon

Nylon was the first commercially successful synthetic polymer. There are two common methods of making nylon for fiber applications. In one approach, molecules with an acid (COOH) group on each end are made to react with molecules containing amine (NH_2) groups on each end. The resulting nylon is named on the basis of the number of carbon atoms separating the two acid groups and the two amines. The product consists of monomers of intermediate molecular weight, which then undergo reaction to form long polymer chains. Nylon fibers are used in many applications, including fabrics, bridal veils, carpets, musical strings, and rope. Solid nylon is used for mechanical parts such as machine screws, gears, and other low-to-medium stress components previously cast in metal. Engineering-grade nylon is processed by extrusion, casting, and injection molding. Nylon can be used as a matrix in composite materials, with reinforcing fibers such as glass or carbon fiber, and has a higher density than pure nylon. Such thermoplastic composites (25% glass fiber) are frequently used in car components next to the engine, such as intake manifolds; good heat-resistance properties of the composites make them feasible competitors to metals.

4. Polyester

Polyesters as thermoplastics may change shape after the application of heat. While combustible at high temperatures, polyesters tend to shrink away from flames and self-extinguish upon ignition. Polyester fibers have high tenacity and elastic modulus as well as low water absorption and minimal shrinkage in comparison to other industrial fibers.

5. Polyphenylene sulfide

Polyphenylene sulfide is an engineering high-performance thermoplastic with a maximum crystallinity of 65%. Polyphenylene sulfide can be molded, extruded, or machined to high tolerances. In its pure solid form, it is opaque white to light tan in color. The maximum service temperature is 218 °C. Polyphenylene sulfide has not been found to dissolve in any solvent at temperatures below approximately 200 °C. Polyphenylene sulfide is one of the most important high-temperature polymers because it exhibits a number of desirable properties. These properties include resistance to heat, acids and alkalis, and to mildew, to bleaches, aging, sunlight, and abrasion. It absorbs only small amounts of solvents and resists dyeing. Polyphenylene sulfide-based composites are used for applications in which great strength and chemical resistance are required at elevated temperatures.

6. Poly(ether ether ketone)

Poly(ether ether ketone) is a semicrystalline high-temperature (up to 500°F) engineering thermoplastic, which is excellent for applications where thermal, chemical, and combustion properties (UL flammability rating of V-0) are critical. At the same time, poly(ether ether ketone) emits little smoke or toxic gas when exposed to flame. This material is tough, strong, and rigid and has superior creep resistance. It also resists radiation and a wide range of solvents. With its resistance to hydrolysis, poly(ether ether ketone) can withstand boiling water and superheated steam used in autoclaves and sterilization equipment at temperatures higher than 482°F. Poly (ether ether ketone) resins are available as unreinforced, 30% glass fiber-reinforced, 30% carbon fiber-reinforced, and HPV-reinforced grades (bearing grade). Typically, the polymer finds applications in automotive, marine, nuclear, oil well, electronics, medical, and aerospace industries.

7. Polyetherimide

Polyetherimide is considered as an advanced thermoplastic with both ether links and imide groups in its polymer chain. Polyetherimide is an amorphous high-performance polymer characterized by excellent thermal properties, good chemical resistance, inherent flame retardancy, and exceptional dimensional stability. Polyetherimide also exhibits high tensile strength without the use of reinforcement, very low smoke emission, and excellent hydrolytic stability. Because of polyetherimide's high stability, its range of processing is wider than that of many other thermoplastics.

8. Polycarbonate

Polycarbonate is a versatile tough plastic used for a variety of applications ranging from bulletproof windows to compact disks (CDs) . The main advantage of polycarbonate over other types of plastics is its unbeatable strength combined with lightweight. While acrylic is 17% stronger than glass, polycarbonate is nearly unbreakable. Bulletproof windows and enclosures seen inside banks or at drive-throughs are often made of polycarbonate. In addition, polycarbonate is merely one-third the weight of acrylic, or one-sixth as heavy as glass, and the only drawback is that it is more expensive than either acrylic or glass. Polycarbonate is also used in the electronics industry. Apple's original iMac featured polycarbonate mixed with clear colors for a transparent computer case. Many cell phones, pagers, and laptops also use clear or opaque polycarbonate in their casings.

Other uses for polycarbonate include greenhouse enclosures, automobile headlights, outdoor fixtures, and medical industry applications, though the list is virtually endless. Somewhat less toxic than polyvinyl chloride to produce, polycarbonate nevertheless requires toxic chemicals in its production phase. However, it is recyclable and environmentally preferable to polyvinyl chloride in applications in which either material can be used.

6.4 Open Molding Process

Open molding, also known as contact molding, open laminating, and wet lay-up, is a method that has been used for the longest time in the polymer–matrix composite industry to make thermoset composite products. It continues to be the preferred process used for the production of a wide range of composite products. It is a simple process with many advantages and uses relatively basic material technology and processing methods. The molding method involves placing reinforcements and liquid resin onto the surface of an open mold (which may or may not be precoated with a gel coat) or onto other substrates as in the case of making a one-off sandwich construction, on-site repairs by applying a reinforcing vacuum-formed acrylic, corrosion-resistant lining on steel, on-site repairs of tanks and pipes, and so on.

The hand lay-up version involves applying the reinforcements and the resin by hand, while the spray-up version uses tailored spray equipment to deposit both the reinforcements and the resin on the mold or an alternative substrate. Open molding is a process typically used for low-to-medium-sized series (from a few to 200–300 parts/y), offering a number of process and product advantages. Producing large complicated shapes as well as smaller and simpler composite products is possible. The hand lay-up process involves low investment costs and little prior working knowledge of the process, while spray-up involves some investment in the tailored spray-up machines and spray guns.

The open mold process is performed in molds made from the same resins and fiberglass reinforcements used in the parts produced in the molds. The molds are created over a simple form; e.g., a professional CNC cut model or a simple shape hand made of a screen covered with Bondo Body Filler. In either case, the mold is created by applying a gel coat over the model, after which sheets of fiberglass wetted with the catalyzed resin are applied in individual layers that are built-up by hand until the desired mold thickness is achieved. Simultaneously, a basic steel or wood frame is fastened to the back of the mold to provide support and to position in an ergonomic manner for producing parts within.

Initially, the open mold process is facilitated by brushing a layer of gel coat into the mold, which creates the actual exterior finish of the final molded part. Then, sheets of fiberglass mat are wetted by brushing catalyzed resin over the mat, and the wetted resin and fiberglass are rolled tight against the gel coat and the additional subsequent layers to ensure that the air within is removed. This is done using a roller similar to a paint roller controlled by the operator manually to roll the wetted glass mat so that the entrapped air is eliminated. Developments in the 1970s provided equipment that can spray the gel coat and the resin with chopped fiberglass into the mold, which significantly improved the application time. However, the fiber and resin still needed to be "rolled" by the operator to eliminate entrapped air and to smooth the backside of the part.

The open mold process is able to cure at room temperature without added heat or pressure, and the actual process will require 4–6 h from start to finish for each molding produced. Well-trained operators and dedicated facilities are required to produce the components and products with high quality. The methods of open mold fabrication include

- Wet lay-up
- Hand lay-up
- Spray-up
- Tape lay-up
- Filament winding
- Autoclave curing.

6.4.1 Wet Lay-Up

In this process, there is usually only one half of the mold and the item to be produced takes the form of this mold. The resin possesses low viscosity without fillers and many additives, and it is spread on the continuous glass fibers stationed in the mold. This technique is widely used; however, health and safety considerations arising because of styrene emissions prompt users to move away from this method. The reinforcement is placed into or against the mold, and the resin is forced between the fibers by using brushes or rollers. There are minimal capital investment costs other than appropriate ventilation, extraction, labor, and tools. Normally, curing occurs at room temperature. This method is particularly suitable for one-off or small number production.

The fiber is positioned in or on the mold by hand. If the mold shows a complex shape, small pieces of mat are cut to fit and then more layers are applied to achieve the required thickness. The liquid resin is poured over the fiber and rolled to ensure complete wetting of the fiber and removal of air bubbles. In general, the resin cures at room temperature with the use of an accelerator and a catalyst. If a hot cure is used, the use of an accelerator is not required. Postcuring of cold cured laminates is recommended (typically 2 h at 80 °C or 16 h at 40 °C depending on the resin type).

Prior to the application of the fiber and resin, the mold cavity is coated with either polyvinyl alcohol or a nonsilicon wax to aid the component release. Release is simply achieved by tapping the wedges between the mold and the component or by the use of compressed air to gently force the pieces apart.

Chopped strand mat is most commonly used, although woven roving is used when stronger and stiffer laminates are required. A gel coat is applied to the mold surface to produce a resin with rich and smooth surface for aesthetic purposes and to protect from the environment. For improved surface finish and corrosion resistance, a surface veil is used, which is applied with an embedded fabric for reinforcement or mixed with resin for smooth surface.

The advantage of wet lay-up is that it is a simple technique requiring low capital investment and no requirement for highly skilled labor. The disadvantages are that it is not suited for strength or weight critical primary structure because the fiber orientation and local resin content cannot be controlled well. The nature of the wet lay-up also has health and safety implications because the vapors from low-molecular-weight monomers can be harmful.

Vacuum-assisted wet lay-up can help prevent the presence of voids caused by air trapping in the laminate. By applying vacuum to the part after reducing lay-up voids, consolidation is improved. The part is sealed within a plastic film (bagging material), and the air is extracted using a vacuum pump.

6.4.2 Hand Lay-Up

Hand lay-up involves open contact molding in one-sided molds and is the most inexpensive and most common process used for making composite products. This is the most common method used for producing composite parts in the US aircraft industry.

In a typical open mold application, the mold is first waxed and sprayed with a gel coat. It may then be cured in a heated oven at approximately 120°F. In the spray-up process, after the gel coat cures, catalyzed resin (usually polyester or vinyl ester with a viscosity of 500–1000 cps) is sprayed into the mold along with chopped carbon fibers. A chopper gun chops the roving directly into the resin spray so that all the materials are simultaneously applied to the mold. Using low-styrene and suppressed-styrene resins, fillers, and high-volume/low-pressure spray guns or pressure-fed resin roller applicators helps reduce emissions of volatile organic compounds.

In the hand lay-up processing, carbon fiber continuous strand mats and/or other fabrics such as woven roving are manually placed in the mold. Each ply is sprayed with the catalyzed resin (with a viscosity of 1000–1500 cps), and the resin is worked into the fiber with brushes and rollers to wet-out and compact the laminate. There are several variations of the hand lay-up technique, as shown in Fig. 6.14.

The fiber content can be increased by up to 50% by curing the part in a vacuum bag using a vacuum of 2–14 psi and curing temperatures under 350°F.

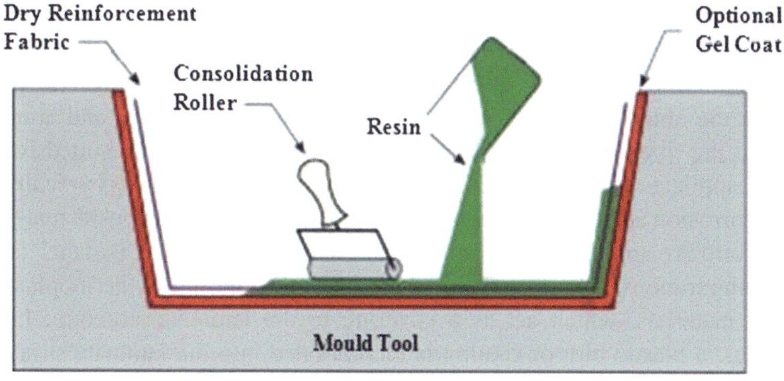

Fig. 6.14 Hand lay-up process

Vacuum-assisted resin transfer molding (VARTM) and infusion molding systems are gaining favor with open mold processors who want to reduce emissions of volatile organic compounds. The applied vacuum compacts the preform and helps the resin to penetrate and wet-out the fiber preform. Fiber contents up to 70% have been reported.

Also, this is the oldest and the simplest method of making polyester composites and is still extensively used. The molds can be made from sheet metal, wood, plaster, or even composites. The first step is to coat the mold with a mold-release agent so that the cured item can be removed from the mold without damage. Fluorocarbons are preferably used for this purpose, although silicones are used when the surface finish is not of primary importance. Silicones leave residues on surfaces that cause problems with subsequent bending or painting. Polyvinylalcohol can also be used as a mold-release agent. The second step is to apply a 0.3–0.4-mm-thick gel coat layer to the surface. This affords a smooth surface finish and good appearance to the product and prevents moisture diffusion, which weakens the resin–fiber interface. After the gel coat has partially cured, an initial coat of the resin is brushed upon the gel coat, followed by manual laying up the reinforcements such as a mat, woven roving, or fabric. The entrapped air is removed by hand rollers, which also ensure complete fiber wetting. Successive layers of reinforcements and resin can be alternatingly added to build up the desired thickness. Curing takes place at room temperature using a proper initiator such as methyl ethyl ketone peroxide.

6.4.2.1 Laminate Materials

There are an infinite number of laminate types that can be developed [59–67]. These materials can be categorized into three basic areas, i.e., core materials, high-strength and stiffness skins, and outer protective layers. Core materials typically serve the function of connecting and spacing the skins to develop stiffness and strength in a sandwich arrangement. The key property of core materials is shear strength, which ensures shear conductivity between the skins; thus, the ability to sustain loads and bending is ensured. Core materials are normally wood, honeycomb, and structural foams. The outer structural layer or skins are traditionally either metals or composites, either in combination with a core material or a multitude of layers of high strength and stiffness. Composite materials offer the widest range of high-strength skins with the ability to change the fiber type (fiberglass, carbon, and aramid) in addition to the fiber volume and orientation. Composites are well suited for large deflection applications requiring high strain capability and fatigue. Typically, they are more corrosion and environmentally resistant than metals. Composite materials in a lamina form are applied in the form of precured prepregs or as "B stage" and wet lay-up configurations. The final group of laminae is composed of thermoplastic and thermoset materials, which act as a covering to the laminate structure. In some applications, a plastic film or coating is incorporated into the laminate structure to protect the structure from impact and environmental effects.

Hand lay-up is also the oldest and simplest method used for producing reinforced plastic laminates. The capital investment for hand lay-up processes is relatively low. The most expensive piece of equipment typically is a spray gun for resin and gel coat application. Some fabricators pour or brush the resin into the molds so that a spray gun is not required for this step. There is virtually no limit to the size of the part that can be made. The molds can be made of wood, sheet metal, plaster, and FRP composites.

In a particular hand lay-up process (otherwise known as wet lay-up), high-solubility resin is sprayed, poured, or brushed into a mold. The reinforcement is then wetted with resin and then placed in the mold. Depending upon the thickness or the density of the reinforcement, the reinforcement may receive additional resin to improve the wetting and allow for better drapeability into the mold surface. The reinforcement is then rolled, brushed, or applied using a squeegee to remove entrapped air and to compact it against the mold surface.

Chopped strand mat is the most inexpensive form of reinforcement used in wet lay-up. It also provides equal reinforcing strength in all directions because of the random orientation of the fibers forming the mat. Woven roving is especially suitable for thick laminates requiring greater strength. Woven fabric and braid can also provide inexpensive reinforcement. Once the reinforcement is thoroughly wetted with the resin, it can be easily formed into complex shapes.

6.4.2.2 Surface Preparation and Bonding

A key step to achieve successful lamination is the bonding process of the layers. There are three basic components [68–70] of the bonding process. First is the surface preparation of the laminae, which improves the ability of the substrate to accept and adhere to an adhesive. Surface preparation varies depending on the material type. Composites use sanding and grinding, surface texturing, or solvent cleaning. The second component is the adhesive itself, including epoxies, urethanes, phenolics, polyesters, solvents, and acrylics. Each adhesive consists of its own attributes depending on the substrate type, in-use requirements, and process constraints. As a general rule, the maximum bond is achieved for a given substrate type when the material itself fails during an ultimate strength test. The maximum lap shear strength of an adhesive is achieved when the adhesive exhibits a cohesive failure in the bond line. The third component of lamination is the process by which the materials are bonded together. This involves a host of parameters, primarily time, heat, pressure, mixture, moisture, and catalysts (initiators). It is important that the three basic components of bonding are properly employed to achieve successful lamination.

6.4.2.3 Laminate Construction

There are three types of laminated construction [71–74]. These include sandwich lamination consisting of at least two high stiffness and strength outer layers connected by a core, all laminated construction consisting of relatively high stiffness and strength layers, and a third type consisting of a structural member reinforced on the tensile or compression or both sides of a flexural beam.

Sandwich lamination constructions are found in many applications ranging from satellite structures to snow skis. Both applications may use a sandwich approach; however, satellites generally require stiffness, strength, and extreme lightweight, while the snow ski requires the laminated beam or composite structure to withstand large deflections and dynamic performance requirements. In addition, the ski structure is integrated with thermoplastic surfaces and metal components. Typically, metal and composite materials are applied to these sandwich structures. Even complex shapes can be achieved by using composite prepregs and wet lay-ups.

All laminate constructions use relatively high strength and stiffness materials. An array of laminate configurations is possible. In aerospace applications, multiple plys are orientated in various directions to provide customized structural strength and stiffness. These principles, although at a less complex level, are used in automotive, industrial, and recreational products in the form of structural members, springs, archery limbs, and bicycles.

6.4.2.4 Multi-ply Construction

The third type of laminated construction is used most often in construction and infrastructure applications [75–77]. Till date, reinforcements have been integrated into glue laminated wood beams (Glu-Lams) and infrastructure components for fabricating columns, beams, and walls. In many of these applications, reinforcement, most typically composites, are used via lamination to the tensile side of a beam or as a wrap on a column. Precured prepreg or wet lay-up composite materials have all been used in these applications. These types of reinforcements improve the strength of laminated wood beams and/or reduce the use of E-rated lumber with less costly new growth wood. Lamination in bridges and buildings provides a low-cost simplified method to revitalize existing structures and increase the load-carrying capability and resistance to seismic events.

In the automotive industry, composites are being combined with metals for performance, weight reduction, and cost advantages. The construction industry is applying composites to wood laminated beams and I-joists.

The possibilities and advantages of laminated materials are significant and provide solutions to product requirements, generally not achievable by using a single material. The advantages in structural performance, reduced weight, and oriented structural properties are some advantages of this approach. In many cases, the result is a simplified and less costly solution to many engineering structural problems.

6.4.3 Spray Lay-Up

Spray lay-up is an open molding composite fabrication process in which the resin and reinforcements are sprayed onto a mold (Fig. 6.15). The resin and glass may be

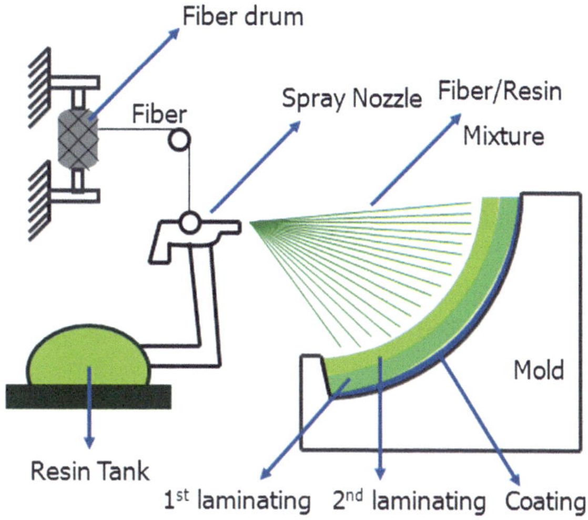

Fig. 6.15 Spray lay-up process

applied separately or simultaneously "chopped" in a combined stream from a chopper gun. Workers roll out the spray lay-up to compact the laminate. Wood, foam, or other core materials may then be added, and a secondary spray lay-up layer imbeds the core between the laminates (sandwich construction). The part is then cured, cooled, and removed from the reusable mold.

Spray lay-up has very little application in aerospace. This technology produces low-specific-strength structures, which usually are not manifested in the end product. Spray lay-up is being used to join back-up structures to composite face sheets on composite tools. Spray lay-up is also in limited use for obtaining fiber-glass splash from transfer tools.

Spray lay-up has been designed to process chopped carbon fibers. A specially designed gun supplies the resin, while chopping the continuous fiber. The resin/fiber mix is then sprayed into a mold. Because of the flow requirements in the gun, high fiber densities are not possible. This technique is an inexpensive and requires only a pneumatic fiber chopping tool and a resin sprayer.

Carbon fiber is used extensively because it is easily chopped into strands, has low static, and fast wet through. Orthophthalic and isophthalic resins such as vinyl ester or polyester are commonly used owing to their high reactivity. Other examples include epoxy and phenolic resins.

One major advantage of this technique is that it permits fast application. Also, the method results in better wetting of the fiber with fewer voids than with hand lay-up. Consequently, parts with better physical properties are obtained. However, the method does produce parts with less uniformity, particularly in thickness in comparison to parts produced by hand lay-up. The laminates tend to be resin rich and hence, are heavier.

Spray-up requires low-molecular-weight resin; the styrene emissions can be high, requiring tight health and safety control. Because the gel and cure result from exothermic chemical reactions, there is a risk of the resulting exotherm being excessive, which may result in the damage of the molded part or mold. It is to be noted that if the ambient temperature is warm, the resin may cure prior to the completion of the spraying application. Examples of applications include manufacture of sanitary ware, caravan bodies, and small dinghies.

6.4.4 *Tape Lay-Up*

In this method, layers of prepreg (reinforcing phase impregnated by liquid resin) tape are applied on the mold surface by a tape application robot.

6.4.5 *Filament Winding*

Filament winding method involves the winding of a continuous filament of reinforcing material onto a rotating mandrel in layers at different layers [78–81]. If a liquid thermosetting resin is applied on the filament prior to winding, the process is called wet filament winding. If the resin is sprayed onto the mandrel with the wound filament, the process is called dry filament winding. This process is primarily used for hollow components with circular or ovoid cross sections, such as pipes and tanks. Fiber tows are passed through a resin bath before being wound onto a mandrel in a variety of orientations, which are controlled by the fiber feeding mechanism and the rate of rotation of the mandrel.

Although the filament winding machine design varies with the part geometry, the basic filament winding process concept is described in Fig. 6.16.

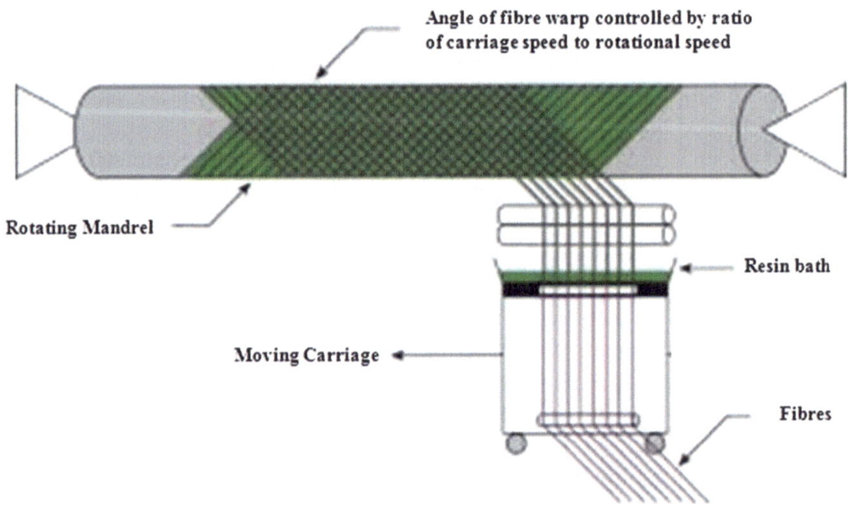

Fig. 6.16 Filament winding process

The fibers are impregnated with a resin (by immersion or by passing over a resin-wetted drum or by injection into the die) before being led to a feed eye, where a controlled bandwidth is set prior to positioning on the mandrel.

Fiber tension is critical to the operation of a filament winding machine. The fibers are supplied on creels, and it is normal to use fiber tensioners (closed loop controlled servo-driven "dancers") in the feed line. The tension required is dependent on the type of the fiber, part diameter, and winding pattern selected. The tension directly affects both the fiber volume fraction and the fraction of voids, which, in turn, influence the strength and stiffness of the composite part.

Once the fiber package is positioned, the resin is taken to full cure, often by heating in an oven. The final stage is the mandrel removal, which leaves the desired hollow component. This may be achieved by hydraulic rams for extracting the steel mandrels. For more complex structures, the mandrel may be composed of low-melting-point materials (e.g., a metal alloy) or a water-soluble salt (leachable plaster), which can be washed out, or a collapsible rubber or nonreusable foam. In some cases, when a liner is required for minimal gas permeability, the liner may also function as the mandrel and hence, need not be removed.

Filament winding has been combined with other fiber application methods such as hand lay-up, pultrusion, and braiding. Classic filament winding involves a spindle with a carriage or carriages to apply hoop and helical fibers. The compaction is carried out by fiber tension. The resin content is now primarily metered. The machines are usually computer controlled, with up to six axes independently monitored. The additional axis comes into play at the fiber turnarounds. The extra head axis allows for better placement of the band and more uniform bandwidth.

Filament winding is typically used in the production of chemical storage tanks and pipelines, gas cylinders, rocket motors, launch tubes, pressure vessels, drive shafts, fishing rods, and missile cases.

6.4.6 Autoclave Curing

Besides conventional curing of the molded parts at room temperature, autoclave curing may be used (Fig. 6.17). Autoclave curing is a method in which a part, molded by one of the open molding methods, is cured by subsequent application of vacuum, heat, and inert gas pressure. The molded part is first placed into a plastic bag from which air is removed by a vacuum pump. This operation removes air

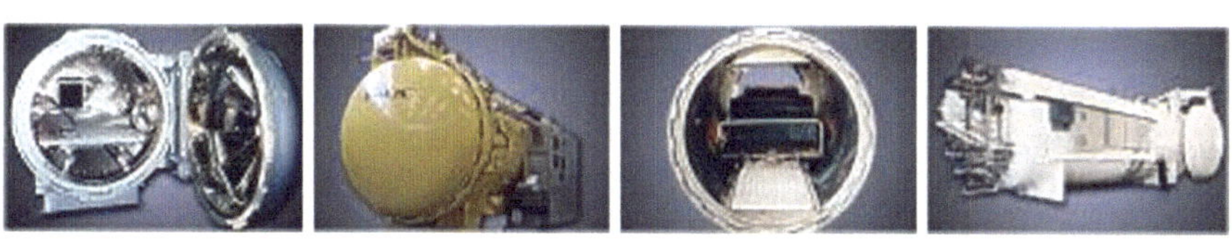

Fig. 6.17 Popular autoclaves used in recent times

inclusions and volatile products from the molded part. Then, heat and inert gas pressure are applied in the autoclave causing curing and densification of the materials.

Autoclave curing enables the fabrication of consistent homogeneous materials. The method is relatively expensive and is used for manufacturing high-quality aerospace products. Autoclaves are also used to process both laminated composites and bonded metallic assemblies such as honeycomb panels. When placed under vacuum or when vented, the heat and pressure can be adjusted to optimize the fiber/resin ratio in composite parts and provide high-strength bonds (using high-strength film adhesives) in bonded assemblies. Autoclaves have been used in the industry for many decades. With progress in technology, autoclave design has also progressed. Figure 6.17 shows popular autoclaves used in recent times.

Benefits of the autoclave cure include

- Achievement of highest strength-to-weight ratio
- Production of high fiber content laminates
- Pressurized curing and reduced voids
- Ability to process large parts or a large number of small parts
- Pressure and heat can be controlled very closely
- Strategic fiber placement
- Flatness.

Typical applications of autoclave curing include the production of precision curing parts, aircraft parts, printed circuit boards, laminates, and thermoset components for the electronics industry.

6.5 Closed Molding Process

Closed molding comprises a broad category of fabrication processes in which the composite part is produced in a mold cavity formed by the joining of two or more tool pieces. There are many variations of the closed molding process, each with its own unique or distinctive aspect. For the purposes of introduction, these variations are categorized into two types: compression molding and liquid injection molding. Compression molding processes uses either sheet molding compound (SMC) or bulk molding compound (BMC). These compounds contain reinforcing fiber, fillers, resin, and curing agents. High filler contents result in a higher part density than is typical for open molded parts. Typical cycle times of a few minutes are achieved with a combination of rapid hydraulic mold closure and curing at elevated temperatures. A large number of parts (thousands to tens of thousands) are required to justify the required capital expenditures. Liquid injection molding processes include RIM, structural reaction injection molding (SRIM), and resin transfer molding (RTM). The liquid resin injection for RIM and SRIM begins with an empty mold cavity. RIM and SRIM typically use highly reactive resins such as urethanes and are well suited for high-rate production. In general, neither compression molding nor RIM/SRIM is a viable alternative to typical open mold production. This is primarily caused by the limited production quantities of typical open molded parts.

The best alternative for close mold production is resin transfer molding. For RTM, the liquid resin injection follows once the mold is first loaded with materials such as the reinforcement fibers and cores. RTM processes produce parts with strengths, stiffness, and weights that most closely resemble typical open mold parts. Cycle times can range from a few minutes for small parts to several hours for large parts. A wide variety of resin chemistries are used according to the design end-use requirements.

6.5.1 *Resin Transfer Molding (RTM)*

RTM is a particularly useful manufacturing process that uses liquid resin to impregnate the stationary fibrous preform [82–86]. During the RTM process, the preform is placed into the mold cavity, the mold is closed, and the resin is injected into the cavity under pressure. The mold with the preform is often placed under vacuum so that the vacuum removes all the entrapped air in the preform and speeds up the RTM process. Once the liquid resin fills the mold cavity, it cures, during which the resin hardens because of the formation of a polymeric network, which forms the matrix of the composites, allowing the part to be demolded. Typically, thermoset polymers such as epoxy, vinyl ester, methyl methacrylate, polyester, or phenolic resin are used with fiberglass, carbon fiber, aramid, and synthetic fiber reinforcements, either individually or in combination with each other. The matrix selection of the polymer and the reinforcement dictates both the molding material cost as well as the molding mechanical and surface finish performance. Along with the polymer and reinforcement, the addition of mineral fillers may be used to enhance fire retardancy, flex modulus, and surface finish.

This technique is well known and has been traditionally applied to moderately large parts in various applications. It allows one to obtain even very complex neat shape parts with good surface finish, in many cases, at reasonable production rates. The fiber architecture, permeability of the preform and fabric crimps, resin viscosity, and temperature of operation influence the wetting of the fabric. Careful process design is required to reproducibly obtain a high-quality product (Fig. 6.18).

Reinforcements are used in their dry form in the mold as binder-bound chopped mat, random-continuous strand mat, stitched "conformable" mat, or woven cloth. The fiber has been either "preformed" to the exact shape of the molding tool in a previous operation or is hand-tailored during the loading process in the molding tool. After the fiber is installed into the mold, a premixed catalyst and resin are injected into the closed mold cavity, encapsulating the fiber within. The primary surface of the molding may be gel coated by spraying the mold surface before installing the fiber. If a gel coat is not required, the exterior finish would be the same on both sides of the molded part. RTM is inherently advantageous in the case

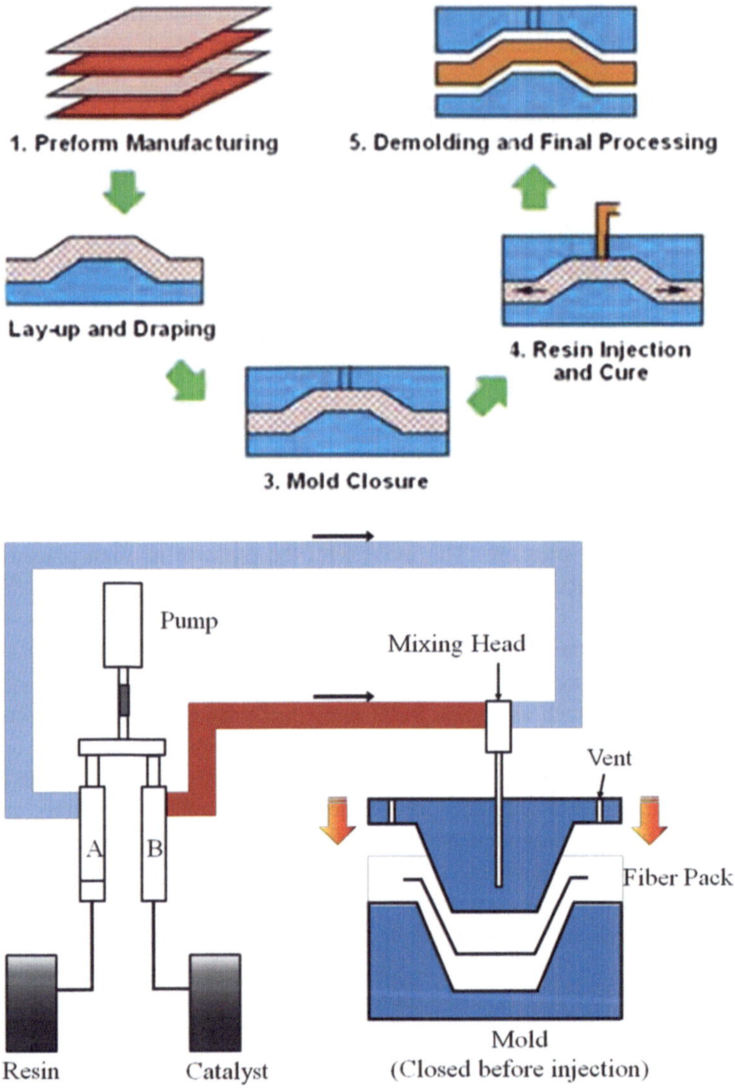

Fig. 6.18 Resin transfer molding process

of low-pressure injection when the injection pressure usually does not exceed 100 psi during the mold-fill process. In most common applications using the RTM process, parts are molded with a cross-sectional thickness of 4 mm with 30 wt% fiber content of the total laminate.

6.5.1.1 Benefits of Using RTM

- Closed molding process is cleaner and healthier, which attracts higher-skilled employees.
- Closed molding operator turnover is dramatically reduced through improved working conditions.
- Closed molding area has organized process flow to maximize throughput efficiency.
- Moldings can be manufactured to close dimensional tolerances.
- Components have good surface finish on both sides.
- Selective reinforcement and accurate fiber management are achievable.
- Ability to build-in fiber volume fraction loadings up to 65%.
- Consistency in thickness and fiber loading, resulting in uniform shrinkage.
- Inserts may be incorporated into moldings.
- Tooling costs are comparatively low compared to other manufacturing processes.
- Uses only low-pressure injection.
- Low-volatile emission during processing.
- Ability to produce near net shape moldings, reducing material wastage.
- Process can be automated, resulting in higher production rates with less scrap.
- Ability to mold complex structural and hollow shapes.
- Low void fraction in the molded components.
- Ability to achieve laminate thicknesses in the range 0.1–90 mm.

6.5.1.2 Current RTM Developments

Three primary factors have a major impact on the use of the RTM process today.

(1) Environmental agencies have recently reduced the allowable styrene (primary volatile organic compound VOC contained in the resins used for SMC, open mold, and RTM) level in the workplace to levels that are very difficult to meet using the open molding process.
(2) The labor force of today is not interested in working in an open mold environment. Thus, the skill level needed to maintain the desired or required product quality and consistency is becoming increasingly rare. The result is an increase in the production cost attributable to the required "reworking" in a postmold operation of the products molded to meet quality standards.
(3) The innovation originating from the suppliers of equipment and tooling have changed the mold design to incorporate the use of vacuum in balance with the injection pressure, enabling lower cost tooling because of the reduction in tooling structure and previously needed external clamping hardware.

6.5.1.3 Background to Current Developments

RTM is facing competition from the open mold and SMC processes, while in most low-volume applications, open mold tooling is the choice for many applications because of low costs. The high-volume applications have chosen the SMC process.

Many tooling methods have been attempted in RTM to both increase tool life and improve surface finish longevity. There have been efforts to use nickel shell, cast aluminum, and cut aluminum. The cut aluminum tooling led to cut steel, and we found that this tooling equals the cost of SMC tooling, and hence, the program can be used for SMC.

All of this tooling effort or change was in reaction to attempts to improve the surface quality of the molded parts by improving the damage resistance of the mold surface. The RTM process began with epoxy tooling surfaces, which led to high-temperature polyester tooling gel coats to improve, release, and reduce wear caused by sticking. The high-temperature polyester was easily damaged, which led to the use of nickel; however, nickel could not be easily repaired, which led to the use of cast aluminum. Cast aluminum contained too much porosity, which led to the use of cut billet aluminum. Cut billet aluminum has an excellent surface; however, for only 20% more cost, cut steel could be used. However, it is excessively off course as far as tooling cost is concerned to make the process competitive with either SMC or open mold.

6.5.2 Light RTM (LRTM)

The Light RTM process (also known as LRTM) for the production of fiber-reinforced composites has recently become the most popular closed molding process for low- to-medium-volume applications. This leading technology has now displaced the former conventional or traditional RTM in most typical marine, automotive, industrial, and medical composite molding applications, requiring two finished sides and close dimensional tolerances.

The recent surge in the acceptance of the LRTM process over the previously used traditional RTM process is in part because of the lower tooling cost of LRTM. In addition, it is also favored because of the ease of converting from the open molding process, which still remains the prominent molding process in the industry.

The LRTM process is founded on what could be considered as minimal tooling structure when compared to traditional RTM or other closed molding processes such as SMC or BMCZ. The tooling structure needed for the LRTM process is similar to that of the open molding process because of the minimal internal cavity pressure. The typical LRTM mold consists of one rigid half, normally for the "A surface" (typically the female cavity) of the mold, with a lightweight see-through semiflexible upper mold half. Each mold half is made from a thermoset polyester or vinyl ester tooling gel coat surface with a fiberglass-reinforced thermoset polymer laminate backing, which is then normally supported by steel box tubing to retain the tooling shape and provide an ergonomic orientation.

LRTM tooling costs are a fraction of the price of equivalent RTM molds; until recently, they were thought to produce less than half the production rate of the higher pressure RTM process. However, with recent innovations in the tool design and process controls, the gap in productivity is closing between the two processes.

Closed molding for either true RTM or LRTM offers far more comfortable and healthy working environments for the molding operators. Hence, the operators are then willing to apply their skills of quality and productivity at a consistently higher level each day. It is true that LRTM will not meet the production rates offered by traditional RTM; however, LRTM will provide a 300–400% increase in per square foot productivity over open mold, with significant improvement of bill of material compliance and lower operator employment turnover.

In LRTM, resin flow rates cannot be accelerated above an optimum level of the greatest flow rate, while maintaining inner-cavity mold pressures below the clamping forces of the external atmospheric pressure. Normally, this limits the injection pressure of the LRTM to less than 1 bar (14 psi), with the cavity held at a constant vacuum level of −0.5 bar (15″ Hg).

As with any composite closed mold production process method, LRTM demands high quality and accurate cross-sectional molds to provide good mold life and consistent production of repeatable parts. This is a key element to successful molding and is commonly overlooked by those venturing into either LRTM or RTM on their own. We strongly encourage those who wish to take advantage of the closed molding processes of RTM or LRTM to allow us to be a major part of achieving success.

Today, the industry is directed toward closed molding, especially LRTM, which is driven by numerous factors such as reduced employee turnover, elimination of VOC emissions, and improved bill of material compliance combined with two finished side moldings that can be engineered to fit into assemblies with far less manual postmold rework. The pending restrictions on VOC emissions now coming to an imminent deadline and the most recent threat by the EPA of listing styrene, the most prominent VOC in the open molding process, as a suspected human carcinogen, has all aided in the recent increased demand in converting many open molders to the closed molding process using RTM, LRTM, and vacuum infusion (VARTM) processes.

6.5.2.1 Comparison of RTM and LRTM Processes

Individual characteristics of each closed molding process can be identified when the fundamental differences in the process methods and related limitations are considered. With traditional RTM, the mold cavity with resin under a moderately high flow rate and pressure is only limited by the structural ability of the molding tool and perimeter clamping or press systems to sustain mold closure. Working within these concerns, RTM tooling and clamping systems can be built with structures sufficient to sustain the flexing caused by the highest expected injection pressure during the molding cycle. Thus, we may choose to use tools made of cored composite with a steel tube backing structure or either CNC machine cut aluminum or steel.

As with any manufacturing process, cost becomes the true limiting factor. Use of tooling (molds) made from fiberglass-reinforced composite materials is generally favored to reduce tooling costs in comparison to the alternative of using machine aluminum or steel materials. Machine cut steel or aluminum tooling is typically limited to very-high-volume applications, which can amortize the tooling cost over far more molding cycles.

The shortcoming of the composite tooling is the service life of the mold surface. Increased tooling surface life has been revolutionized through the use of tooling with replaceable tooling surface "skins." Today, there are a number of skin tooling system designs in the market, which are compatible with RTM and LRTM. One of the most recent designs is the patented ZIP RTM design offered exclusively through JHM Technologies Inc. The ability to replace the perishable mold surface through the use of mold surface "skins" has given composite fiber-reinforced tooling an effective solution in providing long lasting tooling and high-quality moldings.

6.5.3 Vacuum-Assisted Resin Transfer Molding

VARTM is perhaps, the most discussed among low-pressure closed mold-reinforced composite molding techniques. VARTM is a single-sided molding process in which the dry preform (reinforcement or coring materials) is placed into the mold and a cover (or a vacuum bag) is placed over the top to form a vacuum-tight seal (Fig. 6.19). A distribution medium (a mesh) is used and laid on top of the top-release fabric to help maintain an even distribution of the resin and facilitate the flow of resin through the thickness of the panel. The low-viscosity resin typically enters the preform through the resin distribution and vacuum distribution lines with the aid of vacuum. In VARTM, the flow of resin occurs in plane as well as in the transverse directions to the preform. The permeability of the preform, fiber architecture, and fabric crimp has an influence on the wetting of the fabric.

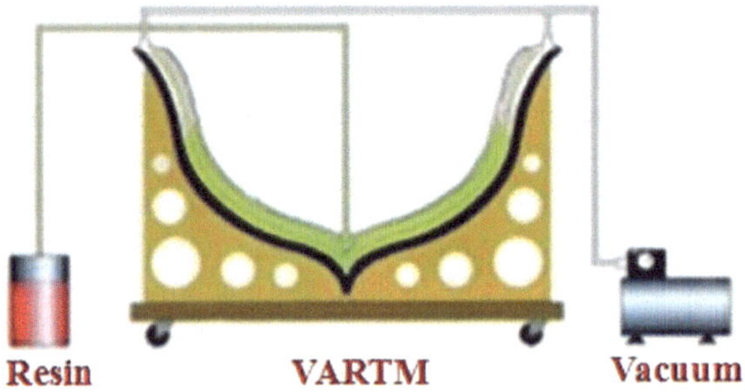

Fig. 6.19 Vacuum-assisted resin transfer molding (VARTM) process

VARTM is most accurately applied to the process of vacuum infusion in which the composites are molded using a rigid mold to provide part geometry and a thin flexible membrane over the fiber, with the outer atmospheric pressure compressing the fiber tight against the rigid mold surface. However, it should be noted that in VARTM, vacuum is added at the exit vent of the molding tool and thus, any form of RTM in which vacuum is applied to the vent would qualify as VARTM. Often, the aerospace industry will have very high-fiber-volume composites molded within a rigid matched mold set in which the fixed cavity mold is clamped closed using a press and the resin injection pressures typically range below 100 psi; although, the pressure could be as high as 600 psi as in the case of filling the mold encapsulating the dry fiber within.

The benefits of the VARTM technology include the following:

- VARTM is applicable to larger, less complex, and lower volume part production of composite parts
- VARTM needs simple and low-cost tooling
- On-site manufacturing and repairing are practical
- High-fiber-volume panels are achievable.

6.5.4 Pultrusion

Pultrusion is a manufacturing process used for producing continuous lengths of FRP structural shapes (Fig. 6.20). Raw materials include a liquid resin mixture (containing resin, fillers, and specialized additives) and reinforcing fibers. The process involves pulling these raw materials (rather than pushing, which is the case in extrusion) through a heated steel-forming die using a continuous pulling device [87–92]. The reinforcement materials are in continuous forms such as rolls of fiberglass mat or doffs of fiberglass roving. As the reinforcements are saturated with the resin mixture (termed as wet-out) in the resin impregnator and pulled through the die, the gelation (or hardening) of the resin is initiated by the heat from the die and a rigid cured profile is formed, which corresponds to the shape of the die.

The design of the pultrusion machine varies with part geometry; however, the basic concept of the pultrusion process is described in Fig. 6.20. The creels position the reinforcements for subsequent feeding into the guides. The reinforcement must be located properly within the composite and controlled by the reinforcement guides.

The resin impregnator saturates (wets out) the reinforcement with a solution containing the resin, fillers, pigment, and catalyst, in addition to any other additives required.

The interior of the resin impregnator is carefully designed to optimize the "wet-out" (complete saturation) of the reinforcements.

On exiting the resin impregnator, the reinforcements are organized and positioned for the eventual placement within the cross-sectional form by the preformer.

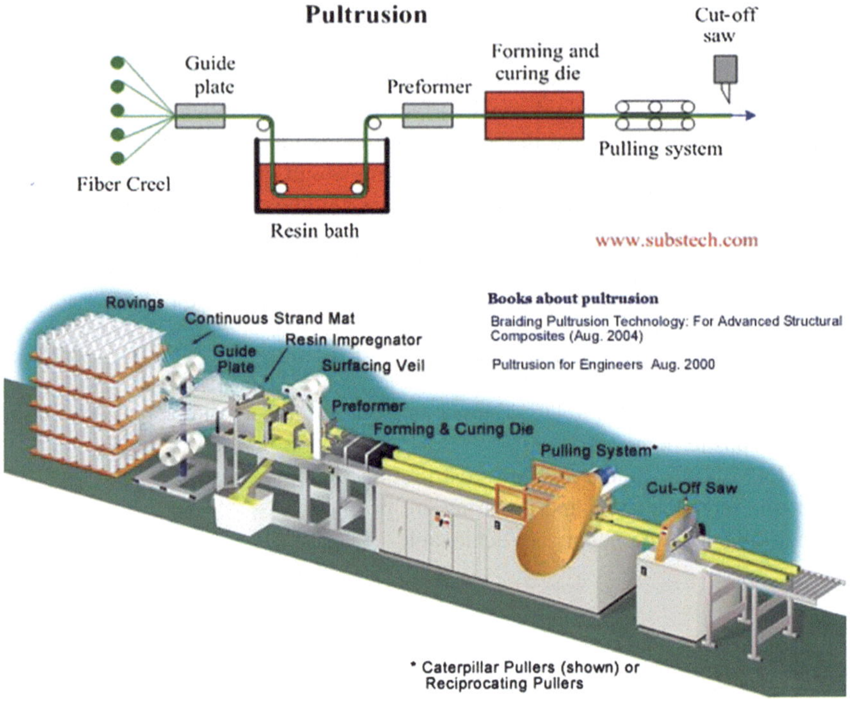

Fig. 6.20 Schematic of pultrusion process

The preformer is an array of toolings, which squeezes away excess resin, while the product is moving forward and gently shapes the materials prior to entering the die. In the die, the thermosetting reaction is heat activated (energy is primarily supplied electrically) and the composite is cured (hardened).

On exiting the die, the cured profile is pulled to the saw for cutting to length. It is necessary to cool the hot part before it is gripped by the pull block (made of durable urethane foam) to prevent cracking and/or deformation by the pull blocks. Two distinct pulling systems are used: a caterpillar counter-rotating type and a hand-over-hand reciprocating type.

In certain applications, a radio frequency (RF) wave generator unit is used to preheat the composite before entering the die. When in use, the RF heater is positioned between the resin impregnator and the preformer. RF is generally only used with an all-roving part.

The initial capital investment for pultrusion is generally higher than open mold or hand lay-up processes. The primary expense for pultrusion manufacturers is the material handling guides and die fabrication costs. The net result is a low-cost process for high volume. The process provides flexibility in the design of the pultruded profiles. Currently, profiles up to 72 inch wide and 21 inch high are possible. Pultrusion can manufacture both simple and complex profiles, eliminating the need for extensive postproduction assembly of components. Because the process is continuous, length variations are limited to shipping capabilities. This process allows for optimized fiber architectures with uniform color, eliminating the need for many painting requirements.

6.5.5 *Thermoforming*

Thermoforming is a molding process used to form sheets of plastic to a mold surface by using heat and vacuum and/or pressure force (Fig. 6.21). This technology appeared to have applications during the prime of thermoplastics. Thermoforming is a downstream plastic process; however, it is often viewed as an alternative to injection molding in numerous markets and end-uses.

Thermoforming is used primarily in low-cost applications for simple embossing of thin plastic sheet and cannot be used to form shapes as rigid as those formed by vacuum forming. Thermoformed plastics are invariably thermoplastics because thermosetting plastics assume their final shape through heat and cannot be molded with this process.

Familiar products manufactured by thermoforming include yoghurt pots and simple trays. The plastics that are used in thermoforming include acrylic, low-density polyethylene (LDPE), and crystalline polyester (CPET).

6.5.5.1 Types of Thermoforming Molds

(1) Machined aluminum molds

Machined aluminum molds are typically built for shallow parts with small draw ratios. Male or female molds and vacuum-form or pressure-form molds can be machined aluminum molds. They can be textured and may offer features such as loose cores, pneumatic cores, and inserts.

Fig. 6.21 Photographs of thermoforming molding and infrared ovens

(2) Cast aluminum molds

Cast aluminum molds are cast at a foundry from a machined pattern from a composite material and are typically built for parts with large draw ratios.

(3) Composite molds

For prototyping and short production runs, cost-efficient composite molds are used for mold construction. These molds produce parts that are to be evaluated for fit, form, and function and may be modified to evaluate possible design changes. These molds are for vacuum forming only and are not temperature controlled. These molds have a limited life.

Benefits of thermoforming technology

- The technology is efficient and very cost-effective
  - Lower machine cost
  - Large parts can be easily formed
- Offers close tolerances, tight specifications, and sharp detail
- Much lower initial project costs
- Inexpensive tooling.

Typical applications of thermoforming are large panels, housings, enclosures, children's swimming pools, small boat hulls, Halloween masks, disposable SOLO cups, ice cube trays, refrigerator door liners, cookie or donut trays, and Glad-Ware food containers.

6.5.6 Vacuum Bagging

This process is basically an extension of the wet lay-up process in which pressure is applied to the laminate once laid-up in order to improve its consolidation (Fig. 6.22). This is achieved by sealing a plastic film over the wet laid-up laminate and onto the tool. The air under the bag is extracted by a vacuum pump, and thus, pressure up to one atmosphere can be applied to the laminate to consolidate it (Table 6.4).

Benefits of this process:

- Large products possible
- Top-quality products through the use of prepregs
- Clean production method
- Low molding costs.

Typical applications are in fields such as the production of large one-off cruising boats, racecar components, and core-bonding in production boats.

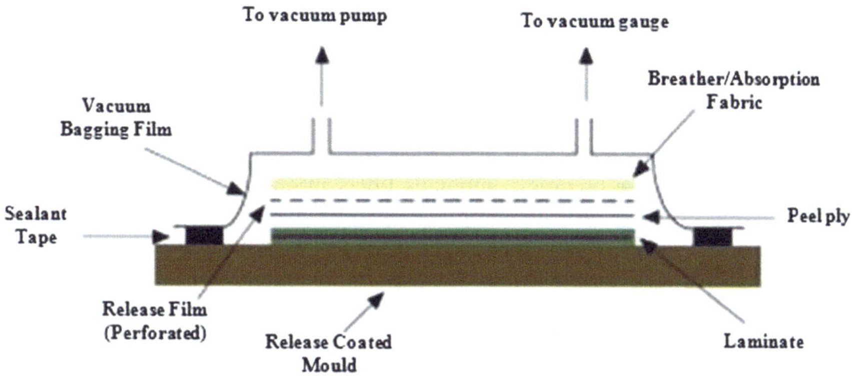

Fig. 6.22 Schematic of vacuum bagging process

Table 6.4 Consumable materials and equipment required for vacuum bagging

Peel-ply	A sacrificial open-weave fiberglass or perforated heat-set nylon ply placed between the laminate and the bleeder/breather to provide the textured and clean surface necessary for further lamination or secondary bonding
Bleeder cloth	A nonstructural fabric designed to absorb excess resin and reactants from the laminate. This may also act as the breather cloth
Breather cloth	A loose-weave or nonwoven porous material used to provide a gas-flow path over the laminate to permit the escape of air, reactants, moisture, and volatiles and to ensure uniform vacuum pressure across the component. This may also act as the bleeder cloth
Release film	A (perforated) sheet of material placed between the laminate and the mold surfaces to prevent adhesion
Edge dams	Profile used to define the edge of the component
Caul plate	A mold or tool placed on top of the laminate inside the bag to define the second surface
Intensifiers	Generally, hard rubber profiles incorporated in the bag to consolidate the laminate at sharp radii
Bagging film	The membrane that permits vacuum to be drawn within the bag
Tacky tape	Adhesive strip used to bond the bag to the tool and provide the vacuum seal
Breach unit	A connector through the bagging film to permit the vacuum to be drawn
Vacuum pipes	The link between the breach unit and the vacuum pump
Resin trap	A container in the vacuum line to collect any excess resin before it can damage the vacuum pump
Vacuum pump	Generally, a high-volume pump (absolute vacuum is rarely required) is suitable for continuous running. For some slow-curing epoxy resins, 24 operations may be needed
Pressure gauges	Generally, clock-type gauges are attached via a breach unit connection

6.5.7 Compression Molding

In compression molding, two matched (usually steel) mold halves are mounted in a press (normally hydraulic) with movement limited to the axis normal to the plane of the mold. This process tends to be associated with a variety of materials, including (but not limited to) the prepreg continuous fibers in epoxy resin and the prepreg short fibers in polyester resin [93–99].

6.5.7.1 Sheet Molding Compound (SMC)

The sheet molding compound (SMC) process is a high-temperature process performed in a matched set of steel molds heated to over 350°F, the SMC compound is laid in a measured volume in the heated mold with the required resin, catalyst, mineral fillers, and reinforcement premixed into the compound in a previous compounding process. Then, the mold is forced closed under greater than 2000 psi of pressure squeezing the compound to fill the entire mold cavity rapidly, enabling curing within 1–3 min. Once cured, the mold is opened and the molding is ejected as a finished part requiring only minimal surface preparation prior to finish painting. One added key advantage of the SMC process is the ability to mold the part to "net" shape, i.e., normally, SMC parts are molded to the final net shape of the part that do not then need postmold trimming. This eliminates the need to trim parts after molding and provides the most consistent part to part trim repeatability.

The molds may be heated or the composite may be preheated and formed in relatively cool molds. A limitation of the process is the uniaxial force applied and hence, while horizontal surfaces experience compression loads, as the component surfaces become vertical, there is little or no component of the force acting to consolidate the material in the vertical plane. Further, near vertical surfaces may be subjected to wrinkling during mold closure. These issues may be resolved by the use of rubber-block molding or hydroforming (pressurized liquid contained in a flexible membrane) wholly or partially substituting the male molds half.

To aid in the processing of flat products, it is helpful to use a sheet molding compound, which can be made in a manner similar to that of a DMC, although higher concentration of reinforcement (20% v/v) and longer staple length (approximately 25 mm) are achieved. The catalyzed resin is deposited on a polyethylene-release film, and the reinforcement is consolidated by adding a top-release film and by passing the sandwich through a series of rollers to yield an SMC of the required thickness. The SMC will have a shelf-life of approximately three to six months at room temperature.

6.5.7.2 Dough Molding Compound (DMC)

Generally used with glass reinforcements, dough is produced using a high-shear mixer by mixing the resin, catalyst, approximately 10% v/v of approximately

10-mm chopped reinforcement, a release agent (such as zinc stearate) together with filler and pigment. After mixing, the dough is formed into a rope or billet. DMC has a lower shelf-life than SMC.

6.5.7.3 Bulk Molding Compound (BMC)

BMC is similar to DMC, although BMC uses an improved resin such as an isophthalic polyester. However, the terminology is restricted to the UK, and in the USA, BMC is identical to DMC.

6.5.7.4 Meeting Demand from the 1970s Through 2012

The automotive industry was served well with the SMC process because their volume justified the capital cost associated with the tooling and the high tonnage presses required. The marine markets could never justify SMC tooling. The varied marine models, the need to react to nearly constant product changes, and the low volume need of each of the moldings make the SMC process very impractical for the marine market and the open molding is a more practical method for their needs. If the volume requires more than one molding per shift, then additional molds are created, each at a very nominal cost; therefore, the open molding process can be very flexible in meeting the needs of the marine or similar volume markets.

6.5.8 Injection Molding

Injection molding is probably the most extensively used method for processing short fiber-reinforced thermoplastics [100–107]. Injection molding is accomplished by large injection molding machines such as that shown in Fig. 6.23. Resin is fed to the machine through the hopper. Colorants are usually fed to the machine directly after the hopper. The resins enter the injection barrel by gravity though the feed throat. Upon entrance into the barrel, the resin is heated to the appropriate melting temperature.

The fiber/resin mixture, whether it is preblended or fed as a physical mixture, is fed into the hopper and transferred into the heated barrel. The material softens by heat transfer from the barrel wall. At the same time, the screw rotates to apply high shear to further heat the material and fill the barrel. The molten material is collected in front of the screw by the rotation of the screw and then injected at high pressure into the mold cavity through the runner and the gate. The mold is cooled below the solidification temperature of the resin in case of thermoplastics composites. The level of automation of this method is the highest among many processing methods. Because of intensive mixing with high shear and passage through a narrow gate, extensive fiber damage occurs; therefore, injection molding of composite materials is suitable only with short fiber-reinforced or particulate-filled polymers.

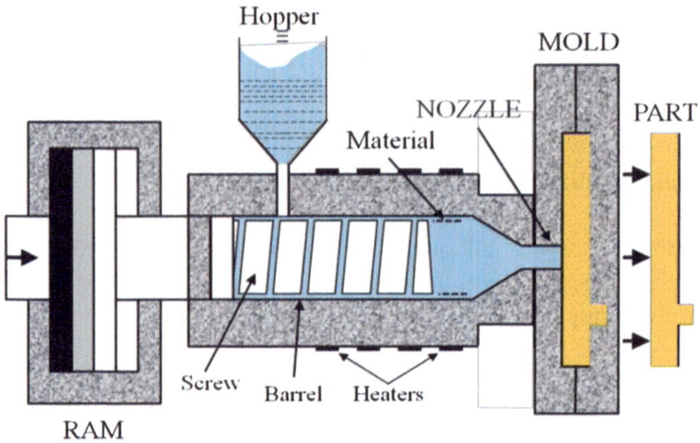

Fig. 6.23 Schematic of injection molding process

There is a critical length of fibers below which the fiber length does not degrade. The critical length is determined by the rheological properties of the composite mold, fiber properties, and instrument factors [108–110]. Less fiber damage occurs when a plunger-type injection molding machine is used rather than a screw-type injection molding machine. Because the plunger-type injection molding machine does not achieve a high degree of mixing in the machine, the raw materials must be thoroughly mixed prior to feeding.

6.5.9 Reaction Injection Molding (RIM)

RIM is widely employed in processing of various thermoset polymers. It usually contains two or more liquids that are accumulated in a high-pressure mixing unit and then injected into a mold, where they are polymerized or cured to form a desired part. Hence, it presents a typical example of a complex nonisothermal and compressible flow of liquid with reaction taking place at high pressure and high temperature. Being mostly motivated by practical applications, many researchers have made many attempts to describe the residual stresses remaining in the injection-molded parts experimentally or theoretically.

RIM is one of the more recent processing methods (Fig. 6.24). Instead of using already polymerized materials such as matrices, highly reactive monomeric or oligomeric ingredients are placed in two tanks, which are then quickly mixed by impingement and injected into the mold cavity. As soon as the two materials are mixed, chemical reaction begins to form a polymeric matrix, which completes typically within 5–30 s. Thus, the major portion of the RIM machine is a high-pressure pump and a metering system. Again, with highly intense shear, only short fibers and fillers can be used as reinforcements. However, RIM uses low-viscosity chemicals,

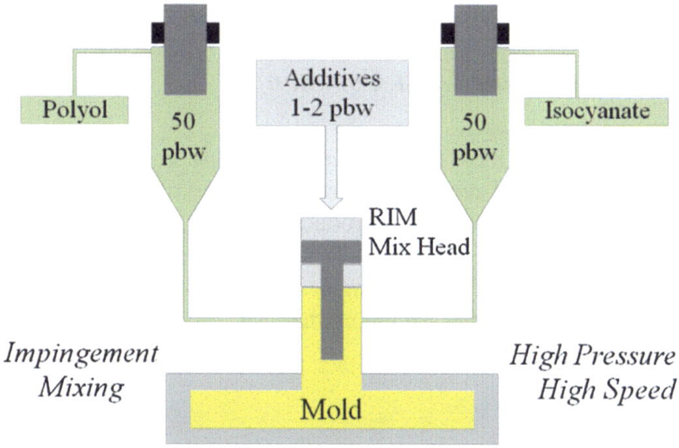

Fig. 6.24 Schematic of reaction injection molding (RIM) process

and this allows the preplacement of continuous fiber-woven fabrics in the mold in a manner similar to resin transfer molding. The distinction between these two methods lies in the preparation of the resin precursor. When the resin formulation is already made, the method is called RTM, while if the resin is prepared in situ by an impingement or static mixer, the method is termed as RIM.

6.5.10 Extrusion

In a thermoplastic process, pellets, granules, or powder are melted and forced through a die under pressure to form a given continuous shape. Typical shapes extruded are flat films and sheets, filaments and fibers, strands for pelletizing, and webs for coating and laminating. The technique is also used for forming composite preformed materials from mixtures of a matrix powder and short fibers and is suitable for metal–matrix composites. The method is widely used for continuous production of films, sheets, tubes, and other profiles and is also used in conjunction with blow molding. A thermoplastic or thermoset molding compound is fed from a hopper to a screw and barrel in which the molding compound is heated to plasticity and then forwarded usually via a rotating screw, through a nozzle possessing the desired cross section. Production lines require input and takeoff equipment that can be complex.

The advantages include low tool cost, the possibility of numerous complex profile shapes, very rapid production rates, the possibility of applying coatings or jacketing to core materials (such as wires). Familiar products manufactured by extrusion include pipes, ingots, guttering, window sills, and insulation on wires.

6.6 Understanding Manufacturing Process of Composites

6.6.1 Focusing on the Most Common

As mentioned above, the most common applications of composite are in automotive, marine, industry, and recreational markets. The volume needs of the automotive industry for OEM production product generally lead to the need for SMC process, yet for low-volume niche vehicle applications. Often, the volume is too low to justify the expenses of steel tooling needed for SMC. Hence, the total overall volume in automotive industries excluding the SMC process is the focus of interest here.

6.6.1.1 Raw Materials

In comparison , each of the focused processes and applications has commonalities in terms of the materials used for composite molding. Each process typically uses a polyester or vinyl ester resin, most often fiberglass reinforcement, with a fiber loading of 25–30 wt% of the total laminate in comparison with the resin. The resin-to-fiber ratio is most interesting and determines the strength and performance of the final molding.

It is to be understood that the corrosion, heat distortion temperature, and general surface finish are driven primarily by the selection of the resin, while the mechanical strength is mostly attributed to the fiber reinforcement. It is also to be noted that the fiber length is a factor that determines the strength and short fibers, typically less than 1″ in length, impart poor strength, in comparison to the same fibers of lengths ranging from 1″ to 3″. The fiber distribution in the cross section of the molded laminate is another critical factor determining the strength and performance. If the fiber is heavily loaded to one side of the laminate, the remaining laminate thickness will be made up of just the resin, which is bound to be unreinforced and prone to cracking if flexed. It is to be noted that a fiber loading of 30 wt% equates to only approximately 17 vol.%, and hence, it is technically for more than 80% of the laminate cross-sectional thickness to be attributed to only the resin, which results in a laminate, which is weak when flexed.

6.6.1.2 Comparison of Process Methods

It can be understood that the most common applications using composites have the same basic resins and reinforcements in common. Hence, in effect, there is no real difference between the materials used in a part molded by open mold or contact molding and the materials used in a part made by RTM, LRTM, or VARTM/vacuum infusion molding. However, the resin and the fiber are the same for each process, and if the fiber-to-resin ratio in the laminate were the same when the cross-sectional fiber distribution is consistent, then each of the process methods would yield a molded part with the same performance in the end. Bearing the commonality of the materials in mind, it is essential to understand the differences in the parts molded using each of the different process methods.

Again, the open mold processes, as well as the RTM and LRTM processes, have the same fiber loading ratio, i.e., typically 30 wt% of fiber and 70 wt% of resin. The variant to the process methods is the VARTM or vacuum infusion process. In these methods, the fiber loading increases to typically 60 wt% and to as high as 70 wt% of the laminate, with the remaining 40–30 wt% attributed to the resin. Hence, if in the case of the VARTM/vacuum infusion process, the much higher fiber volume and the mechanical strength primarily originate from the reinforcement, it is then logical to expect that the performance of the parts obtained by VARTM/vacuum infusion to be different and indeed, marked differences have been observed. Yet, before the differences are considered, it is key to review the differences in the tooling used in VARTM in comparison to RTM or LRTM.

6.6.1.3 Comparison of Actual Mold Cavity of Each Process

In the open molding process, there is only one mold surface and hence, the actual part thickness is totally under the control of the operator. The part thickness is in fact determined by how much resin and fiber is applied to the surface of the mold. The operator can also control the fiber-to-resin ratio by manually adjusting the resin content in relation to the fiber. RTM and LRTM both possess fixed cavity cross sections, which is inherent to the build of their match mold set. Hence, the glass-to-resin ratios are controlled by the type/amount of the fiber loaded and the distribution of the fiber in the laminate cross section. In the VARTM process, there is one rigid mold surface and the opposing mold half is a film on top of the fiber. The actual cavity is determined by the fiber thickness once the outer atmosphere compresses the fiber when a vacuum is created between the rigid mold surface and the film over the fiber.

6.6.1.4 Required Strengths

Bearing in mind that over the years, the vast majority of molded parts have adopted a typical fiber loading of 30 wt% deemed as sufficient for most applications, variations of resins have then enhanced the fiber properties for specific needs of heat or corrosion and surface finish. The other factor that is common to most applications is the need for stiffness or flexure modulus. Because, in the VARTM process there is only one mold surface and the amount of fiber loaded upon the surface sets the part thickness, using the same fiber loading value per square area in the VARTM process as that used in other molding processes, yields a much thinner, and thus, a more flexible laminate.

6.6.2 Distinguishing the Differences

The SMC process produces the highest productivity with the highest surface finish quality. The expected tooling life for a steel tool as that used in the SMC process is greater than 100,000 moldings.

Open molding offers the lowest cost tooling and an exterior finish directly from the mold, which meets the needs of a high-gloss smooth surface, which is normally well suited for most marine and general nonautomotive surface needs. The lower quality surface achieved in comparison to that achieved in SMC can be attributed the mold materials (composites versus steel in SMC). Those in the automotive industry, as well as others who need to meet a "Class A" exterior surface finish in the absence of the volume to justify the use of the SMC process, can use the open molding process, provided the surface of the molding is sanded, primed, and finish painted to meet the finish specifications of the product needs. Additionally, in the open mold process with only a single mold half, the back side of the molding is fully dependent on the skill of the operator to maintain the part thickness and finish smoothness. This requires many industries to grind the individual parts by hand to fit because of the variance, which common to the manufacturing process.

RTM can enjoy the benefits of steel tooling to achieve the true "Class A" surface finish standards; however, the economics would be more viable if the mold is placed in a press to form SMC so that the RTM process is typically performed using composite tooling construction materials, such as the open molding process. The advantage of having two mold halves eliminates the variance in part thickness, which is common in the open mold method, making the RTM part very comparable in dimensional accuracy and repeatability to that obtained by the SMC process.

6.6.3 Why Is Thickness an Issue?

The thickness sets the stiffness when all other factors are identical. For instance, in an "I" beam, the top and bottom rails are separated by a web. The wider the web, the greater is the load the "I" beam can withstand without deflecting. The same holds true in a laminate, i.e., the thicker the laminate is with an equal area volume of fiber, the stiffer will be the laminate. This has been the major stumbling block for the VARTM process from the beginning. By reducing the resin content, the laminate becomes thinner when compared to those obtained in an open mold or in the fixed cavity of the RTM and LRTM processes. This thinner laminate then is far lower in flex modulus or much easier to bend. This issue makes the use of VARTM in the marine market difficult because the fiber is the most costly component of the resin and glass makeup from a volume standpoint, in addition to being the greatest contributor to the final laminate weight. In order to regain the stiffness achieved in the VARTM process, the laminate needs to be "bulked up" with either more fiber or by adding some form of "core" each add cost.

6.6.4 Process Evolution

During the late 1960s through to the early 1970s, the composite industry experienced significant growth, which was driven by the marine industry (as the boats converted fully from previously used wood materials) and the automotive industry (to meet both styling needs and weight reduction of the various exterior body panel components).

To meet the rapid market growth, the suppliers of the primary polymer resins and fiber reinforcements developed their products to support SMC as required by the automotive industry and the open molding process as required by the marine manufactures. It should be noted that generally, the resins and reinforcement for both SMC, open molding, and RTM are nearly identical. Hence, the primary differences lie in the actual molding process methods and not in the materials involved.

6.6.4.1 Orphan Alternative Method: RTM

Throughout the evolution of the open mold and SMC processes, the closed molding process has existed as an alternative. Like the SMC process that uses a matched set of tooling (an upper and lower mold half), the resin transfer molding (RTM) process has been available for use in both the automotive and marine markets.

The RTM process was not accepted by the automotive markets because of their demands of rapid or high-volume production. Also, the marine market also did not accept RTM because of the added cost in tooling because two mold halves are needed, while in the open molding process, only a cavity or single half-mold is needed.

6.7 Summary

The use of composites in space and aerospace applications continues to grow. However, the vast majority of composite molding is used in automotive, marine, industrial, and recreational applications. In the automotive industry, the typical composite molding is supplied through the use of SMC and the group of remaining applications listed all rely on the open molding (spray-up or hand lay-up) process often referred to as contact molding.

The trend toward the increased use of closed molding such as VARTM or RTM (with its most recent innovations such as LRTM) has been influenced in recent years by both an interest in the reduction of VOC emission and to lower the labor content in the production processes, which currently rely on the open molding technology.

At the time of writing this chapter, while there is a strong interest toward the use of low-pressure closed molding methods to avoid excessive VOC emissions and to reduce the need for skilled workers, the majority or the highest volume of parts produced still rely on the open molding methods. However, the future clearly lies with the closed molding processes, although the lack of practical experience on the part of the molders and the added cost of tooling and raw materials used in producing the closed molded parts have continued to slow the conversion from open molding to closed molding over the years.

References

1. J.B. Donnet, R.C. Bansal, *Carbon Fibers*, 2nd edn. (Marcel Dekker, New York, 1990)
2. J.B. Donnet, S.J. Park, Carbon **31**, 434 (1991)
3. P.J. Walsh, vol. 21, ed. D.B. Miracle, S.L. Donaldson, "*Composites*", ASM Handbook, ASM International, Amsterdam (2001)
4. S.J. Park, *Carbon Materials* (Daeyoungsa, Seoul, 2006)
5. S.J. Park, M.K. Seo, *Interface Science and Technology: Interface Science and Composites* (Elsevier, 2011)
6. S.J. Park, M.K. Seo, Carbon **39**, 1229 (2001)
7. W. Ding, J. Chen, J. Liu, Z. Pang, R. Guan, Asian J. Chem. **25**, 7955 (2013)
8. J. Lee, S. Nouranian, G.W. Torres, T.E. Lacy, H. Toghiani Jr., C.U. Pittman, J.L. Dubien, J. Appl. Polym. Sci. **130**, 2087 (2013)
9. X.J. Yang, S.H. Li, X.D. Tang, J.L. Xia, Adv. Mater. Res. **807–809**, 2826 (2013)
10. H.Y. Zhang, H. Xu, C.S. Guan, Adv. Mater. Res. **833**, 322 (2014)
11. X. Zhang, V. Bitaraf, S. Wei, Z. Guo, H.A. Colorado, ALChE J. **60**, 266 (2014)
12. S.J. Park, M.K. Seo, J.R. Lee, Carbon **40**, 835 (2002)
13. W. Dong, H.C. Liu, S.J. Park, F.J. Jin, J. Ind. Eng. Chem. Article in press (2013)
14. A.A. Azeez, K.Y. Rhee, S.J. Park, D. Hui, Comp. Part B Eng. **45**, 308 (2013)
15. J.R. Choi, Y.S. Lee, S.J. Park, J. Ind. Eng. Chem. Article in press (2013)
16. Y. Chen, X. Zhou, X. Yin, Q. Lin, M. Zhu, Int. J. Polym. Mater. Polym. Biomater. **63**, 221 (2014)
17. K. Zhang, Y. Gu, M. Li, Z. Zhang, Mater. Design **54**, 624 (2014)
18. L.L. Hong, A. Moshonov, J.D. Muzzy, Polym. Compos. **12**, 191 (1991)
19. L.T. Drzal, M. Madhukar, J. Mater. Sci. **28**, 569 (1993)
20. M.J. Mathews, S.R. Swanson, Compos. Sci. Technol. **67**, 1489 (2007)
21. S.J. Park, M.K. Seo, J.R. Lee, J. Colloid Interface Sci. **268**, 127 (2003)
22. S.J. Park, M.K. Seo, D.R. Lee, Carbon **41**, 2991 (2003)
23. M. Schwartz, *"Composite Materials", Handbook*, 2nd edn. (McGraw-Hill, New York, 1992)
24. S.W. Tsai, H.T. Hahn, *Introduction to Composite Materials* (Technomic Pub. Co., Lancaster, Pennsylvania, 1980)
25. D.J. Johnson, Structure and properties of carbon fibers, in *Carbon Fibers Filaments and Composites*, ed. by J.L. Figueiredo, C.A. Bernardo, R.T.K. Baker, K.J. Htittinger (Kluwer Academic, Netherlands, 1990)
26. F.T. Wallenberger, *Advanced Inorganic Fibers: Processes, Structures, Properties, Applications* (Kluwer Academic Publishers, Dordrecht, 2000)
27. Q. Guo, J. Liu, L. Chen, K. Wang, Polymer **49**, 1737 (2008)
28. V. Kaushik, J. Raghavan, Compos. Part A **41**, 1210 (2010)
29. R.C. Wetherhold, M. Corjon, P.K. Das, Compos. Sci. Technol. **67**, 2428 (2007)

30. J. Palmer, O.R. Ghita, L. Savage, K.E. Evans, Compos. Part A **40**, 490 (2009)
31. A Market Assessment and Impact Analysis of the Owens Corning Acquisition of Saint-Gobain's Reinforcement and Composites Business (2007)
32. Y. He, G. Tin, M. Pan, D. Chen, Comp. Struct. **109**, 1 (2014)
33. S. Chen, Q. Wang, T. Wang, J. Reinf. Plast. Comp. **32**, 1136 (2013)
34. G. Das, R.D. Kalita, H. Deka, A.K. Buragohain, N. Karak, Prog. Org. Coat. **76**, 1103 (2013)
35. E. Fitzer, R. Kleinholz, H. Tiesler, M.H. Stacey, R.D. Bruyne, I. Lefever, A. Foley, W. Frohs, T. Hauke, M. Heine, H. Jager, S. Sitter, *"Fibers, 5. Synthetic Inorganic", Ullmann's Encyclopedia of Industrial Chemistry* (Wiley-VCH Verlag GmbH & Co., KGaA, Weinheim, Germany, 2008)
36. Y.X. Zhang, H.S. Zhang, Compos. Struct. **92**, 2159 (2010)
37. Y. Bai, Z.H. Huang, F. Kang, J. Mater. Chem. A **1**, 9536 (2013)
38. Y. Bai, Z.H. Huang, F. Kang, Carbon **66**, 705 (2014)
39. M.D. Li, Y. Lu, X. Guo, Appl. Mech. Mater. **470**, 31 (2014)
40. L. Bian, J. Xiao, J. Zheng, S. Xing, C. Yin, A. Jia, Mater. Design **54**, 230 (2014)
41. J.P. Gogoi, N.S. Bhattacharyya, S. Bhattacharyya, Comp. Part B Eng. **58**, 518 (2014)
42. C. Steggall-Murphy, P. Simacek, S.G. Advani, S. Yarlagadda, S. Walsh, Compos. Part A **41**, 93 (2010)
43. S.G. Pantelakis, C.V. Katsiropoulos, G.N. Labeas, H. Sibois, Compos. Part A **40**, 595 (2009)
44. J.A. Ramos, M. Blanco, I.I. Zalakain, I. Mondragon, J. Colloid Interface Sci. **336**, 431 (2009)
45. M. Rico, J. Lopez, C. Ramirez, J. Diez, B. Montero, Polymer **50**, 569 (2009)
46. H.J. Bang, H.Y. Kim, F.L. Jin, S.J. Park, J. Ind. Eng. Chem. **17**, 805 (2011)
47. S.Y. Lee, H.K. Shin, M. Park, K.Y. Rhee, S.J. Park, Carbon **68**, 67 (2014)
48. J. Huang, D. Rodrigue, Mater. Design **55**, 653 (2014)
49. B. Yuan, C. Bao, L. Song, N. Hong, K.M. Liew, Y. Hu, Chem. Eng. J. **237**, 411 (2014)
50. B. Li, H. Yuan, Y. Zhang, Comp. Sci. Technol. **89**, 134 (2013)
51. R. Palazzetti, A. Zucchelli, I. Trendafilova, Comp. Struct. **106**, 661 (2013)
52. M.M. Shokrieh, A. Saeedi, M. Chitsazzadeh, Mater. Design **56**, 274 (2014)
53. Z. Wu, L. Meng, L. Liu, Z. Jiang, L. Xing, D. Jiang, Y. Huang, J. Adhes. Sci. Technol. **28**, 444 (2014)
54. K. Stoeffler, S. Andhelic, N. Legros, J. Roberge, S.B. Schougaard, Comp. Sci. Technol. **84**, 65 (2013)
55. S. Zhou, Q. Zhang, C. Wu, J. Huang, Mater. Design **44**, 493 (2013)
56. H. Wu, L.T. Drzal, J. Appl. Polym. Sci. **130**, 4081 (2013)
57. P.S. Goh, B.C. Ng, A.F. Ismail, M. Aziz, Y. Hayashi, J. Colloid Interface Sci. **386**, 80 (2012)
58. S.H. Hwang, H.W. Park, Y.B. Park, M.K. Um, J.H. Byun, S. Kwon, Comp. Sci. Technol. **89**, 1 (2013)
59. F. Rastellini, S. Oller, O. Salomon, E. Onate, Comput. Struct. **86**, 879 (2008)
60. A.E. Mahi, M. Assarar, Y. Sefrani, J.M. Berthelot, Compos. Part B **39**, 1069 (2008)
61. Y.X. Zhang, H.S. Zhang, Compos. Struct. **92**, 2159 (2010)
62. J. Xu, B.V. Sankar, Compos. Part A **39**, 1625 (2008)
63. S.X. Wang, L.Z. Wu, L. Ma, Mater. Design **31**, 118 (2010)
64. L.J. Deka, S.D. Bartus, U.K. Vaidya, Compos. Sci. Technol. **69**, 725 (2009)
65. P. Feraboli, T. Cleveland, P. Stickler, J. Halpin, Compos. Part A **41**, 557 (2010)
66. D.N. Betts, A.I.T. Salo, C.R. Bowen, H.A. Kim, Compos. Struct. **92**, 1694 (2010)
67. A. Faraz, D. Biermann, K. Weinert, Int. J. Mach. Tools Manuf. **49**, 1185 (2009)
68. H. He, J. Wang, K. Li, J. Wang, J. Gu, Mater. Design **31**, 4631 (2010)
69. R. Jamaati, M.R. Toroghinejad, Mater. Sci. Eng. A. **527**, 4146 (2010)
70. G. Liu, X. Yang, Y. Wang, Polymer **48**, 4385 (2007)
71. T.J.C. Liu, H.C. Wu, Mater. Design **31**, 1971 (2010)
72. S.W. Tsai, H.T. Hahn, *Introduction to Composite Materials* (Technomic Publishing Co., Lancaster, 1980)

73. P. Roca, G. Araiza, Constr. Build. Mater. **24**, 1372 (2010)
74. H. Huang, A.M. Waas, Compos. Sci. Technol. **69**, 2338 (2009)
75. J.J. Nie, Y.D. Xu, L. Zhang, L. Cheng, J.Q. Ma, J. Mater. Process. Technol. **209**, 572 (2009)
76. P.J. Liotier, V. Alain, D. Christine, Compos. Part A **41**, 653 (2010)
77. T.C. Truong, D.S. Ivanov, D.V. Klimshin, S.V. Lomov, I. Verpoest, Compos. Part A **39**, 1380 (2008)
78. G. Li, C. Zhang, Y. Wang, P. Li, Y. Yu, X. Jia, H. Liu, X. Yang, Z. Xue, S.K. Ryu, Compos. Sci. Technol. **68**, 3208 (2008)
79. W. Chen, Y. Yu, P. Li, C. Wang, T. Zhou, X. Yang, Compos. Sci. Technol. **67**, 2261 (2007)
80. A. Zuraida, A.A. Khalid, A.F. Ismail, Mater. Design **28**, 71 (2007)
81. M.T. Arellano, L. Crouzeix, B. Douchin, F. Collombet, H.H. Moreno, J.G. Velazquez, Compos. Struct. **92**, 2457 (2010)
82. P.A. Sreekumar, K. Joseph, G. Unnikrishnan, S. Thomas, Compos. Sci. Technol. **67**, 453 (2007)
83. M. Haider, P. Hubert, L. Lessard, Compos. Sci. Technol. **67**, 3176 (2007)
84. P.A. Sreekumar, S.P. Thomas, J. Saiter, K. Joseph, G. Unnikrishnan, S. Thomas, Compos. Part A **40**, 1777 (2009)
85. H. Salek, P. Trudeau, *SAE Technical. Papers* **7** (2013)
86. Y. Gu, X. Tan, Z. Yang, M. Li, Z. Zhang, Mater. Design **56**, 852 (2014)
87. M.S. Yun, W.I. Lee, Compos. Sci. Technol. **68**, 202 (2008)
88. A. Peled, B. Mobasher, Z. Cohen, Cem. Concr. Compos. **31**, 647 (2009)
89. H.Y. Kim, K.T. Park, J.W. Jeong, Y.H. Lee, Y.K. Hwang, D.H. Kim, Compos. Struct. **91**, 20 (2009)
90. W.F. Ragheb, Thin-Walled Struct. **48**, 709 (2010)
91. H.M. Akil, L.W. Cheng, Z.A.M. Ishak, A.A. Bakar, M.A.A. Rahman, Compos. Sci. Technol. **69**, 1942 (2009)
92. R. Stewart, Reinf. Plast. **53**, 34 (2009)
93. P. Dumont, L. Orgeas, D. Favier, P. Pizette, C. Venet, Compos. Part A **38**, 353 (2007)
94. P. Bhat, J. Merotte, P. Simacek, S.G. Advani, Compos. Part A **40**, 431 (2009)
95. J. Merotte, P. Simacek, S.G. Advani, Compos. Part A **41**, 881 (2010)
96. M.S. Kim, W.I. Lee, W.S. Han, A. Vautrin, C.H. Park, Compos. Part A **40**, 1192 (2009)
97. S.R. Dhakate, R.B. Mathur, B.K. Kakati, T.L. Dhami, Int. J. Hydrogen Energy **32**, 4543 (2007)
98. R. Liu, X. Li, X. Liu, B. Shen, Z. Han, J. Appl. Polym. Sci. **130**, 4032 (2013)
99. K. Senthil Kumar, I. Siva, P. Jeyaraj, J.T. Winowlin Jappes, S.C. Amico, N. Rajini, Mater. Design **56**, 379 (2014)
100. H. Ye, X.Y. Liu, H. Hong, J. Mater. Process. Technol. **200**, 12 (2008)
101. M.A. Khan, J. Ganster, H.P. Fink, Compos. Part A **40**, 846 (2009)
102. S.J. Liu, M.J. Lin, Y.C. Wu, Compos. Sci. Technol. **67**, 1415 (2007)
103. Z.Y. Liu, D. Kent, G.B. Schaffer, Mater. Sci. Eng. A. **513**, 352 (2009)
104. P.W. Ho, Q.F. Li, J.Y.H. Fuh, Mater. Sci. Eng. A. **485**, 657 (2008)
105. J. Li, Z. Chen, X. Wang, T. Liu, Y. Zhou, S. Luo, J. Appl. Polym. Sci. **130**, 4171 (2013)
106. F.J.G. Silva, R.P. Martinho, A.P.M. Baptista, Thin Solid Films **550**, 278 (2014)
107. Y.Q. Wang, J.G. Kim, J.I. Song, Mater. Design **56**, 313 (2014)
108. S.X. Zhang, Z.Y. Ong, T. Li, Q.F. Li, S.F. Pook, Mater. Design **31**, 2897 (2010)
109. G. Morales, M.I. Barrena, J.M. Gomez de Salazar, C. Merino, D. Rodriguez, Compos. Struct. **92**, 1416 (2010)
110. P.A. Santos, M.A.S. Spinace, K.K.G. Fermoselli, M.A.de Paoli, Compos. Part A **38**, 2404 (2007)

Chapter 7
Recent Uses of Carbon Fibers

Abstract In recent decades, the carbon fibers have found wide applications in commercial and civilian aircraft, recreational, industrial, and transportation markets. This chapter will be focused on the applications of virgin carbon fibers and carbon fiber-reinforced composites including adsorbents, molecular sieves, electrodes, catalyst, and biomedical applications. In addition, the applications of the carbon fiber-reinforced composites applied in various industries will be fully covered. Finally, we will discuss the carbon–carbon composites, which have attracted considerable attention in the aerospace and braking systems with excellent thermal behaviors.

7.1 Introduction

Carbon fibers are a new breed of high-strength materials. Carbon fibers contain at least 90% carbon through the controlled pyrolysis of appropriate fibers [1]. The carbon fiber came into being in 1879 when Edison took out a patent for the manufacturing process of carbon filaments for use in electric lamps [2]. However, it was only in the early 1960s when the successful commercial production was started, as the requirements of the aerospace industry, especially for military aircraft, for better and lightweight materials became of paramount significance [3–5].

In recent decades, the carbon fibers have found wide applications in commercial and civilian aircraft, recreational, industrial, and transportation markets. Carbon fibers are used in composites with a lightweight matrix. Carbon fiber composites are ideally suited to applications in which strength, stiffness, low weight, and outstanding fatigue characteristics are critical requirements [6]. They also can be used for applications in which high temperature, chemical inertness, and high damping are important. The suppliers of Advanced Composites Materials Association released the 1997 industry statistics on the worldwide shipments of carbon fibers for composites [7]. However, from 1997 to 1999, there was a global slowing in the demand for carbon fibers. According to Mitsubishi Rayon Co. Ltd. (Tokyo, Japan),

S.-J. Park, *Carbon Fibers*, Springer Series in Materials Science 210,
https://doi.org/10.1007/978-981-13-0538-2_7

a carbon fiber producer, the worldwide consumption for sporting goods is nearly 11 million lb of carbon fibers [8].

As the political situation in the world has changed, the requirement for carbon fibers in military applications has decreased owing to major defense cuts. The emphasis has shifted to commercial applications, which have grown extensively. Therefore, it is not surprising that the carbon fibers are involved in a whole gamut of new applications. Developments do occur very rapidly in the area of composites, and although some of these applications may now have been discontinued or replaced, they serve to illustrate the diverse applications of carbon fibers [9, 10].

This chapter is focused on the various applications of virgin carbon fibers and carbon fiber-reinforced composites. Firstly, the virgin carbon fibers are widely used as adsorbents [11], molecular sieves [12], electrodes [13], catalysts [14], and in biomedical applications [15]. For various applications, carbon fibers are generally utilized as the activated forms through either chemical or physical activation process because of their high specific surface area and superior electrical properties with strong physicochemical properties [16–19]. In this section, we concentrate on the applications of carbon fibers in adsorbents and electrodes. Second, carbon fiber-reinforced composites were recently applied to aircraft, vehicle, sporting goods, and construction. The carbon fiber-reinforced composites can reduce the weight by over 30% while retaining high strength. In this section, we will review the applications of the carbon fiber-reinforced composites applied in various industries. Also, we will discuss the carbon–carbon composites, which have attracted considerable attention in the aerospace and braking systems with excellent thermal properties. At last, the interfacial interaction between the carbon fiber and the composites were briefly introduced.

7.2 Applications of Virgin Carbon Fibers

In general, activated carbons are made by the carbonization of a product such as coconut shells and polymer precursors via the controlled heat treatment and either chemical or physical activation. First, carbon precursors are carbonized up to approximately 800–1000 °C in N_2 atmosphere. Second, carbonized samples are activated by chemical method using either activation agents such as KOH and ZnCl or physical methods such as steam or CO_2 [20–30]. The activation process leads to the formation of the meso- and micropores and can yield a high pore volume of approximately 0.5 cm^3/g.

In 1976, Toho Beslon [31] developed the method for preparing activated fibers from PAN using an Fe salt in air or O_2 at lower temperatures (150–300 °C) for preoxidation. The oxidized fiber was then either activated using a chemical reagent or by physical activation at 700–1000 °C. The product has a high specific surface area of approximately 300–2000 m^2/g [32, 33].

Later, various studies were carried out on the preparation of activated carbon fibers (ACFs) based on PAN, ciscoes rayon, phenolic resin fiber (Kynol), pitch

Table 7.1 Typical properties of activated carbon cloth

Property	Typical value
Construction of cloth	1/1 plain weave
Weight	110 g m^{-2}
Thickness	0.5 mm/layer
Density	0.21 g cm^{-3}
Breaking strength warp	10 N/cm width
Breaking strength weft	15 N/cm width
Apparent internal surface area	1225 m^2/g
External surface area (microscopic)	1 m^2/g
Pore volume (total)	0.5 cm^3/g
Heat of wetting (di-n-butyl phthalate)	25 J/g

fiber, and hybrid resin, i.e., PAN and phenolic fibers [34, 35]. The typical properties of ACC are listed in Table 7.1.

In general, the ACC could be used as adsorbents for many organic molecules, including organic vapors and impurities, i.e., phosphates and nitrates, and bacteria and viruses, and in defense applications such as face masks, respirators, and filters in closed air-conditioning systems, protective clothing, and combat gear [36]. In addition, they are used in industrial applications such as air conditioning, odor control, solvent recovery, liquid purification, enzyme and catalyst support, and medical applications such as bandages, face masks, and blood purification. In this section, we focus on the adsorbents such as hydrogen, carbon dioxide, and organic molecules.

ACFs have unique characteristics compared to those of either granular or powder-activated carbons [37–41]. Thin-fiber shape clearly assures fast intraparticle adsorption kinetics compared to that of either pelletized or granular-activated carbons commonly employed in gas- and aqueous-phase adsorptions [42]. This becomes more important in designing adsorption units in which intraparticle diffusion resistance is the most significant factor, resulting in a considerable decrease in the size of adsorption units. The fibrous form is also favored because of ease in handling when it is used in felt or fabric forms and availability in the consolidated forms by newly developed molding techniques. Because ACFs are considered to have rather defined pore structures and surface characteristics, fundamental studies on adsorption have been tried from various angles.

7.2.1 Activated Carbon Fibers as Adsorbents

The development of the efficient hydrogen storage technology is important for the use hydrogen as future fuel. For application in the industry, the amount of hydrogen storage of 6.5 wt% at desirable conditions is required by DOE (US). Recently, well-known storage systems are compression, liquefaction, or storage as metal hydrides. Nowadays, the use of solid sorbents has been studied as a potential alternative method to store the cryogenic liquids and compressed gases [43–49].

Among various hydrogen storage systems, the metal hydride-based hydrogen storage systems can provide excellent safety with higher volumetric density compared to that afforded by conventional methods. However, this method requires a heavy reservoir to store hydrogen in metal hydrides. Thus, this method is limited to mobile applications.

Hydrogen physisorption is considered as one of the most promising storage technologies to meet the DOE goals. In contrast to the chemisorptions in metal hydrides, the phenomenon of physical adsorption of the undissociated hydrogen molecules on the surface of microporous carbon fibers or particles could improve the situation. This possibility to reach high sorption capacity is because of the ability of carbon to exist in either a finely powdered or fibrous form with highly porous structure, and the specific interactions between the carbon atoms and gas molecules. The total amount of stored hydrogen strongly depends on the pore geometry and pore size distribution as well as the storage pressure and temperature. The sorption of hydrogen usually takes place in the micropores. Macropores have no practical influence on the sorption capacity because they are only important for the gas compression and sorption/desorption reaction rate. Owing to its low density, high surface area and micropore volume, low cost, and accessibility at the industrial level, activated carbons are considered as good candidates for hydrogen storage systems [50].

Previously, Kunowsky et al. [51] reported the effect of the carbonization temperature on the activation of hydrogen storage of carbon fibers. They prepared two carbon fibers carbonized at 973 and 1273 K, and then, the carbon fibers were activated with KOH and NaOH to obtain the optimized pore structures for hydrogen storage.

We found that the highly microporous ACFs were obtained at the lower carbonization temperature and KOH was more effective in developing the micropore structure as an activation agent, as shown in Fig. 7.1.

H_2 adsorption was measured at 298 K up to 20 MPa, and at 77 K up to 4 MPa. The maximum adsorption of hydrogen was 1 wt% at room temperature and 3.8 wt % at 77 K, respectively. The total volumetric storage capacity was 17 g/L at 298 K and 32 g/L at 77 K. Thus, the prepared ACFs showed good hydrogen storage with a well-developed porous structure. Economically, the prepared carbon fibers had the following advantages: (1) low carbonization temperature; (2) low activation agent content with KOH for developing a porous structure; and (3) high carbon yield after carbonization and activation (Fig. 7.2).

Park et al. [52] studied the effect of copper electroplating on high-pressure hydrogen storage of ACFs. In this work, the copper nanoparticles were plated onto the surfaces of ACFs in order to improve the hydrogen storage capacity at 100 atm and 298 K.

The pore structures of Cu/ACFs and total pore volumes were analyzed with reference to N_2 = 77 K adsorption isotherms, and surface morphologies were characterized using scanning electron microscopy (SEM), respectively, as shown in

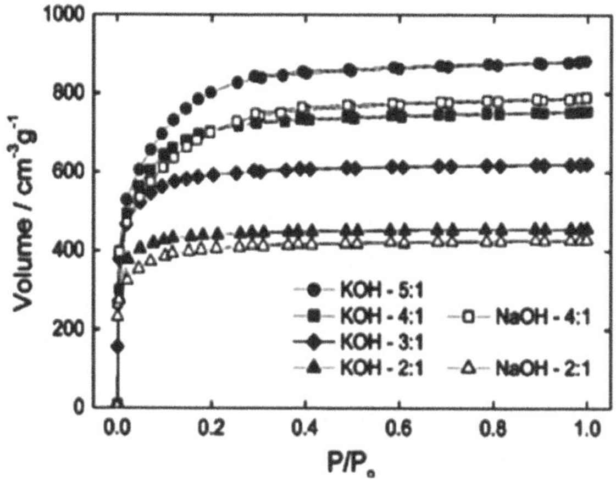

Fig. 7.1 N_2 isotherms of fibers (SL-242, carbonized at 973 K), activated with KOH and NaOH using different hydroxide-to-fiber ratios (w/w) [51]

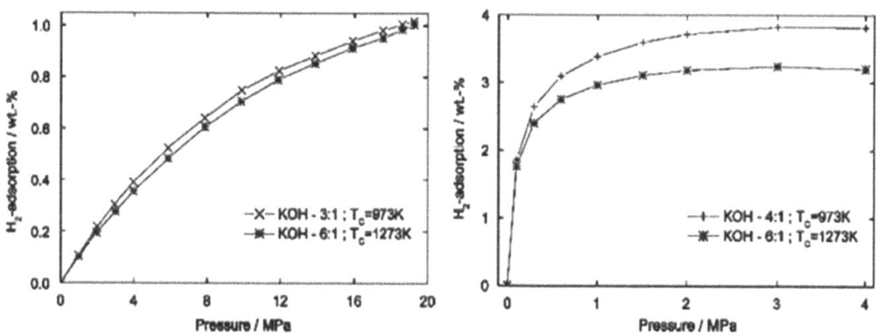

Fig. 7.2 H_2 excess adsorption isotherms measured at room temperature and 77 K for fibers activated using both precursors (S-241 and SL-242, carbonized at 1273 and 973 K, respectively) and different KOH-to-fiber ratios (w/w) [51]

Fig. 7.3. In Fig. 7.3 (left), Cu-10 was partially covered with Cu particles and Cu-30 was fully covered. The sizes of the observed copper particles was over the range 200–300 nm, indicating that some ACF pores could be blocked in the presence of Cu particles, specifically affecting the decreased surface area. As mentioned, the specific surface area and total pore volume decreased with increasing electroplating time (see Fig. 7.3 on the right).

However, the hydrogen storage capacity was enhanced in the presence of Cu nanoparticles for Cu-10 and Cu-30 samples, and then began to decline. These results indicated that the surface properties were weakened owing to the Cu nanoparticles but that the hydrogen storage capacity increased by the modified spillover of hydrogen molecules into the carbon structures (Fig. 7.4).

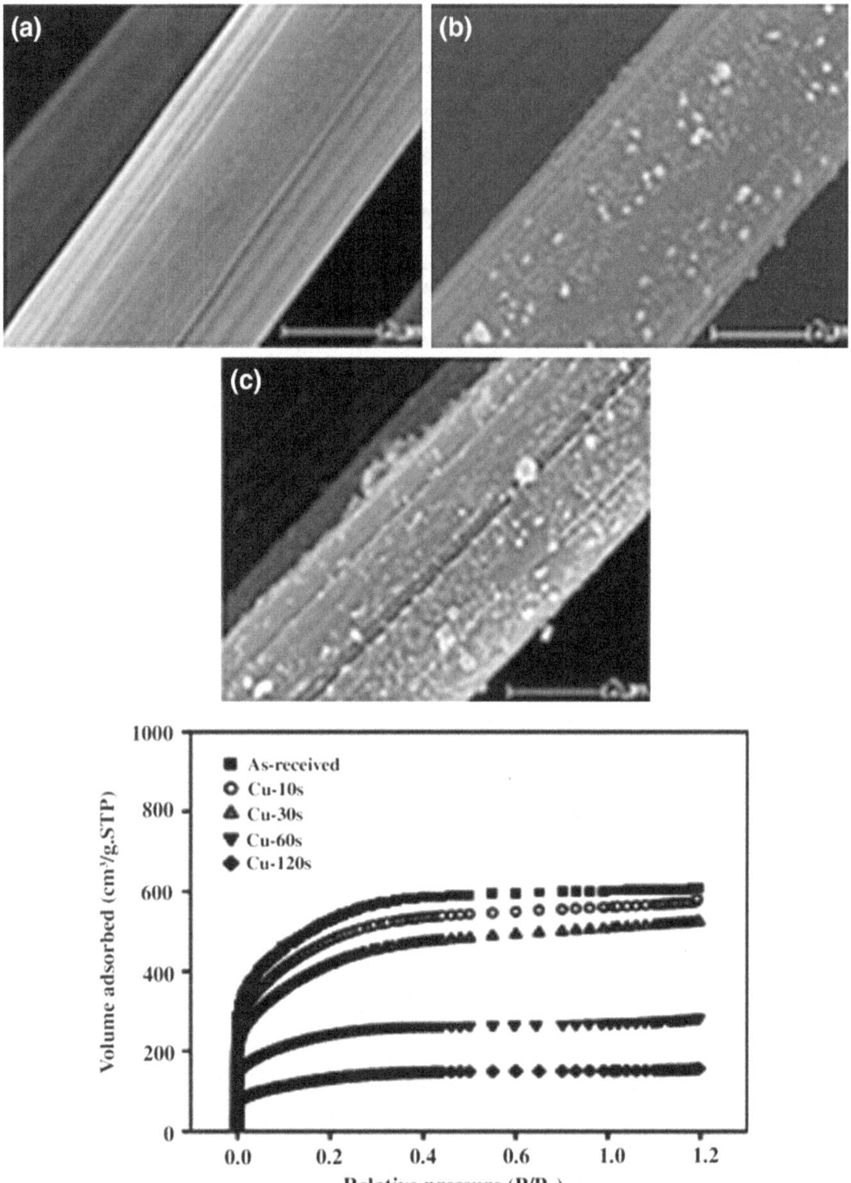

Fig. 7.3 SEM images of as-received and two of the Cu-plated ACFs as a function of plating time. **a** As-received; **b** Cu-10; **c** Cu-30 (*left*) and N_2 = 77 K adsorption isotherms of Cu-plated ACFs as a function of plating time (*right*) [52]

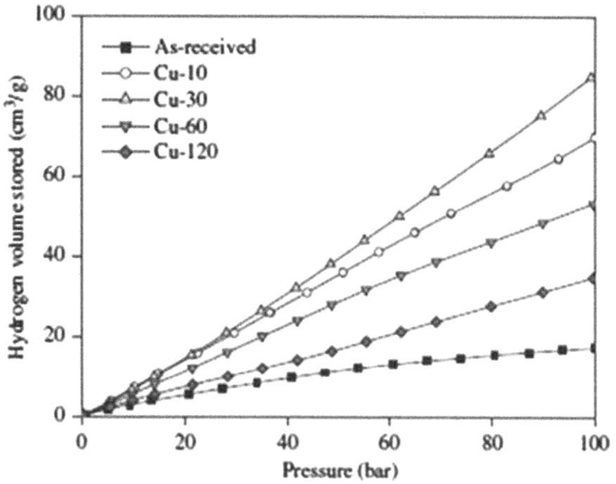

Fig. 7.4 Cu-plated ACF hydrogen storage behaviors at 298 K as a function of plating time [52]

Air we breathe is polluted with contaminants such as toxic gases, dust, and some volatile organic solvents. It is well known that acidic gases such as HCl, SO_x, and NO_x are major air pollutants. HCl gas poses the highest risk when exposed to human body, and it can threaten human life even when present in very small amounts in air.

ACFs are widely used in applications such as air and water purification and solvent recovery. It is recognized that the pore structure and surface functional groups are the most important properties of ACFs, which make them suitable for applications such as adsorption processes. Park et al. studied the effect of oxygen plasma treatment on the removal of hydrogen chloride from the ACFs. In this work, the oxygen plasma treatment of the ACFs was carried out to introduce oxygen containing groups onto the carbon surfaces. The surface properties of the ACFs were determined using X-ray photoelectron spectroscopy (XPS) (Table 7.2), while the SEM images before and after plasma treatment are displayed in Fig. 7.5.

The surface atomic compositions obtained using XPS analysis are presented in Table 7.2. It can be seen that the surface of the ACFs is mainly composed of carbon, oxygen, and nitrogen. The carbon and oxygen contents of the as-received sample are 90.4 and 7.8%, respectively; however, those of P-O-4 are 80.7 and 17.2%. The O_{1S}/C_{1S} ratio of plasma-treated ACFs increased with the treatment time

Table 7.2 Chemical composition of ACFs as a function of treatment time

	C_{1S} (%)	O_{1S} (%)	N_{1S} (%)	O_{1S}/C_{1S} (%)
As-received	90.4	7.8	1.7	8.6
P-O-1	88.5	9.8	1.6	11.1
P-O-2	86.3	11.5	1.7	13.3
P-O-3	83.2	14.7	1.7	17.7
P-O-4	80.7	17.2	1.8	21.3

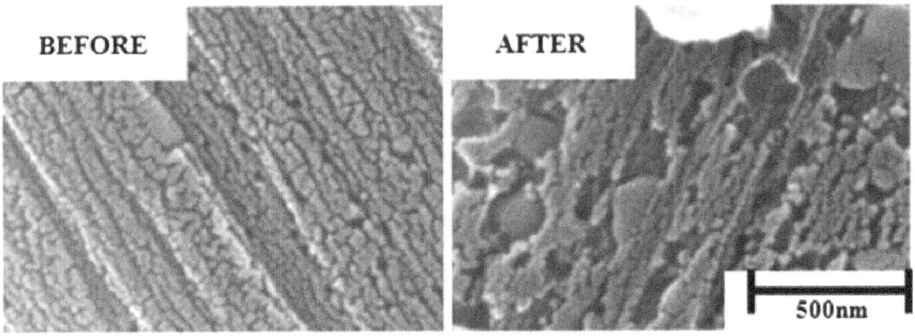

Fig. 7.5 SEM photographs of ACFs *before* (as-received) and *after* (P-O-4) oxygen plasma treatments

owing to the conformation of oxygen-containing surface functional groups on the carbon surfaces.

To study the changes of the surface morphologies of the as-received and P-O-4, SEM images were used at a resolution of ×50,000, as shown in Fig. 7.5. We found that many macropores appeared after the oxygen plasma treatments by the destruction of the carbon surfaces, and it could lead to a decrease in the specific surface area and total pore volume.

N_2/77 K adsorption isotherms were investigated using BET and D–R plot methods to characterize the specific surface area, pore volume, and pore size distribution. The efficiency of hydrochloride removal was confirmed using two types of methods: detecting tubes (range: 1–40 ppm) and gas chromatography. As experimentally found, the hydrochloride-removal efficiency of the ACFs increased with the number of plasma treatment times up to approximately 300%, resulting from the newly formed oxygen-containing functional groups (especially, phenolic and carboxylic) on carbon surfaces, in the decreased specific surface areas or pore volumes. These results indicated that the plasma treatment led to an increase in the hydrochloride removal owing to the improvement of the surface functional groups containing oxygen on the carbon surfaces.

Setoyama et al. [53] tried to determine the micropore distributions of ACFs using a helium adsorption technique and compared the distributions with those determined using the conventional nitrogen adsorption method for three types of ACFs. They excluded the effect of wider pores and assumed the Gaussian distribution for micropore size, and determined the parameters shown in Table 7.3. The portion of wider pores was rather small, and helium was able to better detect the existence of smaller pores compared to nitrogen in ACFs.

Mochida et al. [54, 55] found that the sulfuric acid-treated PAN-based ACF had a high catalytic effect on the reduction of NO by ammonia. The active sites were estimated to be oxygen-containing functional groups on ACF surfaces. They also prepared copper acetate supported on ACF and found that PAN–ACF showed good dispersion ability as a support of copper acetate.

Table 7.3 Micropore properties determined using nitrogen adsorption and helium adsorption by Setoyama et al. [53]

		Cellulose-based KF-1500	Pitch-based A-10	PAN-based FE200
Nitrogen adsorption	Specific surface area (m^2/g)	1147	795	743
	Micropore volume (cm^3/g)	0.25	0.37	0.35
	Micropore width (Å)	10.1	9.4	9.5
Helium adsorption	Narrow pore volume (cm^3/g)	0.54	0.42	0.32
	Wide pore volume (cm^3/g)	0.04	0	0.01
	Micropore width (Å)	9.3	7.6	78.9

Kim et al. [56] studied the adsorption behavior of propylamine on ACFs induced by oxygen functional complexes. The rayon-based ACFs (KF-1500, approximately 1420 m^2/g of specific surface area) were used as the initial materials, and the ACFs were subjected to acid treatment in 1.0 M of HNO_3 solution for 1 h at 373 K. The ACFs before and after the acid treatment were named virgin ACFs (V-ACFs) and acidic ACFs (A-ACFs), respectively, which, after PA adsorption, were renamed PA–V–CFs and PA–A–ACFs.

To facilitate detailed understanding, the O_{1S} peaks before and after the treatments were deconvoluted into surface oxygen complex contributions, as shown in Fig. 7.6, and the specific ratios of the contributions were also provided. The O_{1S} peaks of the ACFs showed four line shapes with binding energies of 530.9, 532.4, 533.8, and 535.2 eV assigned to C=O, C–O, O–C=O, and chemisorbed O_2 (or H_2O), respectively. It was found that the ratios of C=O and O–C=O increased, whereas that of C–O decreased. This result indicated that severe oxidation led to a decrease in the ratio of C–O, resulting in an increase in those of C=O and O–C=O.

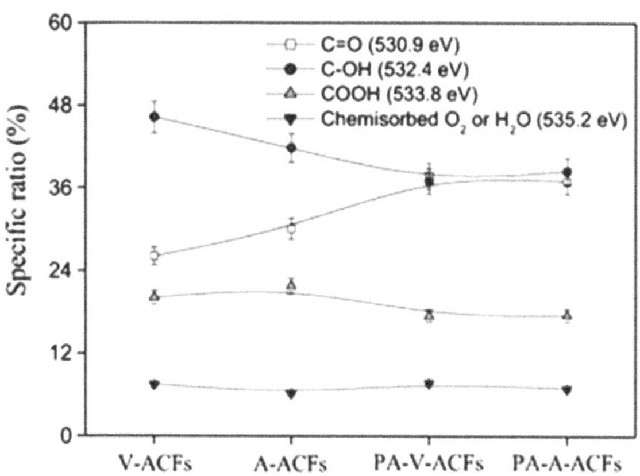

Fig. 7.6 Specific ratios of surface functional groups of ACFs in O_{1S}-XPS results [56]

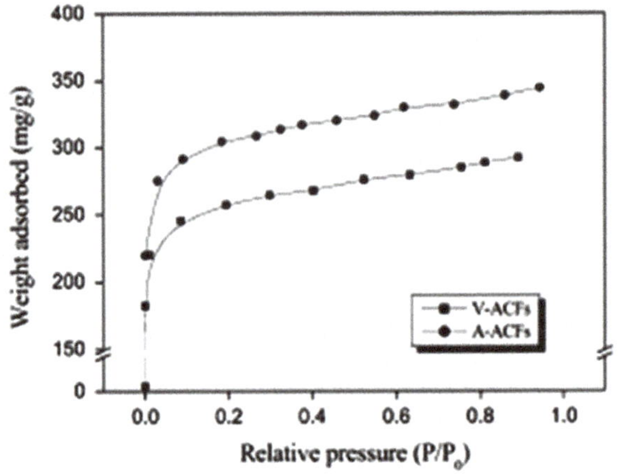

Fig. 7.7 Adsorption isotherms of propylamine on ACFs before and after acid treatment (temperature: 298 K) [56]

Figure 7.7 illustrates the adsorption isotherms of PA on the V–ACFs at 298 K. It was found that the amount of adsorbed PA rapidly increased as a result of the micropore filling effects under 0.1 relative pressure, and then reached a plateau at 1.0 of relative pressure, as representative of a typical Type I isotherm. The PA adsorption curve for the A–ACFs showed a pattern similar to that of the V–ACFs; however, the total PA adsorption amount was larger than that of the V–ACFs. It was noteworthy that the amount of PA adsorbed on the A–ACFs increased, whereas the specific surface area decreased, after the acid treatment. This result clearly showed that the PA adsorption was largely affected not only by the textual properties of adsorbents but also surface oxygen complexes (or functionalities) (Fig. 7.8).

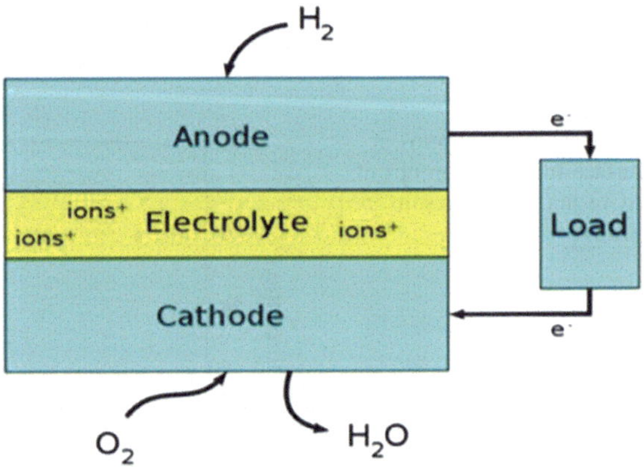

Fig. 7.8 Schematic of a fuel cell

7.2.2 *Carbon Fibers for Energy Storage*

1. Fuel cells

A fuel cell is a device that converts the chemical energy from a fuel into electricity through a chemical reaction with oxygen or another oxidizing agent [57]. Hydrogen is the most common fuel but hydrocarbons such as natural gas and alcohols, including methanol, are sometimes used. Fuel cells are different from batteries in that they require a constant source of fuel and oxygen to run, but they can produce electricity continually for as long as these inputs are supplied [58]. William Grove, a Welsh physicist, developed the first crude fuel cells in 1839 [59]. The first commercial use of fuel cells was in NASA space programs to generate power for probes, satellites, and space capsules. Since then, fuel cells have been used in many other applications. Fuel cells are used for primary and backup power for commercial, industrial, and residential buildings and in remote or inaccessible areas. They are also used to power fuel cell vehicles, including automobiles, buses, forklifts, airplanes, boats, motorcycles, and submarines [60–62].

There are many types of fuel cells, but they all consist of an anode (negative side), a cathode (positive side), and an electrolyte, which allows charges to move between the two sides of the fuel cell. Electrons are drawn from the anode to the cathode through an external circuit, producing direct current. As the main difference among fuel cell types is the electrolyte, fuel cells are classified by the type of electrolyte involved. Fuel cells come in a variety of sizes. Individual fuel cells produce small amounts of electricity, approximately 0.7 V; therefore, cells are "stacked," or placed in series or parallel circuits, to increase the voltage and current output to meet the power generation requirements of applications [63]. In addition to electricity, fuel cells produce water, heat and, depending on the fuel source, small amounts of nitrogen dioxide, and other emissions. The energy efficiency of a fuel cell is generally between 40 and 60%, or up to 85% if the waste heat is captured for use.

Nowadays, many researchers have attempted to improve the efficiency of fuel cells. In this regard, carbon fibers have been widely used as electrode materials, electrical reinforcements in bipolar plates, substrates for metallic catalysts, and so on. In this section, we review certain applications of the carbon fibers for fuel cells.

With the development of electric vehicles, studies on fuel cell and its component materials have been the main focus for researchers all over the world [64]. Materials used as fuel cell electrodes should be light, highly conductive, porous, inert, and stable. In proton exchange membrane fuel cells (PEMFC), carbon fiber paper (CFP) and carbon cloth are now used effectively as electrode substrates to support the catalyst, conduct current, and allow gases to pass through. The properties of the substrate have an important effect on the performance of the fuel cells [65, 66].

CFP, made of carbon fiber through wet paper-making method [67], has better processing properties than that made of pulp and carbon black or graphite powder. However, it is difficult to disperse the carbon fiber evenly in the pulp, and the resistance of the resultant CFP is high. To reduce the resistance, the paper needs to be treated at high temperatures. Zhang and Shen [68] studied the effect of heat treatment temperature on the mechanical and electrical properties of CFP. CFP was prepared using pitch-based carbon felt through impregnation with resin, molding, and heat treatment at different temperatures.

Table 7.4 showed that both specific resistance and tensile strength of CFP greatly decreased with the increasing heat treatment temperatures. However, at temperatures higher than 2273 K, the change decreased. The change of resistance indicated that the removal of noncarbon atoms and rearrangement of carbon network was almost complete up to 2273 K. The smaller the specific resistance, the better the electrical conductivity; this meant that electrical conductivity increased with increasing temperature. However, the amount of resin carbon from resin decomposition at higher temperature decreased; therefore, adhesion between the CFP became weak, leading to lower tensile strength. The SEM images indicated that the carbon fibers were arranged randomly in the paper, which contributed similar conductivity in every direction to the paper [68]. There were many pores between the carbon fibers, which made CFP porous. The sections between the carbon fibers were weaker than the carbon fibers, so they were the key drawbacks of CFP. By keeping the resistance of CFP acceptable, forming adequate resin carbon in between the carbon fibers could improve the tensile properties of CFP. Molding is the key step to determine the thickness of the CFP. The density, specific resistance, and tensile strength of CFP decreased with increasing heat treatment temperature while the electrical conductivity increased. The SEM images indicated that the carbon fibers coated with carbon films were arranged randomly in the paper. CFP with thickness of 0.3 mm, bulk density of 0.4–0.5 g/cm^3, and specific resistance of 200–300 mV m was produced after heat treatment above 2273 K.

Proton electrolyte membrane fuel cells (PEMFCs) are able to efficiently generate high power densities, making the technology potentially attractive for certain mobile and portable applications. Because the bipolar plate is a major part of the PEMFC stack both in weight and volume, the bipolar plate should be developed with its weight and thickness in mind.

Hwang et al. [69] reported the carbon fiber-reinforced epoxy composite-based bipolar plate for PEMFCs. In this study, a bipolar plate for automotive fuel cells was developed using carbon fiber composites by compression molding owing to the fact that carbon/epoxy composites showed not only high electrical and thermal conductivities, but also high specific stiffness and strength. The mechanical and

Table 7.4 Mechanical and electrical properties of CFP [68]

Temperature (K)	Tensile strength (GPa)	Specific resistance (μΩ m)
Green paper	1.53	4700
1273	0.78	1600
2073	0.33	409
2273	0.31	230
2773	0.22	200

thermoelectrical properties of the developed composite bipolar plate were measured to investigate its suitability for the automotive fuel cells. The flexural strengths of the previous composite bipolar plates were primarily significantly lower than the target value of approximately 59 MPa. However, the composite bipolar plate without the corrugated flow channels had a flexural strength of 316 MPa at both room temperature and operating temperature, and the composite bipolar plate with corrugated flow channels had a flexural strength of 195 MPa at room temperature and 180 MPa at operating temperature. In addition, the in-plane electrical conductivity of the cross-ply composite used for this work was 300 S/cm, which was much higher than the DOE target of 100 S cm^{-1}. This was because the pitch based continuous carbon fibers showed high electrical conductivity.

The area-specific resistance is an important property of the fuel cell stack because the energy loss in the fuel cell is proportional to the value of ASR. The values of ASR for the composite bipolar plate were 42 and 58 mΩ cm^2 when the compaction pressures were 1.0 and 0.5 MPa, respectively. Both the high carbon fiber volume fraction and strong bond between the carbon fiber and epoxy resin induced low gas permeability in the composite. The measured permeability was approximately 1.4×10^{-8} m^3 m^{-2} s^{-1}, which was smaller than 2×10^{-8} m^3 m^{-2} s^{-1} for the bipolar plate application.

Kim et al. [70] have studied the effect of nickel coating on improving the electrical conductivity of carbon fibers for applying such a coating on the electrode or bipolar plate of the fuel cell. In this work, nickel/carbon hybrid fibers were prepared by the electrolytic plating on carbon fibers in order to improve the electrical conductivity of the carbon fibers; the effects of nickel content and coating thickness on the electrical conductivity of the fibers were studied. As for the experimental results, it was observed that the electrical conductivity of the nickel/carbon hybrid fibers was dramatically increased in the presence of metallic nickel particles, with the best result observed with a thickness of over 0.75 μm owing to the minimization of the inner pores, as shown in Table 7.5.

Table 7.5 lists the average coating thickness of nickel/carbon hybrid fibers as a function of plating time from 5 to 60 s. The measurement of layer thickness was possible after 10 s of plating time. The thickness increased linearly up to 30 s of plating time and then increased rapidly. In the interval of 10–30 s, the average plating speed was 0.025 μm/s, and then increased to 0.03 μm/s after 60 s of plating time.

To confirm the effects of nickel layer thickness on the electrical resistance of the nickel/carbon hybrid fibers, the four-point probe method was employed; the results

Table 7.5 Average coating thickness of carbon fibers as a function of plating time [70]

	Average coating thickness (μm)
Ni-5	–
Ni-10	0.25
Ni-30	0.75
Ni-60	1.75

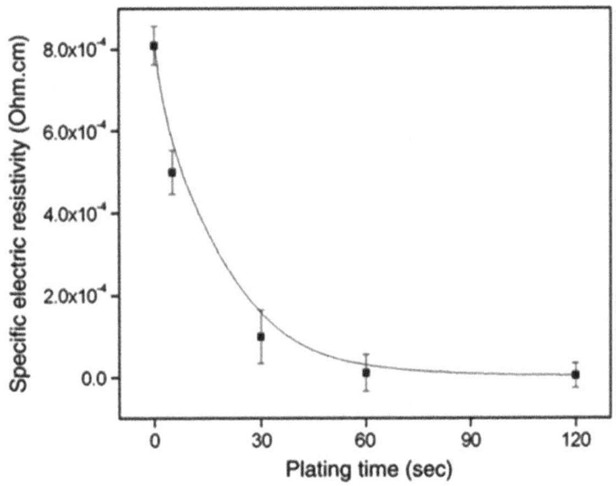

Fig. 7.9 Specific electric resistivity of Ni-plated carbon fibers as a function of plating time [70]

are shown in Fig. 7.9. It was found that the specific resistance decreased with increasing plating time. Also, it was observed that the specific resistivity of the as received sample was over 0.80×10^{-3} Ω cm but that this value decreased dramatically with increasing plating time up to 60 s. From the good linearity of results over the range 5–30 s, it was found that there was a strong correlation between the electrical conductivity and plating time. In the case of electric conductivity after 60 s of plating time, the conductivity was not significantly enhanced. This meant that a perfect Ni layer could be formed over the range 30–60 s of plating time, resulting in a saturation state of electrical conductivity owing to the minimization of inner pores (or contact loss between nickel particles).

2. Supercapacitor

Recently, EDLCs have shown new and promising applications in the capacitive deionization of water and as pulse power sources for digital communication devices and electric vehicles. The popularity of these devices arises from their higher energy density relative to the conventional capacitors as well as their longer cycle life and higher power density relative to batteries [71–75].

In recent years, activated carbon materials, including activated carbons, glassy carbons, carbon nanotubes, graphene, nanodiamond, carbon aerogels, and carbon fibers, have attracted considerable attention as electrode materials for supercapacitors because of their high specific surface areas of up to 500–3000 m^2/g and excellent microstructures and electrochemical properties with low costs. In particular, using fibrous materials yields an additional profit from the construction point of view and thus offers a number of advantages over conventional powdered activated carbons [76–79].

Babel and Jurewicz [80] reported the electrochemical performance of the ACF-based electrode for supercapacitors. Carbon fabrics from viscose fibers activated with KOH have been investigated as possible electrode materials for electrochemical capacitors. The fibers were first pyrolyzed at 400 or 600 °C, then saturated with KOH at C/KOH ratios from 1:3.5 to 1:5 and treated over the temperature range 700–800 °C. The carbon fibers saturated with KOH were analyzed using thermogravimetric and differential thermal analysis in order to obtain information on the temperature dependence of the KOH reaction with carbon. The textural properties of each sample are shown in Table 7.6. The porous features of ACFs were dependent on the KOH activation conditions. A distinct hysteresis of the curves confirmed that the fibers contained considerable volumes of mesopores. The micropores/mesopores ratio slightly exceeded 1, with a high micropore volume of above 0.85 cm^3, which revealed deeply activated carbons. A specific surface area close to 2000 m^2/g was mainly the result of microporosity.

The electrochemical properties of the activated carbons were determined using three-electrode Swagelokw-type capacitors both in 4-M H_2SO_4 and 7-M KOH aqueous electrolytes. Specific capacities of 340 and 270 F/g were determined in acidic and alkaline media, respectively. The electrical capacity for both negative and positive electrodes depended on the treatment method. The capacitance values were discussed taking into account the porous texture, elemental composition, and surface functionality of the ACFs.

Leitner et al. [81] have studied the Nomex-derived ACFs as electrode materials in carbon-based supercapacitors. The SEM micrograph of sample NP1-90 showed that the diameter of the fibers was in the order of 10 μm. The fibers showed a bilobular section, which was more easily appreciated over the high-resolution image of a single fiber [81]. The bilobular morphology was already present in the parent Nomex polymer and maintained along with pyrolysis and activation treatments. Depending on the burn-off owing to the activation, the BET surface area of the carbons was in the order of 1300–2800 m^2/g, thereby providing an extensive network of micropores. The electrochemical characterization was carried out for the electrodes prepared using several ACF samples in an aqueous solution. Their capability as active materials for supercapacitors was evaluated using cyclic voltammetry and impedance spectroscopy. NP1-90 showed the highest capacitance value (175 F/g) in sulfuric acid, indicating that the NP1-90 was the most suitable sample for application in supercapacitors and exhibited excellent capacitive behavior. Furthermore, it was observed that the specific capacitance and performance of the electrode increased significantly with the increasing burn-off degree.

Table 7.6 Textural parameters of activated carbon fibers [80]

Sample	KOH/fiber ratio	Pyrolysis (°C)	Activation (min)	Surface area (m^2/g)	V_{mi}	V_{me}	V_{mi}/V_{me}
AK1	3.5:1	400	45	1978	0.93	0.92	1.01
AK2	4.0:1	400	45	1304	0.50	0.07	7.08
AK3	4.5:1	400	45	1457	0.67	0.43	1.56
AK4	5.0:1	400	45	1977	0.85	0.71	1.20
AK5	4.0:1	400	30	1374	0.57	0.09	6.49
AK8	4.5:1	600	30	1628	0.73	0.21	3.52

This fact could be attributed to an increase in the surface area and porosity with increasing burn-off. Contrary to all the other samples, N-42 fibers showed a lesser increase in capacitance at lower frequencies compared to those of the other samples indicating again that the accessibility of the electrolyte to this carbon was minor. As the burn-off increased, an increase in the capacitance was detectable. Also, an increase in the capacitance shifted to the higher frequency domains, which confirmed that the electrolyte penetration into the pores of the electrode improved at higher burn-off degrees.

3. Lithium battery

Carbonaceous materials have been adopted as the prospective alternate anode for lithium metal in lithium rechargeable batteries [82–86]. Several types of carbon materials show better electrode performance compared to the other intercalation compounds such as transition metal oxides, lithium alloys, and lithium metal; they have high specific capacity, low electrode potential, high coulombic efficiency, long cycle life, and high level of safety because of lack of dendritic lithium metal during charge–discharge cycles [87].

Industrialized mesophase pitch-based carbon fibers have distinctive morphology and several macroscopic transverse textures. Consequently, these materials have attracted much interest and many researches have investigated their electrode characteristics. The pitch-based carbon fibers could also have unique "skin-core" structures consisting of two different crystal structure regions in addition to the morphology and transverse textures [88].

Suzuki et al. [89] used the pitch-based carbon fiber as an anode in lithium rechargeable batteries. Two types of mesophase pitches, A and B, were used for spinning. The physical properties and elemental analysis data of the mesophase pitches used are listed in Table 7.7. The relation between the electrode characteristics and structures or textures of pitch-based carbon fibers controlled by spinning or stabilization conditions was investigated. The graphitized fiber spun at a lower viscosity had a higher degree of graphitization and higher discharge capacity. The graphitized fiber spun at a higher viscosity exhibited a higher rate capability, thereby suggesting that the lithium ions could diffuse more easily in a fiber composed of smaller crystallites compared to the larger ones. The graphitized fiber stabilized at a higher temperature showed a lower irreversible capacity, which might be owing to the relatively less developed graphite structure in the surface region. The graphitized pitch-based carbon fiber prepared with optimum conditions

Table 7.7 Physical properties and elemental analysis data of mesophase pitches [89]

Pitch	SP (°C)	TI (%)	QI (%)	MC (%)	C (wt %)	H (wt %)	N (wt %)	O (wt %)	S (wt %)	H/C
A	297	72	45	100	95	4.6	<0.1	0.2	0.2	0.58
B	293	77	52	100	95	4.8	<0.1	0.2	0.2	0.61

achieved a maximum discharge capacity of 315 mAh/g, an initial irreversible capacity of 10 mAh/g and an initial coulombic efficiency of approximately 97%.

Previously [90], ACF-containing Sn nanoparticles were prepared using impregnation and investigated as a negative electrode material in lithium batteries. The particle size of metallic tin in this composite was 20–40 nm and controlled by choosing porous carbon with an appropriate surface structure. This Sn/ACF composite cycled versus Li metal showed a first discharge capacity as high as 200 mAh/g compared to that of the pristine ACF, which showed only 87 mAh/g. Excellent cyclability with these composites was obtained with ACF BET SSA as large as 2000 m^2/g and 30 wt% Sn. After 20 cycles, the Sn-impregnated OG-7A prepared via aqueous/nonaqueous route showed higher discharge capacities compared to those showed by pristine OG-7A, despite the gradually decreasing capacity owing to the degradation of tin particles especially prepared in a nonaqueous solution. The nonaqueous preparation might provide a larger amount of tin particles, which showed considerable capacity in the first cycle and degraded its reversible capacity during subsequent cycles, on the outer surface of ACF. Tin-impregnated ACF negative electrode was expected to have good cycle capability when many tin nanoparticles were trapped inside the ACF surface structure. Therefore, an ACF with a larger surface area compared to OG-7A would be expected to have a larger pore volume and bring a greater number of tin nanoparticles [90].

7.2.3 Molecular Sieves

Carbon molecular sieves (CMS) are widely used in gas separation processes [91–96]. The unique property of CMS to separate gases based on the different sizes and shapes of molecules has been exploited in commercial applications. The most important feature of CMS is that they have narrow pore size distribution accomplished either by controlled activation or employment of pore narrowing techniques in an inherent pore structure [91]. Freitas and Figueiredo [97] reported that the modification of two activated carbons of different textures by the pyrolysis of benzene could yield the CMS for O_2/N_2 separation. The results showed that this objective could be attained when the carbon precursor had been activated only to a limited extent, and when the carbon deposition was carried out in the proper kinetic regime. Vyas et al. [98] obtained the CMS by methane cracking on bituminous coal and coconut shells, and subsequently, the O_2/N_2 uptake ratio of the best CMS produced was 2.667. Cabrera ct al. [99] described the preparation of CMS for air separation using a two-step hydrocarbon deposition with a single hydrocarbon. They found that the concentration of the carbon-containing compound used in the first step should be larger than that in the second step, so that the pore openings of the micropores of the support narrowed gradually, avoiding pore plugging.

7.2.4 Catalysts

A porous carbon fiber also exhibits catalytic activities [100, 101]. For example, the porous carbon fibers might be used as a catalyst in the oxidation of SO_2 and NO. The catalytic activities are attributed to the surface functional groups such as hydroxyl, carboxylic, and quinone, which are incorporated during the activation stage. Depending upon the type of adsorbate (basic or acidic, anionic or cationic), the surfaces of the porous carbon fibers might also be accordingly functionalized. The surface groups of porous carbon fibers containing electronegative oxygen atoms adversely affected the adsorption of VOCs. On the other hand, the adsorption of SO_2 and NO was found to be primarily controlled by the extent of the surface oxygen functional groups. The results of the surface characterization analysis in relation to those of breakthrough analysis showed that the extent of oxygen functional groups on the surface of porous carbon fibers was less, while the extent of adsorption or oxidation of SO_2 and NO was high [100].

7.3 Applications of Carbon Fiber-Reinforced Composites

The use of fiber-reinforced composites has become an increasingly attractive alternative to the conventional metals for many aircraft components mainly owing to their increased strength, durability, corrosion resistance, resistance to fatigue, and damage tolerance characteristics. Fiber-reinforced composites also provide greater flexibility because the material can be tailored to meet the design requirements and they also offer significant weight advantages. At present, carefully designed individual composite parts are approximately 20–30% lighter than their conventional metal counterparts. Although all-composite airplanes are now available in the world market, advances in the practical use of composite materials should enable further reduction in the structural weight of airplanes. The fiber-reinforced composites used in the aircraft industry are generally either reinforced fibers or filaments embedded in a resin matrix. The most common fibers are carbon and their hybrids. The resin matrix is generally an epoxy-based system requiring curing temperatures between 250 and 350 °F [102–104].

Fiber-reinforced composites used for aircraft structural applications are generally fabricated using sandwich construction having face sheets of either carbon fibers or carbon fibers combined with aramid or glass fibers with a honeycomb core [105]. For interior aircraft applications, composite parts are required to meet the requisite mechanical properties and process abilities. In addition, materials used within the pressurized portion of aircraft must have flammability resistance. Interior parts such as overhead bins, sidewall panels, ceilings, floor boards, galleys, partitions, and cargo floor board liners are made of composite materials, which are generally fiber-reinforced epoxy or phenolic resin. For interior applications, phenolic resin systems are mainly used because of their excellent fire-resistant properties, including low flammability and low smoke and low toxic gas emission [106].

7.3.1 *Aircraft*

The application of fiber-reinforced composites to aircraft can be traced back almost three decades to the F-14 (US Navy) and F-15 (US Air Force) fighters, which used boron/epoxy skins in their empennages [107–110]. Later, the use of carbon fiber-reinforced composites in military and transport aircraft increased resulting from their high performance. The initial applications of carbon fiber-reinforced composites to aircraft structures included secondary structures such as fairings, small doors, and control surfaces. As the technology advanced, the use of carbon fiber-reinforced composites for primary structures such as wings and fuselages increased [107]. The applications of the current aircraft along with the general uses of composite materials in each aircraft are shown in Table 7.8.

The aircraft industry chooses to use carbon fiber-reinforced composites not only to reduce weight, but also because these materials are corrosion and fatigue resistant. The limiting factor in the widespread applications of these materials has been their high cost compared to that of the conventional metals.

1. Fighter aircraft

The trends in the use of fiber-reinforced composites for fighter aircraft are shown in Fig. 7.10. The percentage by weight of composite materials used initially (e.g., F-15E) was small at 2%, but this percentage has since grown to more than 25% for F-22, which is the designated replacement for the F-15E. The F-22 has demonstrated the feasibility and benefits of introducing processes such as resin transfer molding (RTM) to improve the affordability of composite materials in fighter aircraft applications. The use of carbon fiber-reinforced composites in the US Navy's F/A-18E/F equals nearly 20% of its structural weight in flight critical parts, as

Table 7.8 List of aircraft and corresponding composite materials

Fighter aircraft	(US) F-16, F-14, F-18, YF-23, F-22, JSF, UCAV (Europe) Gripen JAS-39, Mirage 2000, Rafael, Eurofighter Typhoon, Lavi, DASA Makoc (Russia) MiG-29, Su series
Transport	(US) KC-135, C-17 (US-Commercial) B-777, B-767, MD-11 (Airbus, European) A-320, A-340, A380, Tu-204, ATR42, Falcon 900, A300-600 ST
General aviation	Piaggio, Starship, Premier 1
Rotary aircraft	V-22, Eurocopter Tiger, Comanche RAH-66, Bell/Agusta BA-609, EH101, Super Lynx 300, S-92

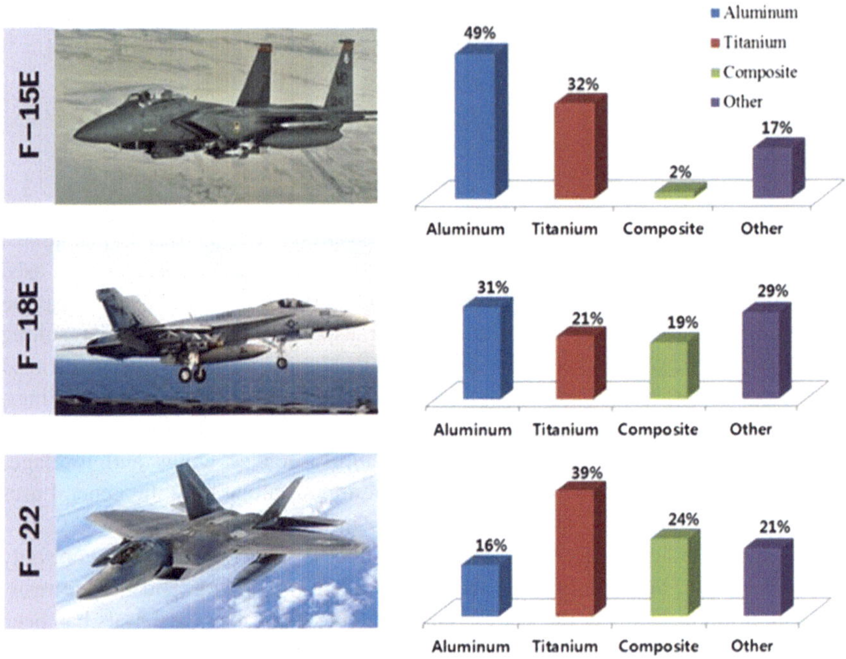

Fig. 7.10 Trends in use of fiber-reinforced composites for fighter aircraft

shown in Fig. 7.11. The choice of composite materials in the F/A-18E/F was dictated by the need to reduce weight and improve strength, reliability, and maintenance in an aircraft carrier environment. The center and aft fuselage skins and other ancillary structures, such as speed brakes and dorsal covers, are all carbon/toughened epoxy construction in the F/A-18E/F. Carbon fibers such as Hexcel's IM7 with improved strength and stiffness properties are used in the wing and tail skins. Although composite materials, in general, are sensitive to impact damage, toughened materials such as Fiberite's 977-3 toughened epoxy system used in the F/A-18E/F have successfully addressed this threat in operations. The AV-8B uses nearly 25% by weight of composite materials in its airframe [111] (Fig. 7.12).

2. Transport aircraft

Airbus industry used advanced composites in the Airbus A300 aircraft, which first flew in 1972. The composite material was used in fin leading edge and other glass fiber fairing panels. In 1979, the in-service evaluation of the Airbus A300 aircraft led to the further use of composite components in the Airbus A300, namely CFRP spoilers and rudders, air brakes, and CFRP landing-gear doors. The use of composites was extended to the Airbus A310 aircraft during 1980–1985, and thereafter, to the Airbus A320 aircraft in 1987. In the Airbus A300B2/B4 aircraft, only glass fiber structures have been used. AFRP and CFRP structures have been

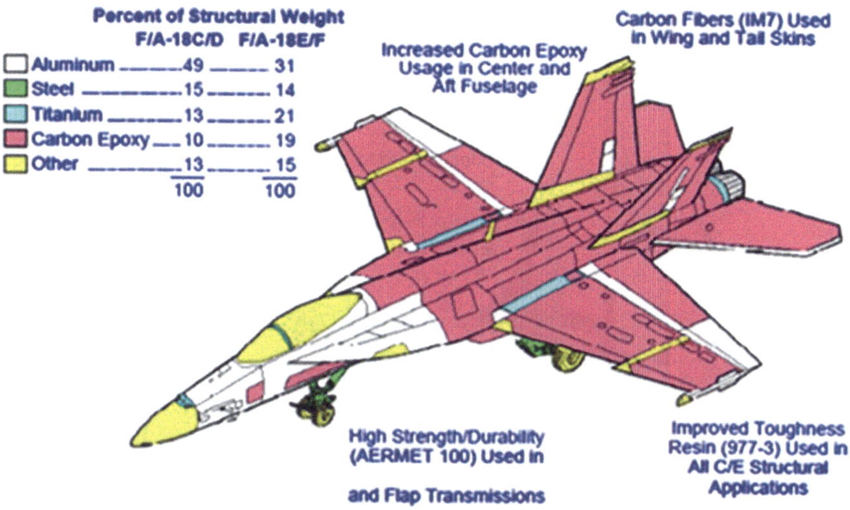

Fig. 7.11 Use of carbon fiber-reinforced composites in the US Navy's F/A-18E/F

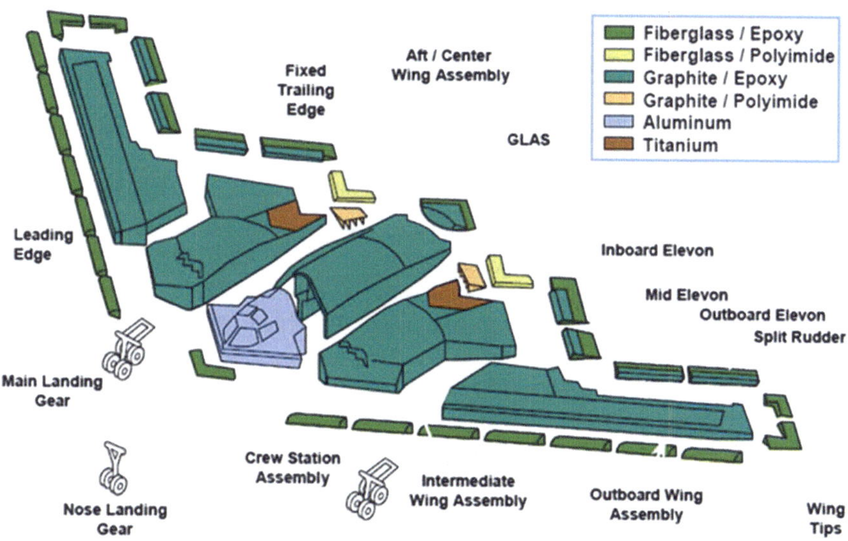

Fig. 7.12 Use of carbon fiber-reinforced composites in B-2 Bomber

used in the Airbus A310-200 aircraft. For the Airbus A320 and Airbus A330/340 aircraft, GFRP, CFRP, and AFRP have been used for the composite structures. The use of composites in the Airbus family of aircraft has consistently increased since the making of the Airbus A300 aircraft. The composites account for approximately 15% of the structure of the Airbus A320 aircraft. Although the total weight of the

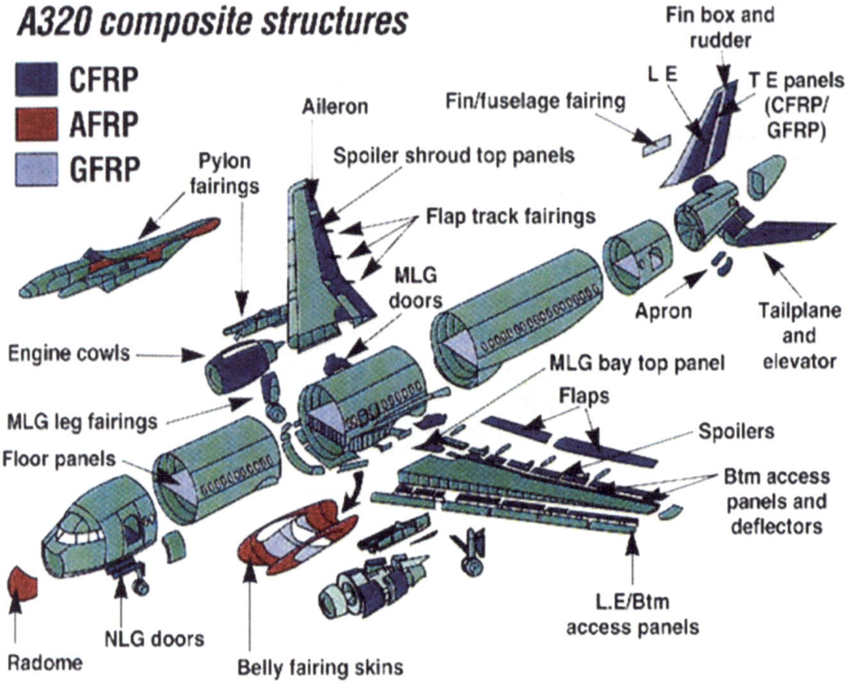

Fig. 7.13 Use of carbon fiber-reinforced composites in the Airbus A320 [111]

composite structures is relatively much higher in the Airbus A330/A340 aircraft, the percentage of weight in relation to the total weight of aircraft is nearly 12%. The use of composites on the A320 aircraft is shown in Fig. 7.13 [111].

The Boeing 777, whose maiden flight was 10 years ago, is approximately 20% composites by weight with the composite materials being used for the wing's fixed leading edge, trailing-edge panels, flaps and flaperons, spoilers, and outboard ailerons. They have also been used for the floor beams, wing-to-body fairing, and landing-gear doors. Using composite materials for the empennage saves approximately 1500 lb in weight. The Boeing 787 will average the extensive use of composite materials (estimates are as high as 50%) in the quest for high efficiency and performance with reduced weight. The composites are also used in major structural elements of modern helicopters, which are approximately 50% composites by weight (Fig. 7.14). The formability of composites has been used to particular advantage in helicopter manufacturing to reduce the number of component parts and therefore, costs [112].

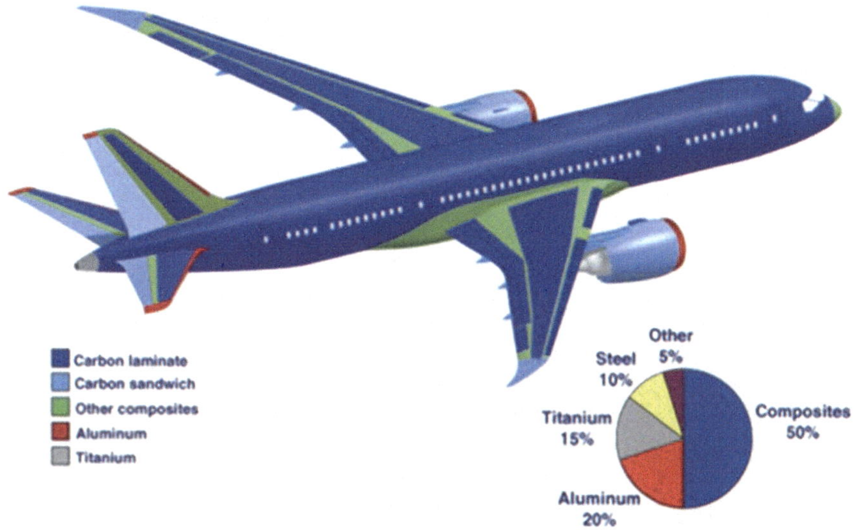

Fig. 7.14 Use of carbon fiber-reinforced composites in the Boeing 787

7.3.2 Automobiles

The automotive companies in the today's world are forced to look for new ways and innovations in manufacturing cars/trucks owing to fierce competition. The cars today should have all the features demanded by the customers at low cost. This has led to the use of composite materials in the construction of parts such as the body, interiors, chassis, hoods, and electrical components. The composite materials have the desired properties to suit the requirements of the car manufacturers. Hence, the composites are more significant today and will continue to be in future in the automobile industry. The recent innovations in the sphere of advanced composites in the automotive manufacturing industry have led to the use of carbon fiber-reinforced plastic (CFRP) with the various thermosets of thermoplastic composites [113] (Figs. 7.15, 7.16, 7.17, 7.18, and 7.19).

In recent years, compressed natural gas (CNG) has become an attractive alternative as a fuel for automobiles such as public buses. Compared to oil-based fuels, the use of CNG reduces the emission of carbon dioxide and nitrogen oxides. Furthermore, the use of carbon fiber-reinforced composites in the design of CNG cylinders has led to attractive lightweight solutions. For the safety of the users, the structural integrity of such composite cylinders must be accurately checked.

Nevertheless, the actual inspection techniques, derived from the techniques used for steel cylinders, might not be well adapted for the composite materials [114].

CNG is an alternative fuel, which is stored under high pressure (3600 psi) in the vehicle [115]. The use of natural gas as an alternative fuel in automotive applications is not widespread primarily because of the high cost and durability of the

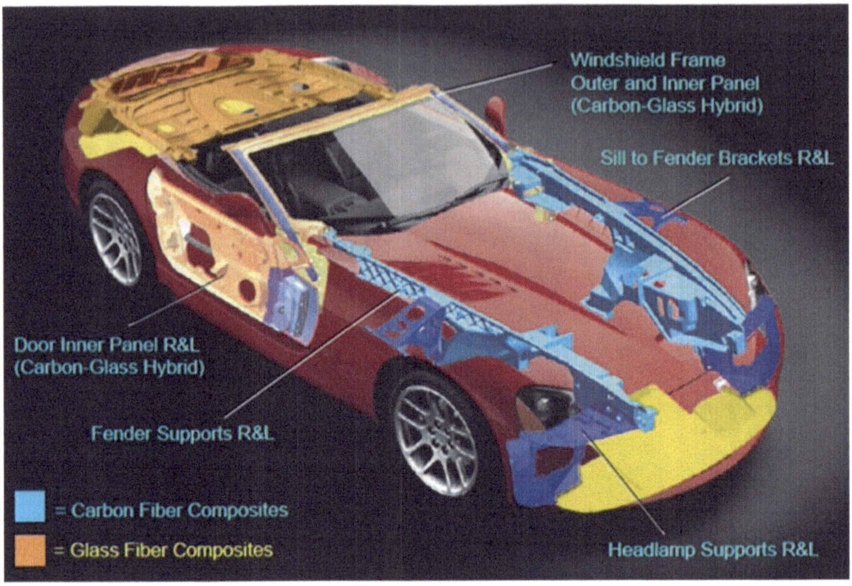

Fig. 7.15 Use of carbon fiber-reinforced composites in automobiles

carbon fiber composite storage tanks. Carbon fiber composites are relatively expensive because of the raw material cost of the carbon fiber, which accounts for approximately 40% of the total tank cost. By introducing large tow-size carbon fibers in the tank design, there exists potential for tremendous cost savings. The cost of large tow-size carbon fiber is approximately one-half the cost of conventional tow-size carbon fiber. However, not all of these savings re realized in the final overall tank cost because of the low fiber strength and low strength translation, which has been demonstrated in large tow-size carbon fiber composite structures [116, 117].

Composite CNG tanks are typically fabricated using the wet-filament winding process. This is a process wherein the fiber tow is passed through a resin bath to impregnate the tow and then wrapped around a mandrel prior to curing in an oven at elevated temperatures [118].

7.3.3 Marine Applications

Recently, the carbon fiber-reinforced composites have been used widely in the marine sector because of their excellent resistance to the marine environment [119]. Complex configurations combined with the advantages of seamless hulls drove the development of CFRP boats. The racing power-boats employed advanced and hybrid composites for high-performance craft and driver safety. Major structural

Fig. 7.16 CNG cylinders manufactured using fiber-reinforced composites

elements, viz. deckhouses, hatch covers, king posts, and bow modules, appeared to be well suited for CFRP construction. Composite applications in the marine segment have been employed in high-speed boats, naval vessels, sailboats, fishing boats, high-capacity trawlers, barges, and other ship components since the beginning of the last decade. The marine industry mainly consumed the carbon fiber-reinforced epoxy as the composite material. The advanced composite

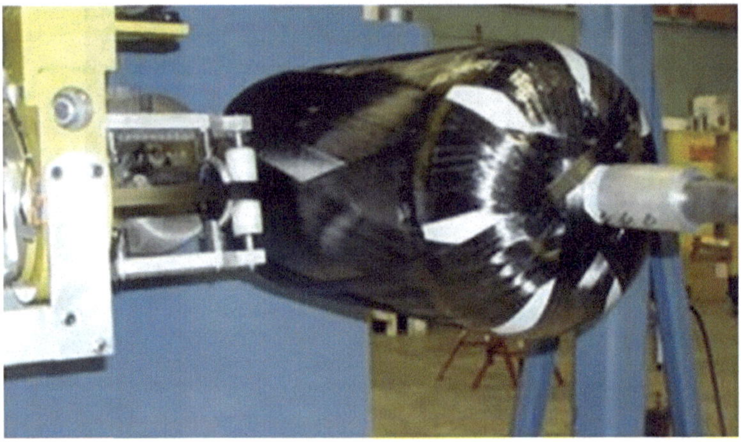

Fig. 7.17 Wet-filament winding process for manufacturing of CNG cylinders

Fig. 7.18 Use of carbon fiber-reinforced composites in racing boats

materials in vessels have the potential to reduce fabrication and maintenance costs, enhance styling, reduce outfit weight, and increase reliability [120, 121].

7.3.4 Wind Turbine Blades

Wind turbine blades convert mechanical energy provided by wind into electrical energy. They are used in machines designed for directing and concentrating energy from the air into a wire that connects the turbine generator to an electrical grid [122]. The cost of energy is dependent on advanced and cost-effective solutions for

Fig. 7.19 Carbon fiber/PEEK/ceramic composite for turbine blades

manufacturing enormous composite structures such as wind blades. The composites are cost-effective materials for large diameter blades and provide the requisite strength and stiffness in structures [123].

Wind turbine blades must be lightweight, rigid, and show fatigue resistance throughout a typical 10-year service life. Such requirements favor advanced composites as the material of choice. A typical wind turbine blade consists of an outer shell supported by a main spar. Traditional materials of the spar used E-glass reinforcing fibers bound in an unsaturated polymer resin matrix. Epoxy as a matrix resin offers better tensile and flexural strength, while the carbon fibers have an extremely high stiffness to density ratio, allowing for longer, lighter turbine blades. Therefore, newer wind turbine blades have been manufactured using carbon fiber-reinforced composites [124, 125].

7.3.5 *Sports Applications*

Sports serve as the test platform for carbon fibers. Carbon fibers are relevant for the sporting goods market as well as commercial applications. With the advanced in the production and applications of technologies, carbon fibers frequently appear in the sporting goods. Over the years, carbon fibers have been used to make golf clubs, hockey sticks, and bicycle stands. Carbon fibers are also found in rackets, fishing rods, snowboards, windsurfing masts, marine hulls, backpacks, tent poles, softballs, and baseball bats [126, 127].

1. Rackets in CFRP

The relative importance of the basic elements of racket constructions, shaft and head-string system, varies with the sport—the head-string system being more dominant wherein the ball momentum is high (tennis) and shaft action being more important, for example, in badminton. High-speed photographic analysis of a ball hitting a squash racket, with a synchronized strain recording on the shaft, has shown that at impact (zero ball velocity), the kinetic energy of the ball was absorbed in the racket and distributed between shaft strain energy and head-string energy in the ratio 1/10. This ratio varied with different shafts, decreasing for more flexible shafts, but it indicated the smaller contribution of the shaft in the energy exchange. In the various racket types, head and shaft need to be appropriately matched. The principal function of the head frame is to provide a stable structural form that will counterbalance the string tension without distortion. Because the stressing will be primarily compressive, high stiffness as well as strength will be important, and hence, CFRP is likely to be advantageous in this region, especially in badminton, wherein a minimum weight is required. The principal problem in fiber design is the holding of the strings without introducing drilled holes into CFRP and this can be achieved in various ways, including the use of material sandwich constructions. Attention must also be paid to the torsion requirements at the shaft-head transition [128].

Because steel-like mechanical properties can be attained in CFRP, it is a tempting starting point simply to replace a tapered high-strength steel tube with an equivalent CFRP element. This approach was used in the racket shown in Fig. 7.20 with shafts formed in unidirectional high-modulus carbon fiber and epoxy resin. These lightweight rackets have satisfactory playing characteristics and endured an 18-month assessment period of severe and varied play. Despite their relatively low torsional rigidity, surprisingly little effect was observed during play [129].

Other types of rackets, particularly badminton and tennis, will undoubtedly benefit from the use of carbon fibers. In both these cases, the latest design developments have involved all-metal tubular construction in an attempt to establish a technically efficient design. In the tennis racket, the modification of wood constructions with CFRP will strengthen and produce lighter designs. The tennis racket manufactured from the carbon fiber-reinforced plastic is shown in Fig. 7.21.

2. Golf shaft

It is generally agreed that it is desirable to aim for minimum-weight shafts that have sufficient stiffness, strength, and impact resistance to fulfill the basic shaft function of accelerating and positioning the club-head; a typical mean value of a medium stiffness shaft is 50 Nm^2. Large weight savings (50% is easily attainable) are possible in CFRP shafts, and this results in a calculable hitting advantage, but it is more likely that the appeal of the CFRP as a shaft material will be in its unique design possibilities with a more modest weight reduction.

These basic requirements have been adequately met by CFRP shafts of various constructions, using both low-modulus and high-modulus carbon fibers, and worthwhile weight reductions gained. The CFRP shafts (Fig. 7.22) have satisfied basic playing needs and indicated that the use of CFRP is feasible and likely to be advantageous. Refinement of design is possible, including accurate matching of head to shaft and controlled variation within sets, because precise material formulations can alter the static and vibrational characteristics of shafts. The balance between the torsional and flexural stiffness, for example, is a design parameter [130, 131].

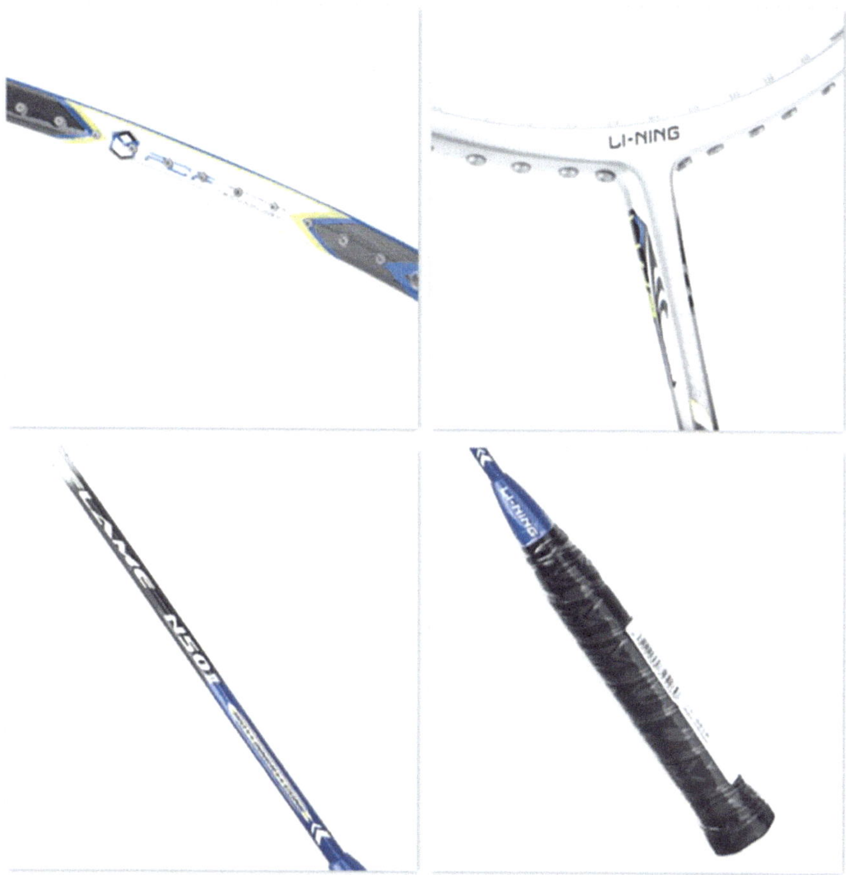

Fig. 7.20 Carbon fiber-reinforced plastic for rackets

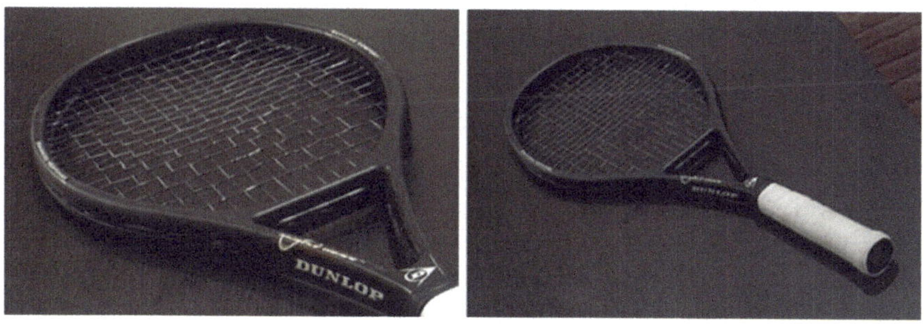

Fig. 7.21 Tennis racket manufactured from carbon fiber-reinforced plastic

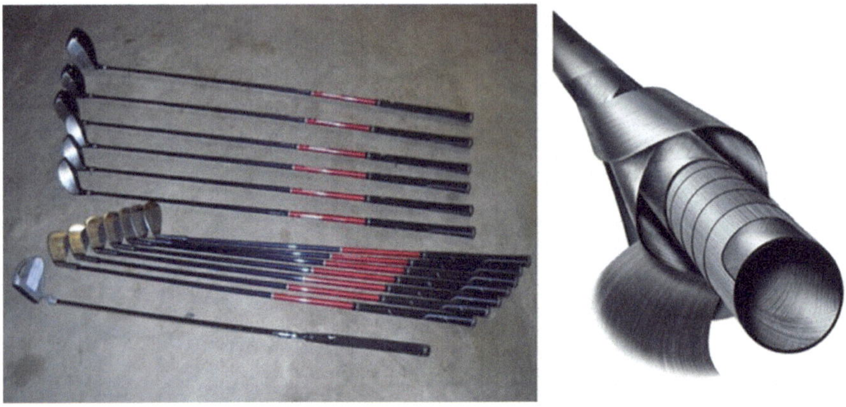

Fig. 7.22 Golf shaft manufactured from carbon fiber-reinforced plastic

7.3.6 Construction

In recent decades, material development in response to a call for more durable infrastructures has led to many exciting advancements. Fiber-reinforced composite designs with unique properties are now being explored in many infrastructural applications. Even concrete and steel are being steadily improved to have better properties and durability.

Fiber-reinforced cement–matrix composites are structural materials, which are gaining importance rapidly owing to the increasing demand of superior structural and functional properties. In particular, the carbon fibers are one of the most common components for reinforcing cementitious materials because of their excellent mechanical properties [132, 133]. When carbon fibers are incorporated into the cementitious matrix, some properties of the carbon fiber-reinforced cement–matrix composites (CFRCCs) are improved, including the tensile strength, flexural strength, compressive strength, toughness, and drying shrinkage [134, 135]. Therefore, the carbon fibers are generally used as a reinforcement material to enhance the mechanical properties of a brittle matrix such as cement paste and especially nonround carbon fibers are more effective for strengthening compared to the round ones [136, 137].

Park et al. [138] used carbon fibers for the reinforcement of cement. In this work, the effect of different cross section types on the mechanical properties of carbon fiber-reinforced cement composites was investigated. To study the effect of shape factor, the mechanical properties of three types of carbon fibers, round, H-shaped, and C-shaped randomly oriented CFRCCs, were investigated. As a result, C-shaped CFRCC showed higher tensile and flexural strength compared to any other shape of carbon fiber-reinforced composites. It was observed that the C-shaped CFRCC presented stronger fiber-matrix interfacial adhesive forces owing to the mechanical anchorage into the matrix. However, the compressive strength of the CFRCC decreased with the increasing aspect ratio and fiber volume fraction. This was probably because of the fact that the amounts of entrained air contents increased during the mixing of each fiber.

Also, both the tensile and flexural strengths of the CFRCC were significantly increased by the additions of fumed silica to the composites. Figure 7.23 shows the flexural strength of CFRCC as a function of the fiber volume fraction. The flexural strength of CFRCC increased with the increasing fiber volume fraction and the CFRCC containing fumed silica showed higher flexural strength compared to that of the CFRCC without fumed silica because of the increasing interfacial areas between the fibers and matrix, resulting from the densification of the microstructure [139, 140]. Also, the C-CFRCC showed higher flexural strengths compared to those of R- or H-CFRCC, regardless of the presence of fumed silica, as shown in Fig. 7.23. In particular, the C-CFRCC containing fumed silica showed the highest flexural strength, approximately 40% higher compared to those of R- and H-CFRCC in a fiber volume fraction of 3%. This was probably because of the fact that the inside space of the C-shaped carbon fibers was compacted by the matrix and the fibers, thus, acted as mechanical anchors, resulting from the C-shaped carbon fibers having a large interface between the fibers and matrix.

The SEM micrographs for C-shaped carbon fibers distributed in the cement matrix are shown in Fig. 7.24. The fibers were distributed homogeneously in the cement matrix and in good contact with each other within an individual cluster, although the clusters were separate.

Cement reinforced with short carbon fibers is attractive owing to its high flexural strength and toughness and low drying shrinkage, in addition to its strain-sensing ability [141–144]. Xu and Chung [144] reported the silane-treated carbon fiber-reinforced cement, which showed improved properties. In this work, the isotropic pitch-based and unsized carbon fibers (length: *5 mm) were used and their properties are listed in Table 7.9.

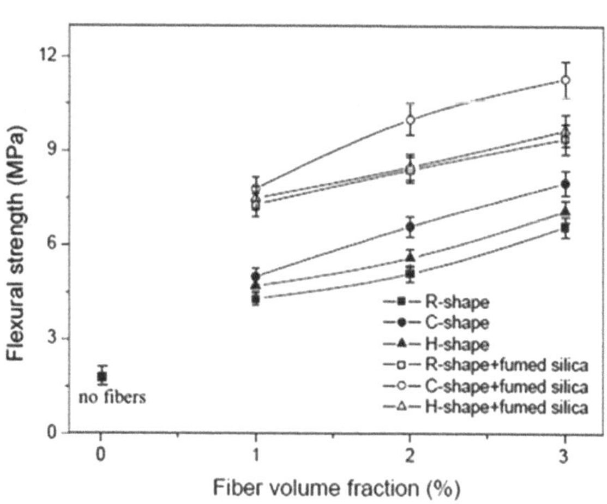

Fig. 7.23 Flexural strength of CFRCC as a function of fiber volume fraction [138]

Fig. 7.24 SEM micrographs of carbon fibers distributed in cement matrix [138]

Table 7.9 Properties of carbon fibers [144]

Filament diameter	15 ± 3 μm
Tensile strength	690 MPa
Tensile modulus	48 GPa
Elongation at break	1.40%
Electrical resistivity	3.0×10^{-3} Ω cm
Specific gravity	1.6 g cm^{-3}
Carbon content	98 wt%

Tables 7.9 and 7.10 show the tensile strength and modulus, respectively, of 12 types of cement pastes. The strength slightly increased while the modulus slightly decreased owing to the addition of methylcellulose and defoamer. However, both strength and modulus increased owing to the addition of fibers. The effectiveness of the fibers in enhancing the strength and modulus increased in the following order: as-received < O_3-treated < dichromate-treated < silane-treated. This trend was indifferent to the use of either as-received or silane-treated silica fumes. For all the formulations, the silane-treated silica fumes yielded substantially higher strength and modulus compared to that yielded by the as-received silica fumes. The highest tensile strength and modulus were exhibited by the cement paste with silane-treated silica fumes and silane-treated fibers; the strength was 56% higher, while the modulus was 39% higher than those of the cement paste with as-received silica fumes and as-received fibers. The strength was 26% higher, while the modulus was 14% higher than those of the cement paste with as-received silica fumes and silane

Table 7.10 Tensile strength (MPa) of cement pastes with and without fibers [144]

Formulation	As-received silica fume	Silane-treated silica fume
A	1.53 ± 0.06	2.04 ± 0.06
A+	1.66 ± 0.07	2.25 ± 0.09
A^+F	2.00 ± 0.09	2.50 ± 0.11
A^+O	2.25 ± 0.07	2.67 ± 0.09
A^+K	2.32 ± 0.08	2.85 ± 0.11
A^+S	2.47 ± 0.11	3.12 ± 0.12

A, cement + water + water reducing agent + silica fume; A^+, A + methylcellulose + defoamer; A^+ F, A^+ +as-received fibers; A^+ O, A^+ + O_3-treated fibers; A^+ K, A^+ + dichromate-treated fibers; A^+ S, A^+ +silane-treated fibers

treated fibers. Hence, the silane treatment with both silica fumes and fibers was nearly equally valuable in enhancing the overall strength.

Conclusively, in composite systems, filler concentration, filler dispersion and the interfacial adhesion between filler and matrix are three of the most crucial factors that can directly dominate the physical and chemical performances of the composites. During the fabrication process of the high-performance materials, these factors should be well considered and analyzed.

References

1. E. Fitzer, D.D. Edie, D.J. Johnson, Carbon fibers-present state and future expectation; Pitch and mesophase fibers; Structure and properties of carbon fibers, in *Carbon Fibers Filaments and Composites*, 1st edn. (Springer, New York, 1989), pp. 3–41, 43–72, 119–146
2. J.B. Donnet, R.C. Bansal, *Carbon Fibers*, 2nd edn. (Marcel Dekker, New York, 1990), pp. 1–145
3. A. Shindo, Japan Patent Publication No. (1962)4405 (1962), Patent Application No. 195928287
4. Committee on High Performance Synthetic Fibers for Composites, Commission on Engineering and Technical Systems, National Research Council, *High-Performance Synthetic Fibers for Composites* (National Academy Press, Washington, DC, 1992) pp. 23, 56–64, 86
5. O.P. Bahl, J.B. Donnet, T.K. Wang, S. Rebouillat, J.C.M. Peng, *Carbon Fibers*, 3rd edn. (Marcel Dekker, New York, 1998), pp. 1–84 (Chap. 1)
6. T. Roberts, *The Carbon Fiber Industry*: *Global Strategic Market Evaluation 2006–2010* (Materials Technology Publications, Watford, 2006), pp. 10, 93–177, 237
7. SACMA Releases, *Carbon Fiber Industry Statistics*, Composites News, No. 1 (1998)
8. Carbon fibers Seen as Having Big Long Term Growth Infrastructure is Next Big Trend Driver, *Advanced Materials and Composites*, News, No. 3 (1999)
9. P. Morgan, *Carbon Fibers and Their Composites* (CRC Press, Boca Raton, 2005)
10. D.L. Chung, *Carbon Fiber Composites* (Butterworth-Heinemann, Boston, 1994), pp. 3–65
11. W. Feng, S. Kwon, E. Borguet, R. Vidic, Environ. Sci. Technol. **39**, 9744 (2005)
12. M. Teng, J. Qiao, F. Li, P.K. Bera, Carbon **50**, 2877 (2012)
13. R. Elazari, G. Salitra, A. Garsuch, A. Panchenko, D. Aurbach, Adv. Mater. **23**, 5641 (2011)

14. F. Yang, K. Cheng, Y. Mo, L. Yu, J. Yin, G. Wang, D. Cao, J. Power Sources **217**, 562 (2012)
15. V. Leung, F. Ko, Polym. Adv. Technol. **22**, 350 (2011)
16. Z. Sedláková, G. Clarizia, P. Bernardo, J.C. Jansen, P. Slobodian, P. Svoboda, M. Kárászová, K. Friess, P. Izak, Membranes **4**, 20 (2014)
17. S. Andjelic, J. Roberge, N. Legros, L. Khoun, S.B. Schougaard, *SAE Technical Papers* 7
18. J. Shi, L. Bao, K. Kemmochi, Polymer Comp. **35**, 1494 (2013)
19. J.A. Onwudili, N. Insura, P.T. Williams, J. Energy Inst. **86**, 227 (2013)
20. Y. Zhang, S.J. Park, J. Polym. Sci., Part B: Polym. Phys. **55**, 1890 (2017)
21. Y. Zhang, M. Park, H.Y. Kim, B. Ding, S.-J. Park, Sci. Rep. **7**, 1 (2017)
22. Y. Zhang, J.R. Choi, S.-J. Park, Compos. Part A Appl. Sci. Manuf. **101**, 227 (2017)
23. L.Y. Meng, S.J. Park, Mater. Chem. Phys. **137**, 91 (2012)
24. S.Y. Lee, S.J. Park, J. Solid State Chem. **207**, 158 (2013)
25. J.H. Lee, I.J. Kim, S.J. Park, Electrochim. Acta **113**, 23 (2013)
26. G.Y. Heo, S.J. Park, Power Technol. **239**, 94 (2013)
27. K.S. Kim, S.J. Park, J. Power Sourc. **244**, 792 (2013)
28. S.Y. Lee, B.J. Kim, S.J. Park, J. Solid State Chem. **199**, 258 (2013)
29. S.Y. Lee, S.J. Park, J. Colloid Interface Sci. **389**, 230 (2013)
30. Y. Zhang, M. Park, H.Y. Kim, S.J. Park, J. Alloys Compd. **686**, 106 (2016)
31. K. Shimazaki, M. Hirai, Nippon Kagaku Kaishi **7**, 739 (1992)
32. Toho Belson Co., UK Patent Publication No. 2062599 A (1981)
33. Toho Belson Co., Japan Patent Publication No. 132,193 (1976)
34. K. Shimazaki, Nippon Kagaku Kaishi **7**, 807 (1993)
35. Kuraray Chemicals Co., Japan Patent Publication No. 7,583 (1980)
36. K. Kawazoe, T. Osawa, Rep. Spec. Proj. Environ. Sci. (1982)
37. Y. Zhang, S.J. Park, J. Catal. **355**, 1 (2017)
38. J.K. Zhao, Y. Zhu, J.F. Yu, Adv. Mater. Res. **811**, 370 (2013)
39. S.I. Su, W.C. Shih, C.M. Wang, Y.S. Liu, S.P. Wu, Fibers Polym. **14**, 1808 (2013)
40. C. Tang, W. Sun, W. Yan, RSC Adv. **4**, 523 (2014)
41. L. Sun, Y. Yao, L. Wang, Y. Mao, Z. Huang, D. Yao, W. Lu, W. Chen, Chem. Eng. J. **240**, 413 (2014)
42. M. Suzuki, Carbon **32**, 577 (1994)
43. Y. He, S. Yang, H. Liu, Q. Shao, Q. Chen, C. Lu, Y. Jiang, C. Liu, Z. Guo, J. Colloid Interface Sci. **517**, 40 (2018)
44. S.Y. Lee, S.J. Park, J. Colloid Interface Sci. **384**, 116 (2012)
45. S. Shrestha, G. Son, S.H. Lee, T.G. Lee, Chemosphere **92**, 1053 (2013)
46. H.A. Al-Aoh, M.J. Maah, R. Yahya, M.R.B. Abas, Asian J. Chem. **25**, 9573 (2013)
47. Z. Sun, Y. Yu, S. Pang, D. Du, Appl. Surf. Sci. **284**, 100 (2013)
48. J.S. Bae, S. Su, Int. J. Greenhouse Gas Contol. **19**, 174 (2013)
49. Y. Yao, V. Velpari, J. Economy, Fuel **116**, 560 (2014)
50. C. Carpetis, Int. J. Hydrogen Energy **5**, 423 (1980)
51. M. Kunowsky, B. Weinberger, F. Lamari Darkrim, F. Suárez-García, D. Cazorla-Amorós, A. Linares-Solano, Int. J. Hydrogen Energy **33**, 3091 (2008)
52. S.J. Park, B.J. Kim, Y.S. Lee, M.J. Cho, Int. J. Hydrogen Energy **33**, 1706 (2008)
53. N. Setovama, T. Suzuki, K. Kaneko, Ultramicropore Characterization by He Adsorption. Charact. Porous Solids III (1993)
54. I. Mochida, Y.-N. Sun, H. Fujitsu, S. Kisamori, S. Kawano, Nippon Kugaku Kuishi **6**, 885 (1991)
55. Y.-C. Chiang, W.-L. Hsu, S.-Y. Lin, R.-S. Juang, Materials. **10**, 511 (2017)
56. B.K. Kim, S.K. Ryu, B.J. Kim, S.J. Park, J. Colloid Interface Sci. **302**, 695 (2006)
57. R.S. Khurmi, R.S. Sedha, *Materials Science* (S. Chand & Company Ltd, India, 2010)
58. C. Rayment, S. Sherwin, *Introduction to Fuel Cell Technology*, University of Notre Dame, department of Aerospace and Mechanical Engineering (2003)
59. F.M. Amatucci1, S. Coleman, *Entrepreneurship Theory Pract* **31**, 971 (2007)

60. S. Basu, *Recent Trends in Fuel Cell Science and Technology* (Springer, New Delhi, 2007)
61. M. Cohen, *Presented at Intelec 2008, San Diego, California. Published in IEEE Intelec 2008 Proceedings* (2008)
62. P. Qian, H. Zhang, J. Chen, Y. Wen, Q. Luo, Z. Liu, D. You, B. Yi, J. Power Sourc. **175**, 613 (2008)
63. N. Karim, S. Jonathan, *How Fuel Cells Work: Polymer Exchange Membrane Fuel Cells*, How Stuff Works. Accessed 4 Aug 2011
64. N. Mahato, A. Banerjee, A. Gupta, S. Omar, K. Balani, Prog. Mater Sci. **72**, 141 (2015)
65. J. Denton, J.M. Gascoyne, R.J. Potter, US Patent 5,865,968 (1999)
66. M.S. Wilson, J.A. Valerio, S. Gottesfeld, Electrochim. Acta **40**, 355 (1995)
67. N.J. Walker, *Carbon Fibers: Technology, Uses and Prospects* (The Plastics and Rubber Institute, 1986)
68. X. Zhang, Z. Shen, Fuel **81**, 2199 (2002)
69. I.U. Hwang, H.N. Yu, S.S. Kim, D.G. Lee, J.D. Suh, S.H. Lee, B.K. Ahn, S.H. Kim, T.W. Lim, J. Power Sources **184**, 90 (2008)
70. B.J. Kim, W.K. Choi, M.K. Um, S.J. Park, Surf. Coat. Technol. **205**, 3416 (2011)
71. G. Gryglewicz, J. Machinowski, E. Lorenc-Grabowska, G. Lota, E. Frakowiak, Electrochim. Acta **50**, 1197 (2005)
72. H. Oda, A. Yamashita, S. Minoura, M. Okamoto, T. Morimoto, J. Power Sourc. **158**, 1510 (2006)
73. J.S. Im, S.W. Woo, M.J. Jung, Y.S. Lee, J. Colloid Interface Sci. **327**, 115 (2008)
74. C.L. Liu, W. Dong, G. Cao, J. Song, L. Liu, Y. Yang, J. Electrochem. Soc. **155**, F1 (2008)
75. A. Castro-Muñiz, F. Suárez-García, A. Martínez-Alonso, J.M.D. Tascõn, T. Kyotani, Chemsuschem **6**, 1406 (2013)
76. Y. Zhang, K.Y. Rhee, D. Hui, S.J. Park, Compos. Part B Eng. **143**, 19 (2018)
77. A. Braun, M. Bärtsch, B. Schnyder, R. Kötz, O. Haas, A. Wokaun, Carbon **40**, 375 (2002)
78. Z. Ryu, H. Rong, J. Zheng, M. Wang, B. Zhang, Carbon **40**, 1144 (2002)
79. Y. Zhang, M. Park, H. Yong, B. Ding, S. Park, Appl. Surf. Sci. **384**, 192 (2016)
80. K. Babel, K. Jurewicz, J. Phys. Chem. Solids **65**, 275 (2004)
81. K. Leitner, A. Lerf, M. Winter, J.O. Besenhard, S. Villar-Rodil, F. Suarez-Garcıa, A. Martınez-Alonso, J.M.D. Tascon, J. Power Sources **153**, 419 (2006)
82. T.H. Nguyen, A. Fraiwan, S. Choi, Biosens. Bioelectron. **54**, 640 (2014)
83. G.H. Waller, S.Y. Lai, B.H. Rainwater, M. Liu, J. Power Sources **251**, 85 (2014)
84. M.S. Park, B.O. Jeong, T.J. Kim, S. Kim, K.J. Kim, J.S. Yu, Y. Jung, Y.J. Kim, Carbon **68**, 265 (2014)
85. M. Alaf, H. Akbulut, J. Power Sources **247**, 692 (2014)
86. S.H. Chung, A. Manthiram, Electrochem. Comm. **38**, 91 (2014)
87. T. Nagaura, in *4th International Rechargeable Battery Seminar*, Deerfield Beach, FL, USA (1990)
88. S.C. Bennett, D.J. Johnson, Carbon **17**, 25 (1979)
89. K. Suzuki, T. Iijima, M. Wakihara, Electrochim. Acta **44**, 2185 (1999)
90. M. Egashira, H. Takatsuji, S. Okada, J.I. Yamaki, J. Power Sources **107**, 56 (2002)
91. V. Gomes, M. Hassan, Sep. Purif. Technol. **24**, 189 (2001)
92. T.J. Giesy, M.D. LeVan, Chem. Eng. Sci. **90**, 250 (2013)
93. M. Rungta, L. Xu, W.J. Koros, Carbon **50**, 1488 (2012)
94. J. Rother, T. Fieback, Adsorption **19**, 1065 (2013)
95. T. He, Q. Li, Y. Ju, J. Chem. Eng. Japan **46**, 811 (2013)
96. X. Ning, W.J. Koros, Carbon **66**, 511 (2014)
97. M. Freitas, J. Figueiredo, Fuel **80**, 1 (2001)
98. S. Vyas, S. Patwardhan, S. Vijayalakshmi, B. Gangadhar, Fuel **72**, 551 (1993)
99. A. Cabrera, J. Zehner, C. Coe, T. Gaffney, T. Farris, Carbon **31**, 969 (1993)
100. S. Adapa, V. Gaur, N. Verma, Chem. Eng. J. **116**, 25 (2006)
101. J. Yun, H.I. Kim, Y.S. Lee, J. Mater. Sci. **48**, 8320 (2013)
102. C. Dong, Compos. A **42**, 419 (2011)

103. P.K. Mallick, *Fiber-Reinforced Composites: Materials, Manufacturing, and Design* (Marcel Dekker, New York, 1993)
104. D.H. Middleton, *Composite Materials in Aircraft Structure* (Longman Scientific and technical, Harlow, 1990)
105. http://www.universalmetaltek.com/honeycomb_panel.htm
106. A. Knop, L.A. Pilato, *Phenolic Resins: Chemistry, Applications and Performance, Future Directions* (Springer, Berlin, 1985)
107. A.R. Bunsell, J. Renard, J. Renard, *Fundamentals of Fibre Reinforced Composite Materials* (CRC Press, Boca Raton, 2005)
108. H. Fukagawa, T. Hirogaki, T. Kato, A. Kato, M. Seki, Key Eng. Mater. **523–524**, 226 (2012)
109. Y. Zhang, K.Y. Rhee, S. Park, Compos. Part B Eng. **114**, 111 (2017)
110. Z. Kapidžić, L. Nilsson, H. Ansell, Comp. Struct. **109**, 198 (2014)
111. R.B. Deo, J.H. Starnes, R.C. Holzwarth, *Low-Cost Composite Materials and Structures for Aircraft Applications*, RTO-MP-069 (II) (2001)
112. D.R. Tenney, J.G. Davis, R. Pipes, J.N. Byron, *NASA Composite Materials Development: Lessons Learned and Future Challenges* (NASA, Langley Research Center, 2009)
113. S.J. Park, M.K. Seo, *Interface Science and Composites* (Academic Press, London, 2011)
114. A.R. Bunsell, Reinf. Plast. **50**, 38 (2006)
115. J.M. Miller, *Propulsion Systems for Hybrid Vehicles, Power and Energy Series 45*, IET (2004)
116. D.L. Beshears, J.M. Starbuck, Smart Onboard Inspection of High Pressure Gas Fuel Cylinders. Conference: *1st International SAMPE Automotive Conference*, Detroit, MI, 27–29 Sept 1999
117. J. Starbuck, L. Cataquiz, *Evaluation of Large Tow-Size Carbon Fiber for Reducing the Cost of CNG Storage Tanks*, SAE Technical Paper 2000-01-1526 (2000)
118. S.T. Peters, *Composites Materials and Processes* (Digital Engineering Library By McGrawHill, Mountain View, California, 2004)
119. Y. Miyano, M. Nakada, J. Ichimura, E. Hayakawa, Compos Part B **39**, 5 (2008)
120. S. Biswas, A. Mittal, G. Srikanth, Composites: A Vision for the Future. http://www.tifac.org.in/index.php?option=com_content&view=article&id=539
121. I. Kimpara, Marine Struct **4**, 117 (1991)
122. K. Grogg, Harvesting the Wind: The Physics of Wind Turbines (Carleton College, 2005)
123. P. Brøndsted, H. Lilholt, A. Lystrup, Annu. Rev. Mater. Res. **35**, 505 (2005)
124. D. Walczyk, *An Overview of Composite Wind Turbine Blade Manufacturing* (Rensselaer Polytechnic Institute 2011)
125. H.G. Willett, Reinf. Plast. **56**, 34 (2012)
126. L.N. Sun, Z. Deng, Adv. Mater. Res. **341–342**, 173 (2011)
127. J.X. Liang, Y.T. Cai, Appl. Mech. Mater. **217–219**, 66 (2012)
128. W. Paton, Composites **1**, 221 (1970)
129. K. N. Lo, US Patent 4,360,202 (1982)
130. M.P. Huntley, C.L. Davis, M. Strangwood, S.R. Otto, in *Proceedings of the Institution of Mechanical Engineers Part L-Journal of Materials-Design and Applications*, **220**, 229 (2006)
131. A. Kojima, H. Horii, Effect of torsional properties of CFRP golf club shafts on the speed and the directional stability of the ball, ed. by Z. Maekawa, E. Nakata and Y. Sakatani, *Proceedings of the 4th Japan international SAMPE symposium* (pp. 1328–1334). Yokohama: SAMPE (1995)
132. E. Fitzer, M. Heine, *Fibre Reinforced Composite Materials*, vol. 2 (Elsevier, Amsterdam, 1988), pp. 73–148
133. P. Tsotra, K. Friedrich, Compos. Part A **34**, 75 (2003)
134. T.O. Mason, M.A. Campo, A.D. Hixson, L.Y. Woo, Cem. Concr. Compos. **24**, 457 (2002)
135. Y. Xu, D.D.L. Chung, Carbon **39**, 1995 (2001)
136. S.J. Park, J.B. Donnet, J. Colloid Interface Sci. **206**, 29 (1998)
137. H.A. Toutanji, T. El-Korchi, R.N. Katz, Cem. Concr. Res. **23**, 618 (1993)

138. S.J. Park, M.K. Seo, H.B. Shim, K.Y. Rhee, Mater. Sci. Eng., A **366**, 348 (2004)
139. S.J. Park, M.S. Cho, Carbon **38**, 1053 (2000)
140. H.N. Garden, L.C. Hollaway, Compos. Part B **29**, 411 (1998)
141. Y. Zhang, S.J. Park, Carbon **122**, 287 (2017)
142. C.E.M. Gomes, Adv. Mater. Res. **753–755**, 616 (2013)
143. O. Galao, F.J. Baeza, E. Zornoza, P. Garcés, Cement Concrete Comp. **46**, 90 (2014)
144. Y. Xu, D.D.L. Chung, Cem. Concr. Res. **29**, 773 (1999)

Chapter 8
Carbon/Carbon Composites

Abstract Carbon fiber-reinforced carbon matrix composites (C/C composites) have excellent properties such as specific strength and specific elastic modulus, along with excellent fatigue resistance, thermal shock resistance, ablation resistance, frictional properties, lightweight, heat/thermal conductivity, and dimensional stability. C/C composites are the materials that exhibit almost no decline in mechanical properties, even at 2000–3000 °C, when protected from oxidation. A method of directly coating the surface of the fiber, to form an oxide film on the surface of the produced composite material, and method for loading directly into the filler when producing the resin have been proposed as the method for imparting antioxidation properties typical. This chapter is a detailed review on the manufacturing methods, antioxidation properties, and applications of C/C composites.

8.1 Manufacturing of Carbon/Carbon Composites

Recently, carbon/carbon composites (C/C composites) have been expanding in application fields such as space, aviation, various kinds of ultrahigh-temperature heat-resistant materials, and carbon brake disks [1–15]. Thermal stabilities and mechanical properties of C/C composites are reduced due to the formation of many cracks and pores in the interior due to the gas released by pyrolysis of the matrix during the carbonization process during manufacturing [16–25]. In order to compensate for this problem, it is necessary to repeat the process of re-impregnation and re-carbonization several times to increase the density. One of the major concerns is achieving densification within a faster process time. This is directly related to the reduction of the process cost, so research is being conducted on the production of C/C composites through a single manufacturing process [25–42].

In addition, a method of adding a ceramic powder having a specific physical property to a matrix has been reported. Among them, $MoSi_2$ is known to have a high melting point (2100 °C) and excellent high-temperature oxidation resistance, and brittle–ductile transition (BDT) properties are reported. In the case of C/C

S.-J. Park, *Carbon Fibers*, Springer Series in Materials Science 210,
https://doi.org/10.1007/978-981-13-0538-2_8

composites containing $MoSi_2$ as an additive, it has been reported to have BDT properties at about 900 °C [43–55].

The microstructure of the matrix plays an important role in determining the final physical properties and in particular the mechanical properties of C/C composites. From this point of view, much research has been conducted on the degree of graphitization and microstructure control of the matrix to obtain the desired performance. The degree of graphitization of the matrix is greatly influenced by the heat treatment temperature. One of strategies for efficient graphitization, catalysts are used as a method to enhance the degree of graphitization of the matrix at this final heat treatment temperature, thereby enabling graphitization at lower temperatures. Typical catalysts are inorganic catalysts such as iron oxide and nickel oxide [56–70].

A typical example of a brittle matrix composite is that the breaking strain of the matrix is higher than the breaking strain of the fiber and that the nature of the fiber–matrix bond is the most important to control the fracture behavior. One of the biggest issues of C/C composites is that oxidation reduces the characteristics of carbon and graphite [75–90]. Currently, studies for improving the oxidation inhibiting effect due to the mechanical properties of such C/C composites are widely carried out. In addition, C/C composites have a high anisotropic graphite crystal structure and form a strong covalent bond along the basal plate to show the unique creep resistance of the graphitized carbon. However, it is known that when the temperature is higher than about 500 °C, it is very sensitive to the oxidation reaction, so that it may be limited in the temperature range of use due to the rapid collapse of the composite when exposed to high temperatures in the air. Therefore, these materials are required to have improved thermal stability at high temperatures in the air [75–90].

In this respect, various thermal designs can be made, such as having a value of more than copper along the direction of the reinforced carbon fiber and insulating property in the direction perpendicular to the fiber while maintaining the advantages of graphite and carbon material. Due to these excellent properties, it is attracting attention as a heat-resistant and abrasion-resistant material for aerospace such as brake disk for aircraft, leading edge of space shuttle, nozzle of rocket, and nose cone.

Figure 8.1 shows general C/C composites manufacturing process, and the process includes liquid impregnation method and chemical vapor infiltration method depending on how matrix carbon is formed.

8.1.1 Liquid Impregnation Method

In the liquid impregnation process, the composites or fiber preforms are impregnated with an organic matrix precursor under vacuum. Impregnated preforms are carbonized and pyrolyzed at a heat treatment temperature of 800 °C or higher after curing in an autoclave. During pyrolysis, the preform forms multiple layers of carbon

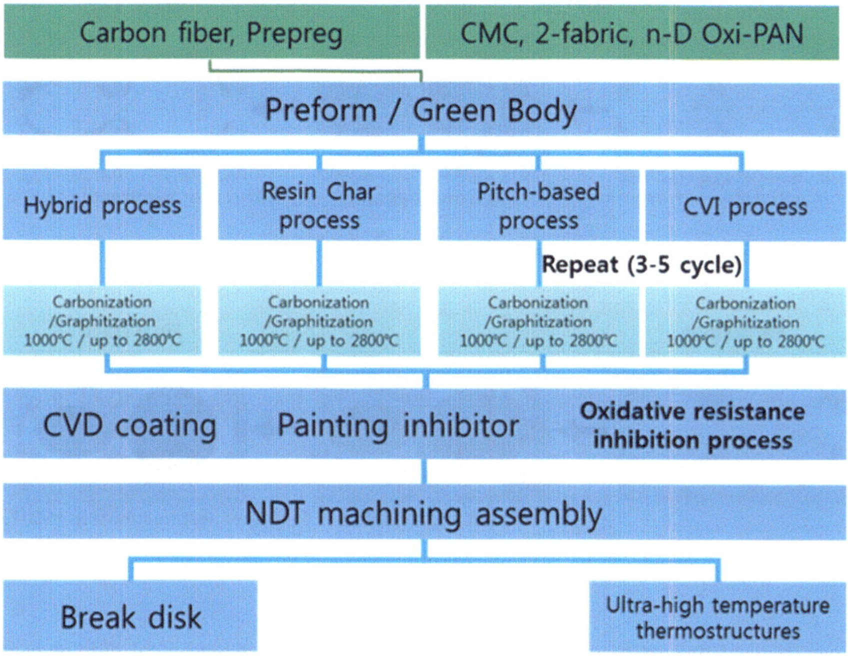

Fig. 8.1 Use of carbon fiber-reinforced composites in automobiles

matrices deposited from hydrocarbon gases such as methane, propane at elevated temperatures above 1000 °C. This impregnation and carbonization processes are repeated several times to obtain the density and physical properties of the desired state while filling the pores vacated. The number of densification steps is also closely related to the carbonization rate of the matrix and the efficiency of the selected impregnation material (resin or pitch). The manufacturing of C/C composites proceeds a graphitization treatment at a very high temperature of 2200–2800 °C, to have ordered crystalline structure close to the density of graphite.

Liquid-phase impregnation method is divided into a method of using a thermosetting resin such as an epoxy, a phenol, and a furan resin, and a method of using tar or pitch depending on the type of the matrix. As compared with the case where the thermosetting resin undergoes the solid-phase carbonization process, pitch is advantageous because of going through a liquid-phase carbonization process, and the orientation of the aromatic molecules increased after carbonization, such as small points are uniformly generated in accordance with the thermal conductivity is excellent and shrinkage.

Figures 8.2 and 8.3 show schematically the impregnation mechanisms of C/C composite using pitch and thermosetting resin. In the case of pitch, much voids and gaps are left due to volume reduction due to volatilization after carbonization or graphitization. In order to solve this problem, the method of increasing the density by repeating the impregnation and carbonization processes of the pitch several

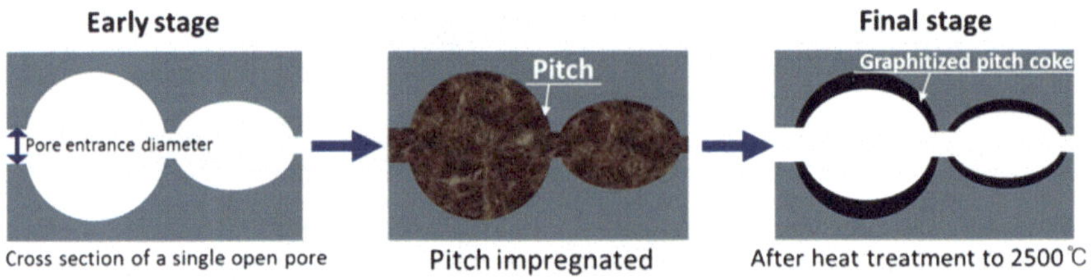

Fig. 8.2 Liquid-phase impregnation–pyrolysis repetition method of the pitch

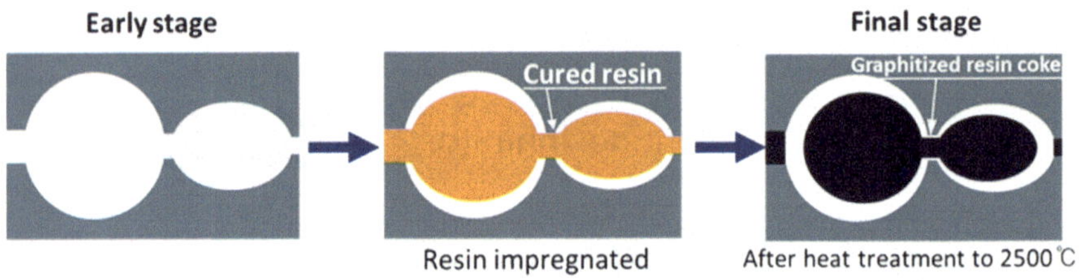

Fig. 8.3 Liquid-phase impregnation–pyrolysis repetition method of the thermosetting resin

times is used, but as the number of steps is repeated, the economical efficiency is getting lowered. In this respect, it is currently being manufactured by mixing or modifying various other techniques than simply repetition of impregnation/carbonization. Generally, when the pitch is carbonized at atmospheric pressure, the carbonization yield is very low. However, when the carbonization is performed at high pressure, the carbonization yield is significantly increased. Therefore, the density increase rate per impregnation/carbonization is much larger, and the densification can be performed even fewer times.

Figure 8.2 shows the pore closing mechanism which occurs in the case of the liquid-phase impregnation–pyrolysis repetition method of the pitch. As the impregnated pitch is carbonized, it is escaped from the pores and the matrix is contracted, so that the entrance of the pores is less likely to be clogged. However, there is a disadvantage that pressurized carbonization is required for efficient high density.

Figure 8.3 shows the pore closing mechanism exhibited in the liquid-phase impregnation–pyrolysis repetition method of the thermosetting resin. This is an appropriate method for filling the micropores because pressurized carbonization is unnecessary, but the liquid resin shrinks and the pore opening is clogged, making it difficult to increase the density.

8.1.2 Chemical Vapor Infiltration Method

Chemical vapor infiltration method is a method of forming a matrix by allowing pyrolytic carbon generated from pyrolysis of hydrocarbons to be deposited on the fiber surface inside the preform while hydrocarbon is spoken to a preform having many pores made of carbon fibers. Compared with the liquid impregnation method, it is advantageous to fabricate a relatively dense matrix. Figure 8.4 shows the chemical vapor impregnation method, and a matrix having good adhesion to carbon fibers can be obtained, but a phenomenon occurs in which the inlet of the pores is clogged under some deposition conditions. An important problem to be solved in the production of composite materials by chemical vapor deposition is to prevent clogging of the surface pores.

Various methods such as isothermal method, thermal gradient method, pressure gradient method, and pulse method have been developed as chemical vapor infiltration method to date. These classification criteria are divided into whether the preform has a uniform temperature, temperature, or a pressure gradient when the pyrolysis carbon is deposited, and whether the reaction gas is in contact with the preform or flows through the preform. Figure 8.5 shows various methods of chemical vapor infiltration methods schematically.

Figure 8.5a shows the isothermal method. In course of the reaction gas flowing around the preform heated to a certain temperature in the furnace, the gas diffuses into the preform to deposit matrix materials. Since the temperature of the surface and inside of the substrate is kept uniform, the deposition rate of the surface and the inside is relatively uniform. From this reason, the isothermal method has the advantage of simultaneously densifying preforms of different sizes and shapes. However, since the surface and the interior are deposited at the same time, the surface is clogged before the interior is filled. Therefore, in order to obtain a high density, the removal of the surface layer of the preform and the infiltration process must be repeated, which requires a long time of deposition. However, it is a commercially attractive method because it is possible to simultaneously deposit many parts of various sizes and complex shapes in one reactor.

Figure 8.5b represents the thermal gradient method. When the susceptor is preformed and the susceptor is induction heated, the temperature of the surface in contact with the susceptor is higher than the temperature of the noncontacted

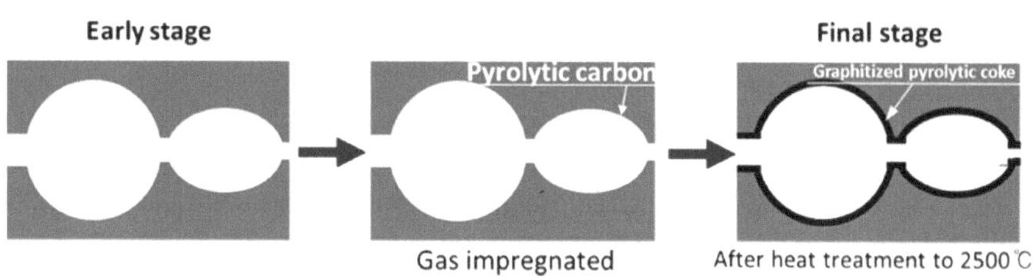

Fig. 8.4 Chemical vapor impregnation method

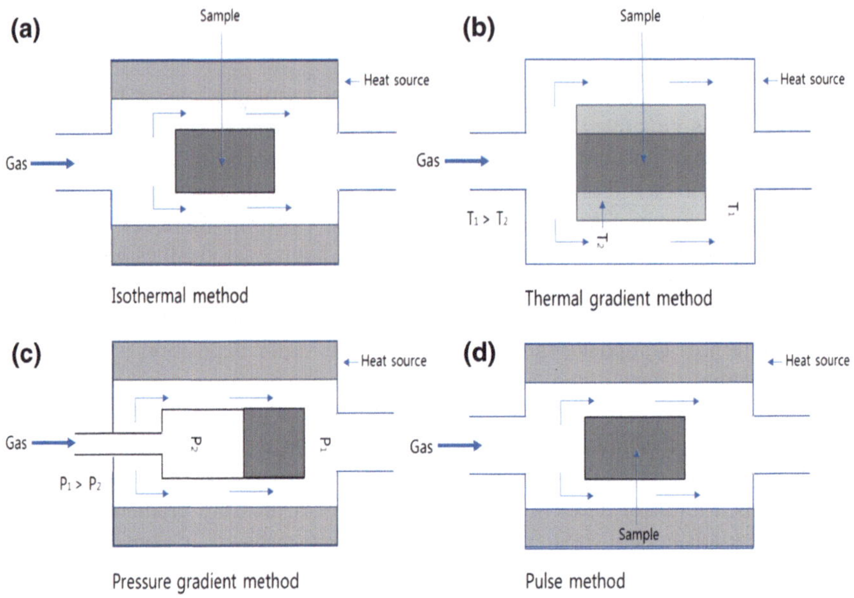

Fig. 8.5 Various methods of chemical vapor infiltration methods

surface, so that a temperature gradient is formed inside the preform. Thus, there is a difference in the deposition rate due to the temperature gradient on the surface and inside of the preform, where the deposition starts from the interior with higher temperature and proceeds toward the surface. Since the thermal gradient method depends on diffusion, it is relatively slow, and it is possible to produce a composite material having a small pore size by this method. However, since it is difficult to maintain a uniform deposition condition during the deposition process, it is difficult to obtain a uniform microstructure of the deposition layer.

In the pressure gradient method (Fig. 8.5c), the preform is located inside one end of the tube into which the raw material gas is injected. When the raw material gas is injected, a pressure difference is formed between the upper and lower portions of the preform, and this pressure difference acts as a driving force for penetration. In this method, the deposition rate is high, but the density gradient of the deposition material is produced, and the shape of the preform is limited. The advantage of the pressure gradient method is that it can reduce the deposition time because of the higher deposition temperature and faster mass transfer rate than the isothermal method.

The pulse method is a method in which vacuum is applied to the chamber and the reaction gas is injected repeatedly to densify the preform (Fig. 8.5d). In the case of the isothermal method, the surface is clogged before the inside of the preform is filled, but the pressure pulse method can compensate this disadvantage. When vacuum is applied to the chamber, the interior of the preform is evacuated. At this time, the reactive gas is easily injected into the preform and deposited. Thus, uniform deposition occurs over the surface and inside of the preform. An advantage of this method is that the preform can be densified uniformly in a short time, and there is no great limitation on the shape. The disadvantage is that the equipment is expensive and the efficiency of the raw material gas is low, which increases the operation cost. The infiltration rate depends on the deposition temperature, the concentration of the starting gas, the gas holding time, and the number of pulses.

8.2 Antioxidations of Carbon/Carbon Composites

Carbon oxidizes readily at temperatures above 400 °C. Because the C/C composites are very sensitive to oxidation at high temperatures, when exposed to high temperatures midair, the composites undergo a rapid decay, and therefore require effective antioxidation measures at high-temperature atmosphere, which is essential for the manufacture and applications of C/C composites. In other words, the antioxidation properties of common carbon/plastic composites are also important along with their mechanical and interfacial properties, because they can significantly influence the final material properties and hence the performance and lifetime of the composites. However, difficulties in studying the antioxidation of C/C composites caused by cracks triggered by the difference in the thermal expansion coefficient, the adhesion between the oxidation-resistant barrier and C/C composites, etc., impede the fabrication of oxidation-resistant barriers that are stable over a wide range of workable temperatures [45–86].

Therefore, numerous investigations attempted to protect the C/C composites from oxidation. General approaches for enhancing the thermal stability and antioxidation by preventing oxidation include (i) direct coating of the carbon fiber with an oxygen barrier layer (or oxidation-resistant barrier) for the antioxidation, protection of the carbon fiber during the manufacture of the composites, and adjusting the interfacial properties between the fiber and the branches to achieve the desired mechanical properties of the C/C composites; and (ii) directly adding an oxidation inhibitor to the matrix during the manufacture of C/C composites to suppress the oxidation [55–58, 86–92].

8.2.1 Oxygen Barrier Layer

Coating is an efficient method for protecting C/C composites at high temperature. Many ceramics are promising candidate materials for oxidation protective applications on C/C composites. Ceramic materials are widely used as oxygen barrier layers to provide protection against oxidation owing to its excellent antioxidation properties and good compatibility with C/C composites. It is well known that refractory can improve ablation and erosion resistance and control ablation

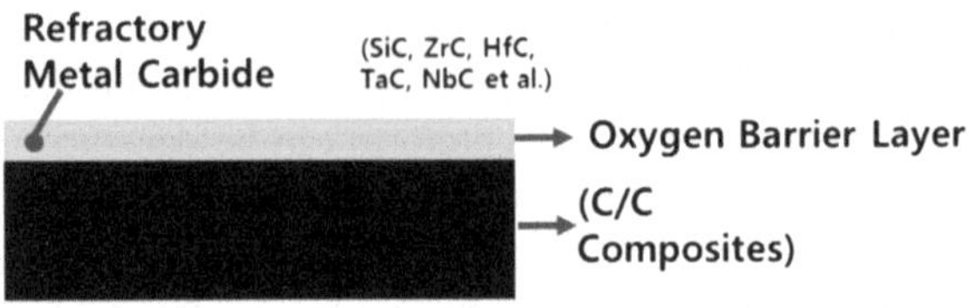

Fig. 8.6 Oxygen barrier method for antioxidation of C/C composites

appearance effectively by being doped into the matrix or forming coating on the materials. Figure 8.6 represents the oxygen barrier method for antioxidation of C/C composites.

Among the refractory, SiC was widely used as the bonding and buffer layer such as SiC/mullite and SiC/borosilicate glass multilayer coatings because of its good compatibility with C/C composites. The reactions occurring during the oxidation process were as follows:

$$2SiC(s) + 3O_2(g) \rightarrow 2SiO_2(l) + 2CO(g) \tag{8.1}$$

$$SiC(s) + 2O_2(g) \rightarrow SiO_2(l) + CO_2(g) \tag{8.2}$$

$$SiO_2(s)(l) \rightarrow SiO_2(g) \tag{8.3}$$

$$C(s) + O_2(g) \rightarrow CO_2(g) \tag{8.4}$$

$$2C(s) + O_2(g) \rightarrow 2CO(g) \tag{8.5}$$

In addition, many studies showing excellent antioxidation effects by coating SiC single material as well as other ceramic materials have been reported. Huang et al. reported yttrium silicate oxidation protective coating for SiC-coated C/C composites (Fig. 8.7).

Among the promising candidates, ZrC possesses a melting point as high as 3540 °C and is one of the most promising candidates for ultrahigh-temperature applications due to its excellent high-temperature performance and its high melting point over 2700 °C of ZrO_2, which has a lower vapor pressure and could form protective layer on surface to reduce the diffusion speed of oxygen. The reactions occurring during the oxidation process were as follows:

$$ZrC(s) + 2O_2(g) = ZrO_2(s) + CO_2(g) \tag{8.6}$$

$$2ZrC(s) + 3O_2(g) = 2ZrO_2(s) + 2CO(g) \tag{8.7}$$

$$ZrO_2(s) = ZrO_2(l) \tag{8.8}$$

Some research groups reported ZrC-doped C/C composites by liquid impregnation and elucidated the ablation mechanism in an arc jet heater. Shen et al. reported C/C composites doped with ZrC by a three-step process and researched its ablation mechanism in an oxyacetylene torch. Tong et al. prepared pitch-derived

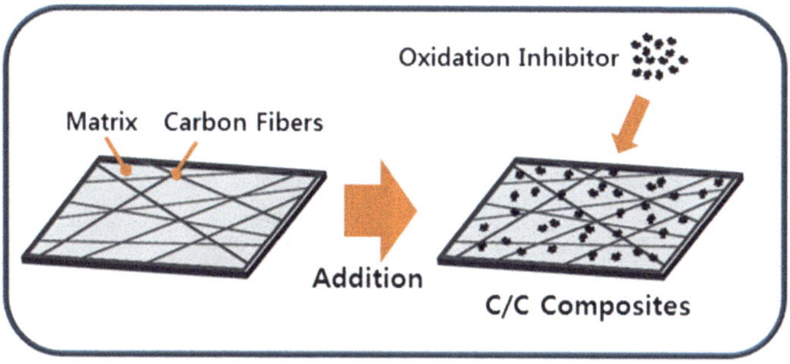

Fig. 8.7 Cross-sectional SEM pictures of yttrium silicate coating. **a** $SiC/SiO_2{\cdot}Y_2O_3$, **b** $SiC/1.5SiO_2{\cdot}Y_2O_3$, **c** $SiC/1.5SiO_2{\cdot}Y_2O_3/SiO_2{\cdot}Y_2O_3$, **d** $SiC/2SiO_2{\cdot}Y_2O_3/1.5SiO_2{\cdot}Y_2O_3/SiO_2{\cdot}Y_2O_3$. [ref]

ZrC/C composites using Zr-containing pitch by hot-pressed technology and reported that a dense and uniform ZrO_2 film acting as an effective oxygen barrier could be formed which reduced the ablation rates of the composites greatly. In order to improve the matching ability between ZrC coating and C/C composites, SiC and ZrC are also used together owing to its good compatibility with C/C composites. Liu et al. prepared a two-layer SiC–ZrC coating by CVD and investigated its laser ablation behaviors. Some research groups developed the thermodynamic description for the C–Si–Zr system based on its constituent binaries and critically reviewed experimental data. However, the above-mentioned composites by hot-pressing or molding are not suitable to the manufacture of complex-shape or large-size components, and the cost of hot-pressing and CVD process is high as well. Therefore, it is necessary to find a high-efficiency and low-cost approach to improve the ablation properties of C/C composites [82–92].

8.2.2 *Oxidation Inhibitor*

In this section, the direct addition of oxidation inhibitor into the matrix was suggested with details. When manufacturing C/C composites, in order to grant thermal stability and oxidation resistance, the ceramic-based SiC, SiO_2, $MoSi_2$, ZrB_2, SiB_4, TEOS, and Si_3N_4 have been used, and the study regarding the thermal stability of these materials has been constantly conducted by the thermogravimetry. Among the ceramic materials, $MoSi_2$ has high fusing point (2100 °C) and excellent high-temperature oxidation resistance, along with brittle–ductile transition at around 900 °C. It was added to PAN-based carbon fiber and phenol resin, a precursor of carbonizing matrix. The C/C composites were manufactured by unidirectional singe process, which is rid of densification through numerous re-immersion and re-carbonization steps. Also, the advantages of TEOS, that are nontoxic, liquid, and chemically stable (low reactivity and corrosion), which are not

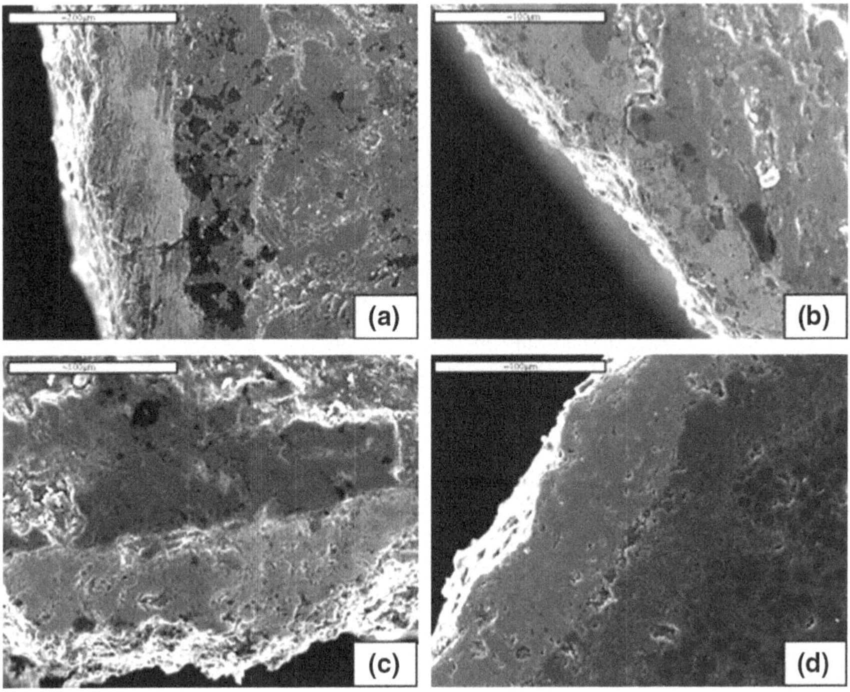

Fig. 8.8 Schematic representation of addition of oxidation inhibitor on C/C composites

possessed by $MoSi_2$ used in the manufacture, were utilized for the manufacture of C/C composites, for further discussions [30, 56–63] (Fig. 8.8).

Generally, the methods for enhancing the thermal stability and oxidation resistance are as follows: directly coating the carbon fiber with antioxidation film to prevent and protect the fiber from oxidation as well as granting desired mechanical properties by adjusting the interfacial properties between the fiber and branch materials; directly adding an oxidation inhibitor to the matrix when manufacturing C/C composites to grant oxidation resistance to prevent the oxidation; and finally, coating the entire surface of C/C composites with an antioxidation film which can hardly be penetrated by oxygen.

Park and Seo reported the effect of oxidation inhibitors, i.e., $MoSi_2$, on the oxidation behavior of C/C composites. PAN-based carbon fibers impregnated with phenolic resin used as a precursor of the carbonized matrix were modified by the addition of molybdenum disilicide ($MoSi_2$) at different concentrations, i.e., 0, 4, 12, and 20% by weight for improving the antioxidation properties of the composites (Fig. 8.9). Oxidation behavior of C/C composites made using $MoSi_2$ as an oxidation inhibitor in flowing air has been studied over the temperature range 600–1000 °C. The results showed that the oxidation behavior of the composites made without the use of $MoSi_2$ differed from those of the composites made using $MoSi_2$ and involved a degradation

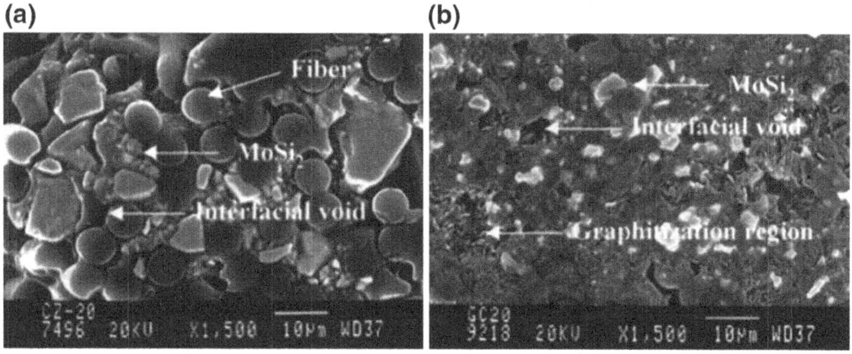

Fig. 8.9 SEM pictures of C/C composites made with 20 wt% $MoSi_2$. **a** Carbonization and **b** graphitization

mechanism of its own, resulting in the highest oxidation rate during the initial phase of the oxidation.

The composites made using $MoSi_2$ led to a significantly improved oxidation resistance owing to both the reduction of the crack in composites and the formation of a mobile diffusion barrier for oxygen when compared to those made without using $MoSi_2$. Hence, the carbon active sites were blocked, and the oxidation of carbon was limited. This was probably owing to the effect of the inherent $MoSi_2$ properties, resulted from the formation of the protective layer against oxygen attack above 800 °C.

The activation energy of containing $MoSi_2$ in C/C composites is shown in Fig. 8.10 [30]. In case of $MoSi_2$ in C/C composites, the activation energies are increased up to 83 kJ/mol for carbonization and 93 kJ/mol for graphitization. This is probably attributed to the inhibition effect of $MoSi_2$ to the blockage of active sites

Fig. 8.10 Activation energies of C/C composites as a function of $MoSi_2$ content with carbonization and graphitization [30]

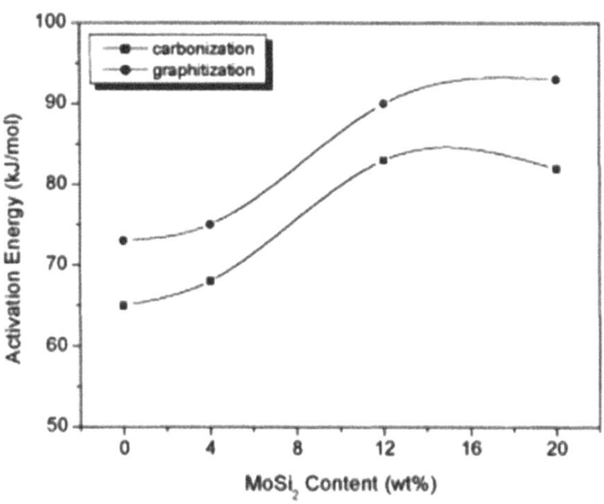

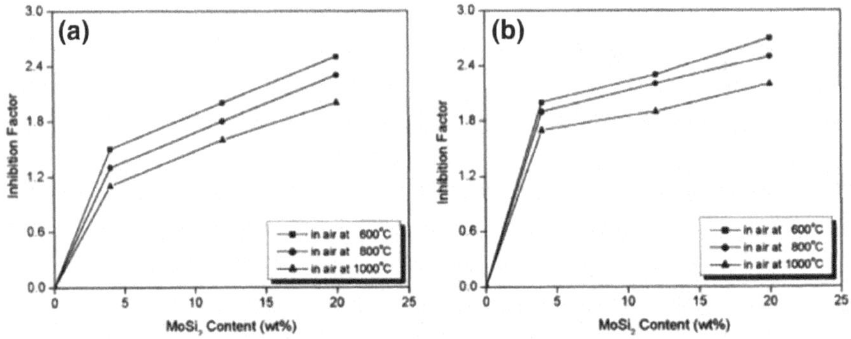

Fig. 8.11 Variation of the inhibition factor (I_f) of C/C composites measured at 10% level of burn-off as a function of $MoSi_2$ content. **a** Carbonization and **b** graphitization

on the composite surface, resulting in improving the antioxidation properties and in producing the presence of BDT properties above 900 °C that can grow the degree of adhesion at interfaces between fibers and matrix. The composites consisting of different contents of the oxidation inhibitor $MoSi_2$ displayed an increase of the work of adhesion between the fibers and matrix, which improved both the fracture toughness and impact properties of the composites. The 12 wt% $MoSi_2$ composites exhibited the strongest mechanical and mechanical interfacial properties. This was probably because of the improvement of the London dispersive L component, WLA, of the work of adhesion, resulting in an increase in the interfacial adhesion force among the fibers, filler, and matrix in this system [30].

As experimental results, the effect of the amount of $MoSi_2$ on the inhibition factor (I_f) measured of the carbonized and graphitized composites at different temperatures in a given burn-off level of 10% by weight is shown in Fig. 8.11. For the relatively small presence of $MoSi_2$ in the composites, i.e., below 4% by weight, I_f values increase very sharply with the $MoSi_2$ contents. Thereafter, with increasing of the $MoSi_2$ contents, I_f values were increased slowly. These results mean that the additions of $MoSi_2$ are significant effects on oxidation rate of composites in all heat treatment temperatures, though their antioxidation rates are somewhat different.

Figure 8.12 exhibits an interesting result of the C/C composites in the different heat treatment temperatures in a given retention time of 20 h. These results indicate that the specimen prepared with the 12 wt% $MoSi_2$ received at 1000 °C shows intensity spectra of SiO_2, while the same specimen prepared at 600 °C identifies the $MoSi_2$ spectra. Therefore, it is indicated that a new formation of oxygen protection layer, SiO_2, of C/C composites during pyrolysis plays an important role in a mobile diffusion barrier at high temperatures, resulting in increasing antioxidation of C/C composites. The excellent oxidation resistance of $MoSi_2$ derives from a protective layer of silica (SiO_2) that forms upon exposure to an oxidizing atmosphere at elevated temperature. The occurrence of either reaction is controlled mainly by the kinetics of the reaction, and the high-temperature oxidation behavior of $MoSi_2$

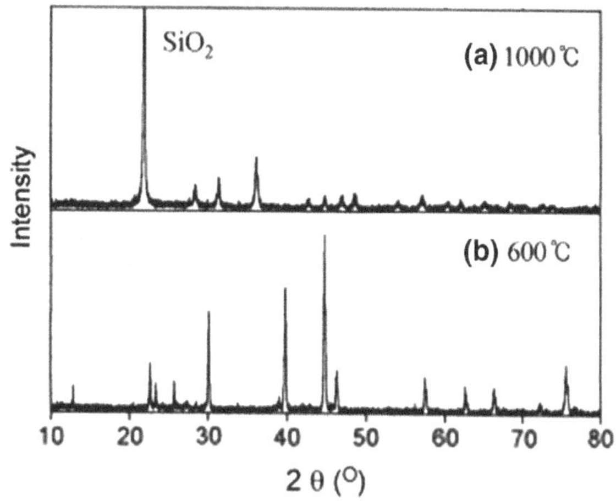

Fig. 8.12 XRD patterns of the C/C composites in a given retention time of 20 h. **a** 1000 °C and **b** 600 °C

(above 800 °C) is governed by the reaction of (8.9), while the low-temperature oxidation (below 800 °C) is governed by the reaction of (8.10). Above 800 °C, the oxidation of silicon is accompanied by the formation of the silicide, Mo_5Si_3, between the $MoSi_2$ and outer SiO_2 layers [30].

$$5MoSi_2 + 7O_2 \rightarrow Mo_5Si_3 + 7SiO_2 \quad (8.9)$$

$$2MoSi_2 + 7O_2 \rightarrow 2MoO_3 + 4SiO_2 \quad (8.10)$$

There is a strong evidence that the formation of a transient Mo–Si–O surface oxide layer (metastable phase or more likely mixed oxide of MoO_3 and SiO_2) plays a crucial role in the kinetics of $MoSi_2$ oxidation and, in addition to thermodynamics, there is also a kinetic factor involved [21, 22]. This SiO_2 has been detected even after 20 h of oxidation at 1000 °C as seen in Fig. 8.12.

8.3 Applications of Carbon/Carbon Composites

Unlike many high-temperature alloys and ceramics, C/C composites do not undergo degradation in strength above 2000 °C in inert environments. Figure 8.13 shows the application of C/C composites to various industrial fields. They have excellent thermal stability coupled with high mechanical performance either in an inert atmosphere or vacuum. However, the main drawback of carbon-based materials is low oxidation resistance. Although the high specific strength and specific stiffness, and strength retention at elevated temperatures make C/C composites an attractive structural material, the issue of carbon oxidation limits their use. The development of a series of innovative aerospace technologies such as hypersonic and space shuttles, which are emerging as twenty-first century aerospace and space

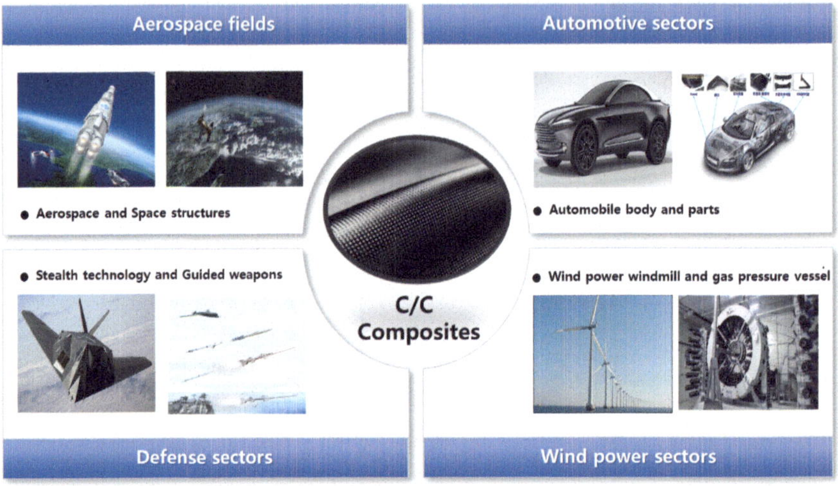

Fig. 8.13 Applications of C/C composites

transportation vehicles, has become possible with the advent of ultralight and ultraheat-resistant C/C composites along with the development of traditional aviation technologies. In the 1990s, carbon fiber-reinforced composite materials, which have been applied in the lightweight aspect of the gas industry since the 1970s, have created new areas as a structural material for aircraft stealth combined with electromagnetic wave absorbing coating materials in addition to these lightweight structural characteristics. Rockets for space flight C/C composites are the only solution now, because they involve a rapid thermal shock. In addition to the application of these structural materials for the space shuttle, it is also applied to heat-resistant structural components of intercontinental ballistic missiles in advanced military powers. Advanced composite materials technologies for aerospace and space are rapidly spreading from sports to leisure, automobile industry, general industrial and civil engineering structures, making it possible to predict the emergence of large-scale composite materials market that will replace the existing metal materials market.

References

1. S.S. Yao, F.L. Jin, K.Y. Rhee, D. Hui, S.J. Park, Compos. Pt. B-Eng. **142**, 241 (2018)
2. H. Deng, K. Li, H. Cui, H. Li, Y. He, J. Zheng, G. Song, Tribol. Int. **121**, 231 (2018)
3. Q. Shen, H. Li, W. Li, Q. Song, J. Alloy. Compd. **738**, 49 (2018)
4. J.W. Shim, S.J. Park, S.K. Ryu, Carbon **391**, 1635 (2001)
5. C. Shao, H.Y. Kim, J. Gong, B. Ding, D.R. Lee, S.J. Park, Mater. Lett. **57**, 1579 (2003)
6. M.K. Seo, S.J. Park, Chem. Phys. Lett. **395**, 44 (2004)
7. S. Qingliang, L. Hejun, Z. Fengling, S. Qiang, F. Qiangang, Corros. Sci. **132**, 204 (2018)

8. G. Mittal, V. Dhand, K.Y. Rhee, S.J. Park, W.R. Lee, J. Ind. Eng. Chem. **21**, 11 (2015)
9. M.K. Seo, J.R. Lee, S.J. Park, Mater. Sci. Eng. A **404**, 79 (2005)
10. S.J. Park, Y.S. Jang, J. Colloid Interf. Sci. **261**, 238 (2003)
11. J.B. Donnet, S.J. Park, H. Balard, Chromatographia **31**, 434 (1991)
12. S.J. Park, B.J. Kim, Mater. Sci. Eng. A **408**, 269 (2005)
13. X. Chao, L. Qi, W. Tian, X. Hou, W. Ma, H. Li, Compos. Pt. B-Eng. **136**, 72 (2018)
14. T. Li, Y. Zhang, J. Sun, J. Ren, P. Zhang, Y. Fu, L. Zhang, J. Am. Ceram. Soc. **101**, 1371 (2018)
15. C. Xu, L. Song, H. Zhu, S. Meng, W. Xie, H. Jin, Compos. Struct. **188**, 7 (2018)
16. Y. Zhang, H. Wang, T. Li, Y. Fu, J. Ren, Ceram. Int. **44**, 3056 (2018)
17. B.J. Kim, Y.S. Lee, S.J. Park, Int. J. Hydrogen Energy **33**, 4112 (2008)
18. S.J. Park, K.D. Kim, Carbon **39**, 1741 (2001)
19. S.J. Park, J.B. Donnet, J. Colloid Interf. Sci. **206**, 29 (1998)
20. K. Li, X. Zhou, G. Shi, S. Farhan, Z. Zhao, Ceram. Int. **44**, 1964 (2018)
21. C. Wang, K. Li, Q. He, C. Huo, X. Shi, J. Alloy. Compd. **741**, 937 (2018)
22. H.B. Zhao, J.B. Cheng, Y.Z. Wang, J. Alloy. Compd. **736**, 71 (2018)
23. B. Zhang, X. Li, Appl. Compos. Mater. **25**, 191 (2018)
24. X. Li, Y. Tang, J. Song, W. Yang, M. Wang, C. Zhu, Y. Lin, Carbon **129**, 236 (2018)
25. L. Feng, K. Li, B. Xue, Q. Fu, L. Zhang, Mater. Des. **113**, 9 (2017)
26. L. Feng, K.Z. Li, B. Xue, Q. Song, X.R. Song, Q.G. Fu, Mater. Lett. **187**, 158 (2017)
27. L. Ma, L. J. He, C. S. Mo, L. B. Zhang, M. S. Pan, Y. F feng Huang, Adv. Compos. Lett. **26**, 101 (2017)
28. X. Zhang, X. Li, G. Yuan, Z. Dong, G. Ma, B. Rand, Carbon **114**, 59 (2017)
29. S.J. Park, M.K. Seo, Carbon **39**, 1229 (2001)
30. P. Chowdhury, H. Sehitoglu, R. Rateick, Carbon **126**, 382 (2018)
31. Y. Jia, H. Li, Q. Fu, Z. Zhao, J. Sun, Corros. Sci. **123**, 40 (2017)
32. C. Wang, K. Li, X. Shi, J. Sun, Q. He, C. Huo, Compos. Pt. B-Eng. **125**, 181 (2017)
33. L. Zhou, J. Huang, H. Ouyang, L. Cao, W. Hao, X. Wang, J. Alloy. Compd. **712**, 288 (2017)
34. X. Jiao, T. Li, Y. Li, Z. Zhang, F. Dai, Z. Feng, Ceram. Int. **43**, 8208 (2017)
35. C. Wang, X. Jin, H. Cheng, C. Hong, X. Zhang, Mater. Des. **131**, 177 (2017)
36. L. Cao, J. Liu, J. Huang, L. Zhou, X. Yong, X. Shen, C. Yao, Ceram. Int. **43**, 16512 (2017)
37. L.Z. Xue, K.Z. Li, Y. Jia, J.J. Ren, S.Y. Zhang, Compos. Struct. **178**, 210 (2017)
38. D.S. Li, H.W. Duan, W. Wang, D.Y. Ge, L. Jiang, Q.Q. Yao, Compos. Struct. **172**, 229 (2017)
39. Y. Jia, H. Li, Q. Fu, J.A. Sun, Surf. Coat. Technol. **309**, 545 (2017)
40. X. Yong, L. Cao, J. Huang, W. Kong, J. Su, C. Li, J. Liu, Surf. Coat. Technol. **311**, 63 (2017)
41. Y. Jia, H. Li, J. Sun, L. Li, Q. Fu, Int. J. Appl. Ceram. Technol. **14**, 331 (2017)
42. K.Z. Li, T. Duan, J.P. Zhang, N.K. Liu, M.Y. Zhang, J. Mater. Sci. Technol. **33**, 71 (2017)
43. M. Cakici, R.R. Kakarla, F. Alonso-Marroquin, Chem. Eng. J. **309**, 151 (2017)
44. S. Wang, F. Zeng, Y. Li, Y. Gu, F. Zhang, X. Xiong, Mater. Res. Bull. **91**, 189 (2017)
45. X. Jin, X. Fan, C. Lu, T. Wang, J. Eur. Ceram. Soc. **38**, 1 (2017)
46. B. Du, C. Hong, Q. Qu, S. Zhou, C. Liu, X. Zhang, Ceram. Int. **43**, 9531 (2017)
47. B. Du, C. Hong, X. Zhang, A. Wang, Y. Sun, Ceram. Int. **44**, 3505 (2018)
48. M. Jafari, J. Mostaghimi, H. Monajatizadeh, M. Rafiei, Surf. Eng. **34**, 220 (2018)
49. L. Wang, Q. Fu, F. Zhao, Intermetallics **94**, 106 (2018)
50. J.Y. Jing, Q.G. Fu, R.M. Yuan, Surf. Eng. **34**, 47 (2018)
51. X.F. Zhang, K.S. Zhou, M. Liu, C.M. Deng, C.G. Deng, S.P. Niu, S.M. Xu, Ceram. Int. **43**, 13075 (2017)
52. X. Xi, H. Xiong, A. Shui, M. Huang, S. Xiao, H.T. Lin, J. Eur. Ceram. Soc. **37**, 5044 (2017)
53. Z. He, P. Lian, Y. Song, Z. Liu, J. Song, J. Zhang, W. Liu, J. Eur. Ceram. Soc. **38**, 453 (2018)
54. F. Zhao, Q. Fu, L. Wang, S. He, Ceram. Int. **44**, 5440 (2017)
55. Y. Jia, H. Li, X. Yao, Z. Zhao, J. Sun, Ceram. Int. **43**, 3601 (2017)
56. S.J. Park, B.J. Kim, J. Colloid Interf. Sci. **275**, 590 (2004)
57. Y. Jia, H. Li, X. Yao, J. Sun, Z. Zhao, J. Eur. Ceram. Soc. **38**, 1046 (2017)

58. D. Chen, H. Gu, A. Huang, Y. Deng, Ceram. Int. **44**, 6089 (2018)
59. B. Xu, Y. An, P. Wang, X. Jin, G. Zhao, Ceram. Int. **43**, 783 (2017)
60. C. Wang, K.Z. Li, Y. Lu, L. Peng, B. Li, Z. Zhao, J. Alloy. Compd. **745**, 569 (2018)
61. C. Xue, H. Zhou, J. Hu, H. Wang, J. Xu, S. Dong, J. Adv. Ceram. **7**, 66 (2018)
62. C. Zhou, P. Wang, Z. Chen, C. Wei, W. Han, Ceram. Int. **44**, 2496 (2018)
63. T. Feng, H.J. Li, M.H. Hu, H.J. Lin, L. Li, Surf. Eng. **34**, 40 (2018)
64. S. Zhou, Y. Qi, P. Wang, Y. Cheng, W. Han, Adv. Appl. Ceram. **116**, 462 (2017)
65. A. Vinci, L. Zoli, E. Landi, D. Sciti, Corros. Sci. **123**, 129 (2017)
66. Y. Jiang, D. Feng, H. Ru, W. Wang, C. Zhang, Surf. Coat. Technol. **339**, 91 (2018)
67. F. Liu, H. Li, S. Gu, X. Yao, Q. Fu, Ceram. Int. **44**, 6619 (2018)
68. S. Ariharan, P. Sengupta, A. Nisar, A. Agnihotri, N. Balaji, S.T. Aruna, K. Balani, J. Therm. Spray Technol. **26**, 417 (2017)
69. S.J. Park, Y.S. Jang, J.W. Shim, S.K. Ryu, J. Colloid Interf. Sci. **260**, 259 (2003)
70. S.J. Park, M.H. Kim, J.R. Lee, S. Choi, J. Colloid Interf. Sci **228**, 287 (2000)
71. S.J. Park, M.S. Cho, Carbon **38**, 1053 (2000)
72. L.Y. Meng, S.J. Park, J. Colloid Interf. Sci. **352**, 498 (2010)
73. J.B. Donnet, S.J. Park, Carbon **29**, 955 (1991)
74. S.J. Park, M.K. Seo, Y.S. Lee, Carbon **41**, 723 (2003)
75. S.K. Ryu, B.J. Park, S.J. Park, J. Colloid Interf. Sci. **215**, 167 (1999)
76. M.J. Jung, J.W. Kim, J.S. Im, S.J. Park, Y.S. Lee, J. Ind. Eng. Chem. **15**, 410 (2009)
77. B. Zou, Y. Hui, W. Huang, S. Zhao, X. Chen, J. Xu, X. Cao, J. Eur. Ceram. Soc. **35**, 2017 (2015)
78. S.J. Park, Y.S. Jang, K.Y. Rhee, J. Colloid Interf. Sci. **245**, 383 (2002)
79. L. Li, H. Li, X. Yin, Y. Chu, X. Chen, Q. Fu, J. Alloy. Compd. **645**, 164 (2015)
80. W. ZhongLiu, X. Peng, L. Zhuan, H. Wen, L. Heng, Y. XiaoYu, C. WenBo, J. Eur. Ceram. Soc. **35**, 3789 (2015)
81. Y. Zhang, Z. Hu, B. Yang, J. Ren, H. Li, Ceram. Int. **41**, 2582 (2015)
82. X. Ren, H. Li, Y. Chu, Q. Fu, K. Li, Int. J. Appl. Ceram. Technol. **12**, 560 (2015)
83. B. Zhang, J. Huang, H. Ouyang, L. Cao, & Li. C. Ceram. Int. **42**, 17932 (2016)
84. D. Wang, Y. Zeng, X. Xiong, G. Li, Z. Chen, W. Sun, Y. Wang, Ceram. Int. **41**, 7677 (2015)
85. S.J. Park, M.S. Cho, J. Mater. Sci. **35**, 3525 (2000)
86. S.J. Park, B.J. Kim, J. Colloid Interf. Sci. **291**, 597 (2005)
87. Y. Chu, H. Li, H. Luo, L. Li, L. Qi, Corros. Sci. **92**, 272 (2015)
88. L. Li, H. Li, Y. Li, X. Yin, Q. Shen, Q. Fu, Appl. Surf. Sci. **349**, 465 (2015)
89. Y.L. Zhang, J. Ren, S. Tian, H. Li, X. Ren, Z. Hu, Corros. Sci. **90**, 554 (2015)
90. T. Feng, H.J. Li, M.H. Hu, H.J. Lin, L. Li, Ceram. Int. **42**, 270 (2016)
91. X. Ren, H. Li, Y. Chu, K. Li, Q. Fu, Ceram. Int. **40**, 7171 (2014)
92. X. Ren, H. Li, Y. Chu, Q. Fu, K. Li, Surf. Coat. Technol. **247**, 61 (2014)

Chapter 9
Novel Carbon Fibers and Their Composites

Abstract Traditionally, the application of carbon fibers has been limited to very special fields such as aerospace and military because of their high cost. However, various techniques for low-cost carbon fibers are under development in terms of using low-cost precursors, low-cost manufacturing processes, and even functional coating methods. Moreover, future applications of carbon fibers can widen not only as structural reinforcements but also in information technology-based applications such as housings for electric devices, smart cloths, and healthcare items. In this chapter, we will cover the carbon fibers and their composites in recent various applications. In particular, these are classified as a low-cost production technique of carbon fibers for general industries, thin carbon fibers for extreme industries, and functional carbon fibers for smart composites.

9.1 Novel Low-Cost Production Technique of Carbon Fibers for General Industries

Carbon fibers are ultralight and show high conductivity and high thermal resistance. It is used as a core material in composites in aerospace, sports, automotive, defense, and semiconductor industries. However, commercial carbon fibers are excessively expensive ($33/kg) to be widely used in the automotive and construction industries. Recently, there have been extensive efforts to produce low-cost (<$11/kg) carbon fibers with excellent properties.

Figure 9.1 shows the price structure of carbon fibers. As shown in Fig. 9.1, the precursors contribute to the largest proportion of cost (51%), while the oxidative stabilization step accounts for 18% of the cost. Therefore, low-cost precursors and oxidative stabilization for low energy consumption are essential (Fig. 9.2).

In this chapter, we explain the use of low-cost precursors for the production of biomass-based low-cost carbon fibers by low-cost manufacturing processes.

Lignin is the most abundant natural polymer and exists in the walls of plant cells as a main ingredient [2–13]. It is readily available and relatively cheap and is structurally appealing as a precursor for carbon fibers. It is a high molecular weight

S.-J. Park, *Carbon Fibers*, Springer Series in Materials Science 210,
https://doi.org/10.1007/978-981-13-0538-2_9

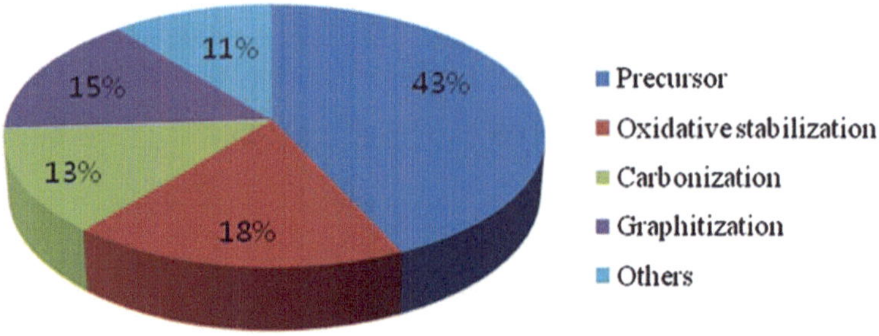

Fig. 9.1 Price structure of polyacrylonitrile (PAN)-based carbon fiber production [1]

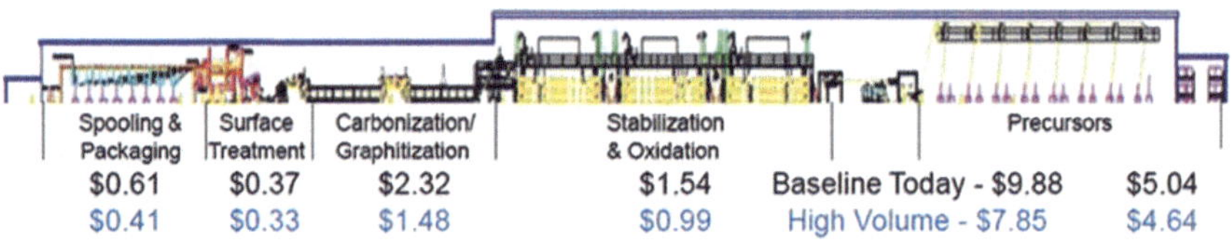

Fig. 9.2 Production cost of PAN-based carbon fibers [1]

polyaromatic macromolecule with a reported total worldwide production of approximately 26 million ton/year [2]. Kadla et al. [3] reported the production of the first lignin-based carbon fibers from commercial kraft lignin using a thermal extrusion process. Fusible lignin was used to form a fine filament of carbon fiber by a thermal pretreatment under vacuum. Figure 9.3 shows the carbon fibers produced form hardwood kraft lignin . The carbon fibers were produced at an overall yield of 45%. The tensile strength and modulus increased with decrease in the fiber diameter, and the values are comparable to those of much smaller-diameter carbon fibers produced from phenolated exploded lignin. The fibers show a promising tensile strength of 400–550 MPa and a modulus of 30–60 GPa, and hence, kraft lignin should be further investigated as a precursor for general-grade carbon fibers (Table 9.1).

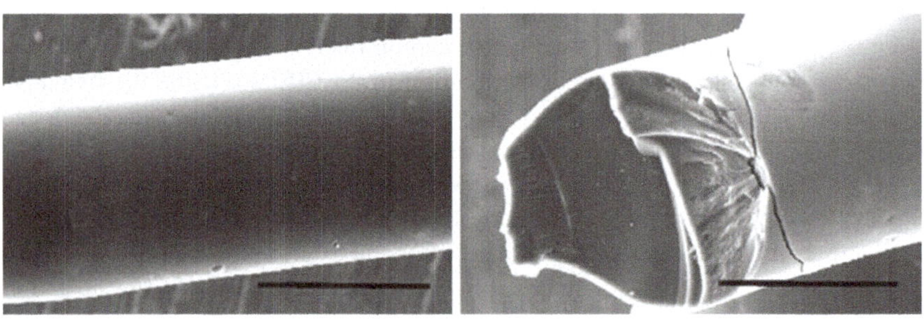

Fig. 9.3 Scanning electron microscopy images of carbon fibers produced from hardwood kraft lignin (×1000 magnification, scale bar = 25 μm) [3]

Table 9.1 Mechanical properties of HWKL and HWKL/PEO carbon fibers [3]

	Diameter (μm)	Tensile strength (MPa)	Modulus (GPa)	Elongation (%)
HWKL	46 ± 8	422 ± 80	40 ± 11	1.12 ± 0.22
HWKL–PEO (100 K) 97-3	34 ± 4	448 ± 70	51 ± 13	0.92 ± 0.21
HWKL–PEO (100 K) 95-5	46 ± 3	396 ± 47	38 ± 5	1.06 ± 0.14
HWKL–PEO (300 K) 95-5	44 ± 4	392 ± 89	44 ± 12	0.91 ± 0.16
HWKL–PEO (600 K) 97-3	33 ± 2	458 ± 97	59 ± 8	0.79 ± 0.21
HWKL–PEO (600 K) 95-3	63 ± 7	339 ± 53	33 ± 2	1.25 ± 0.26

Qin et al. [4] investigated the effect of organoclay on lignin-based carbon fibers. Table 9.2 shows the textural properties of carbon fibers with and without organoclay reinforcement. The tensile strength of pyrolytic lignin-based carbon fibers was improved by organoclay reinforcement at loadings below 1.0 wt%. Increasing the organoclay content above 1.0 wt% resulted in a drop in the tensile strength of carbon fibers, although this was also accompanied by an increase in the fiber diameter, which is known to decrease the tensile strength. The addition of organoclay to pyrolytic lignin can improve the mechanical properties of the lignin-based carbon fibers.

Maradur et al. [5] have made attempts to develop lignin-based carbon fibers made of a copolymer of lignin with acrylonitrile. Using a wet-spinning process, the PAN–lignin copolymers were then spun into fibers. The carbon fibers were prepared by the subsequent thermal treatment of the spun fibers. A variety of physical changes were observed to occur during the thermostabilization process, which cause the initial changes in the color of the spun fibers from pale yellow to dark brown (Fig. 9.4).

The scanning electron microscopy (SEM) images of the as-spun, stabilized, and carbonized fibers obtained at low and high magnifications are shown in Fig. 9.5.

Table 9.2 Textural properties of carbon fibers with and without organoclay reinforcement [4]

Organoclay loading (type)	Diameter (μm)	Young's modulus (GPa)	Tensile strength (MPa)
0 wt% clay	49 ± 2	36 ± 1	370 ± 38
1.0 wt% (cloisite 20A)	45 ± 1	32 ± 1	422 ± 24
2.0 wt% (cloisite 20A)	50 ± 2	29 ± 1	307 ± 25
5.0 wt% (cloisite 20A)	55 ± 2	26 ± 1	240 ± 17
1.0 wt% (cloisite 30B)	47 ± 1	32 ± 1	438 ± 24
2.0 wt% (cloisite 30B)	50 ± 1	30 ± 1	357 ± 29
5.0 wt% (cloisite 30B)	57 ± 2	30 ± 1	255 ± 25

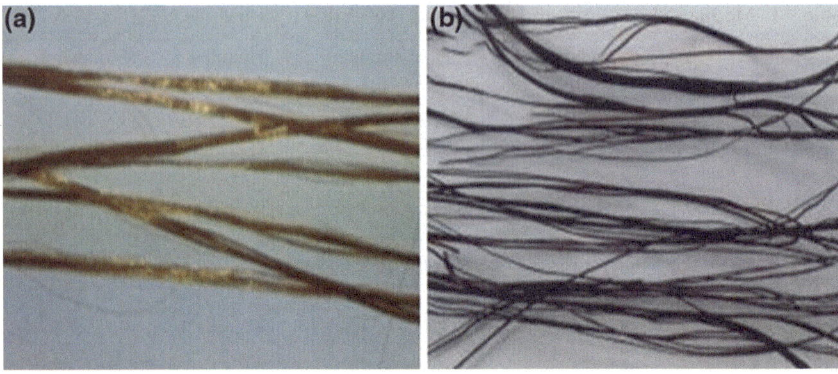

Fig. 9.4 Photographs **a** of as-spun fibers and **b** of stabilized fibers [5]

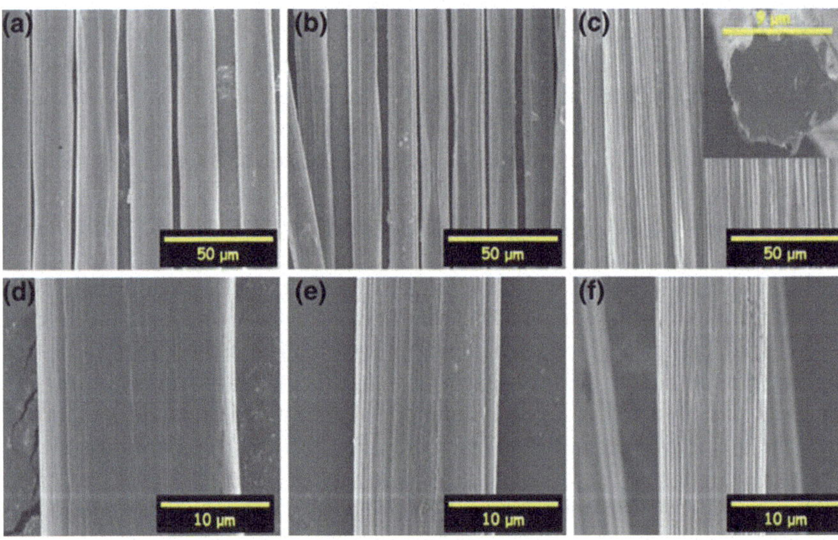

Fig. 9.5 Field emission (FE)—SEM images **a**, **d** of as-spun, **b**, **e** of stabilized, and **c**, **f** of carbonized fibers. Inset shows cross-sectional FE–SEM image [5]

The average diameter of the fibers decreases from 20 to 15 to 11 µm with increase in heat treatment. The drastic volumetric changes involved complex physical and chemical reactions, such as carbon densification and gas evolution. At high magnification, the surface of the carbonized fibers became rougher and developed striped patterns, and the carbonized fibers appeared more porous than the fibers stabilized through the pyrolysis process at high temperature.

An increased awareness that nonrenewable resources are becoming scarce and our dependence on renewable resources has increased. This century could be called the cellulosic century because more and more renewable plant resources for products are being discovered. It has been generally stated that natural fibers are renewable and sustainable. The living plants from which the natural fibers are taken are renewable and sustainable, while the fibers themselves are not renewable. Table 9.3 shows the main fibers used commercially in composites, which are now produced worldwide [6]. Also, the natural fiber types are listed in Table 9.4 [7].

If carbon fibers can be prepared from biomass-based natural fibers, which are abundant in nature and less expensive, it would be desirable and useful, although the fibers are not continuous, limited in fiber length, and show relatively low mechanical properties in comparison with conventional rayon-based and PAN-based carbon fibers.

Cho et al. [8] investigated the effect of alkali pretreatment of jute-based carbon fibers. Figure 9.6 displays the SEM images of the cross sections of jute fibers before and after carbonization. Raw jute fibers consist of a number of cells, typically found in cellulose-based natural fibers. After carbonization, clearly the inner structure (struts between the cells) of the fiber became dense and showed the presence of

Table 9.3 Commercial sources of major fibers [6]

Fiber source	Global production (10^3 ton)
Bamboo	30,000
Jute	2300
Kenaf	970
Flax	830
Sisal	378
Hemp	214
Coir	100
Ramie	100
Abaca	70
Sugarcane bagasse	75,000
Grass	700

Table 9.4 Types of natural fibers [7]

Bast fibers	Flax, hemp, kenaf, jute, mesta, ramie, *Urena*, roselle
Leaf fibers	Pineapple, banana, sisal, screw pine, abaca, curaua, *Agave*, cabuja, henequen, date palm, African palm
Seed fibers	Cotton, kapok
Fruit fibers	Coconut, coir
Wood fibers	Hardwoods, softwoods: numerous types (*10,000 varieties)
Grasses and reeds	Wheat, oat, barley, rice, bamboo, bagasse, reed, corn, rape, rye, esparto, elephant grass, canary grass

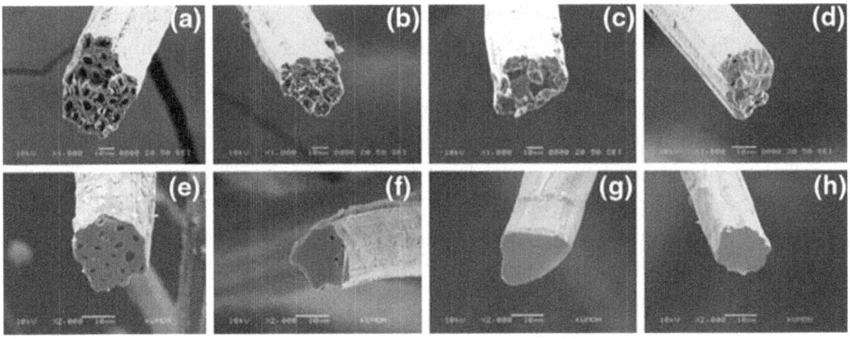

Fig. 9.6 SEM images showing cross sections **a** of raw, **b** of 5 wt% NaOH, **c** of 10 wt% NaOH, and **d** of 15 wt% NaOH-treated jute fibers. Micrographs **e–h** are cross-sectional images of (**a–d**), respectively, after carbonization [8]

smaller cells with different sizes. It seemed that some cells were fused or combined together, indicating that reorganization of the inner structure may have occurred during carbonization.

9.2 Novel Thin Carbon Fibers for Extreme Industries

In Sect. 9.1, the low-cost techniques of production of carbon fibers were overviewed for applications in general industries. In this section, carbon fibers with remarkable physical performances will be focused on for applications in extreme industries. Normally, T-1000-grade carbon fibers from Toray Co. Ltd are recognized as high-performance carbon fibers. However, in this section, super-performance carbon fibers, which can be used in space or deep seas or those that can be applied in parts where normal high-performance carbon fibers cannot be applied, will be discussed. Practically, after the proposal of building a space elevator made of carbon nanotubes, many researchers have been trying to evaluate a variety of materials for their suitableness. However, the limitation of normal carbon fibers with respect to physical strengths for extreme applications has urged many scientists to research the production of super carbon fibers. Among many research topics, high-strength carbon fibers with nanometric diameters have gained much attention.

Carbon fibers with nanometric diameters can be typically divided into two types. The carbon fibers can be manufactured by electrospinning. In this process, polymeric long fibers or nanofiber webs are slit, twisted, oriented, and finally carbonized. The other types of carbon fibers can be produced by wet or dry yarning of carbon nanotubes. The carbon fibers made by these methods cannot be produced at suitable costs and failed to show the physical performance as super carbon fibers, although it may be possible to produce super carbon fibers for application in extreme industries by dry or wet yarning.

9.2.1 Novel Continuous Carbon Nanofibers

9.2.1.1 Submicron Carbon Fibers

The diameter of commercial carbon fibers is in the range 6–12 μm [9]. The fiber diameter determines the specific surface area of the fiber and the fiber–resin interfaces. Normally, mechanical and interfacial properties of composites reinforced with carbon fibers depend on the interfacial adhesion between the fiber surface and the matrices [10]. Therefore, smaller fiber diameter can lead to higher interfacial adhesion because of the increase in the interface of the filler matrix. Hence, several carbon fiber manufacturers specifically focus on submicron carbon fibers.

In submicron carbon fibers, the fiber diameter ranges from several hundred nanometers to three micrometers. These carbon fibers can be produced from very fine polymer fibers obtained by high orientation process. However, it is challenging to apply these thin carbon fibers to general industries because of their high manufacturing cost, low production scales, low carbonization yields, and large fraction of worsted fibers [11–15].

However, various methods (some of which are listed below) have been developed for precursor treatment.

1. Bicomponent spinning

The formation of nanofibers by bicomponent spinning is a two-step process, which involves (1) spinning islands-in-the sea (INS) or segmented pie bicomponent fibers and (2) the removal of one component [16, 17]. INS fibers, first created by Toray Industries, typically consist of a large number of individual "islands" of one polymer surrounded by a "sea" or matrix of another polymer [18–22]. The sea polymer is removed by dissolution with a solvent or by melting to leave ultrafine filaments. The most frequently investigated approach is to spin INS fibers using melt-spinning processes. Figure 9.7 shows nanofibers obtained by bicomponent spinning. Several hundreds of nanofibers are observed [23].

2. Melt-blowing technique

Today, melt blowing [24] is the primary production technique used to obtain small-diameter fibers. The melt-blowing technique developed in the 1950s [25] at the Naval Research Laboratory is a nonwoven manufacturing process, in which a polymer melt is extruded through dies, attenuated by heated high-velocity airstreams (typically at the same temperature as the molten polymer) and spun into fibers. Nonwoven webs composed of 500 nm fibers were produced in a melt-blowing process using different combinations of primary plates and secondary plates in a modular die plate assembly. Polypropylene nanofibers with an average diameter of 250 nm were melt-blown using a special die. Melt blowing may be used to produce INS fibers containing more than 600 island fibrils with diameters as low as 50 nm [26].

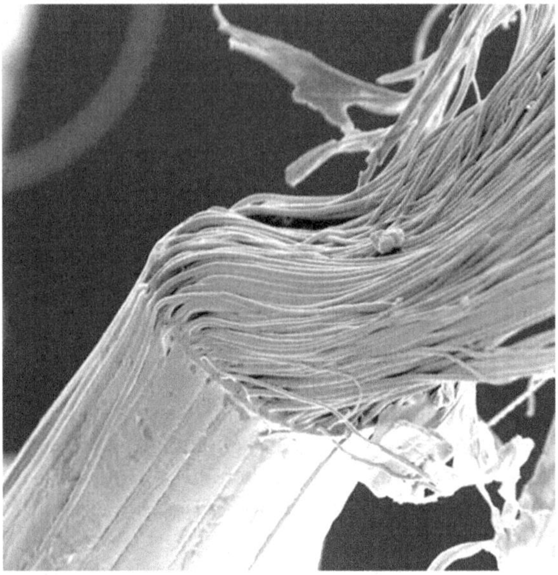

Fig. 9.7 Nanofibers obtained by bicomponent spinning [23]

3. Flash-spinning technique

Flash spinning is a modified version of the spunbonded technology in which a polymer such as polypropylene is dissolved and extruded and the solvent is made to rapidly evaporate at the spinneret, which causes the individual filaments to be disrupted into a highly fibrillar form, which collected on a moving screen to form a web. As a further development to the melt-blowing process, flash spinning has been used to produce plexifilamentary film-fibril strands consisting of ultrafine fibers from a spin fluid. Flash-spinning processes [27] developed to date do not produce fibrous webs composed completely of nanofibers, though a polyolefin nonwoven fibrous structure comprising both micrometer fibers and submicrometer fibers is available.

4. Electrospinning

Electrospinning [28–32] has received much attention as a technique to produce various ultrafine polymeric fibers, which are otherwise difficult to fabricate by other methods. It has been demonstrated to be a simple and viable method to produce fibers with diameters in the range of nanometers to submicrometers from polymers, ceramics, composite materials , and metals in the form of solution or melt [30].

9.2.1.2 Long Carbon Fibers by Electrospinning

Several types of methods such as vapor-grown method, laser ablation, arc plasma, and electrospinning followed by carbonization have been developed for the preparation of carbon nanofibers [33–35]. In addition to these, the electrospinning method can produce polymeric nanofibers of various sizes ranging from ten nanometers to a few micrometers. Moreover, essential equipment is less expensive than that used in other methods. In electrospinning, the applied electric field and solution conductivity are important parameters, which influence the fiber diameter during spinning in addition to parameters such as the surface tension of solution, jet length, solution viscosity, surrounding temperature and humidity, solution feeding rate, collector materials, and collection speed. Figure 9.8 shows the electrospinning apparatus and mechanism [36].

Various polymers such as polyacrylonitrile (PAN) [37, 38], polyvinyl alcohol (PVA) [39, 40], and polyimide (PI) [41, 42] can be electrospun when the viscosity is controlled by heat or solvent. To make continuous carbon fibers, electrospun polymeric nanofibers are carbonized at temperatures exceeding 1000 °C. Continuous carbon nanofiber is a long thin strand of material with a diameter of approximately 5–200 nm, composed mostly of carbon atoms bonded together in microscopic crystals and aligned parallel to the long axis of the fiber. Among various polymers, PAN is a predominant candidate to obtain carbon materials because of possibility of obtaining a high carbon yield after the carbonization process.

The preparation of continuous carbon fibers from PAN nanofibers involves three main stages, i.e., electrospinning, oxidative stabilization, and carbonization [43]. Suitable stabilization of the electrospun PAN nanofibers is essential for high carbon yields and good physical strength. Stabilization involves heating PAN nanofibers in an air atmosphere. During the stabilization process, cyano groups (–C≡N) undergo cyclization and N-containing benzene-like structures are formed (as shown in Fig. 9.9). This is why PAN is used as a carbon precursor [44]. This process prevents melting during subsequent carbonization, resulting in high carbon yield and good fiber shapes. In the next step, stabilized nanofibers are carbonized in an inert atmosphere to increase the carbon content by removing noncarbonized components in the form of gases such as H_2O, NH_3, CO, HCN, CO_2, and N_2 with a yield of approximately 50–60% [45, 46].

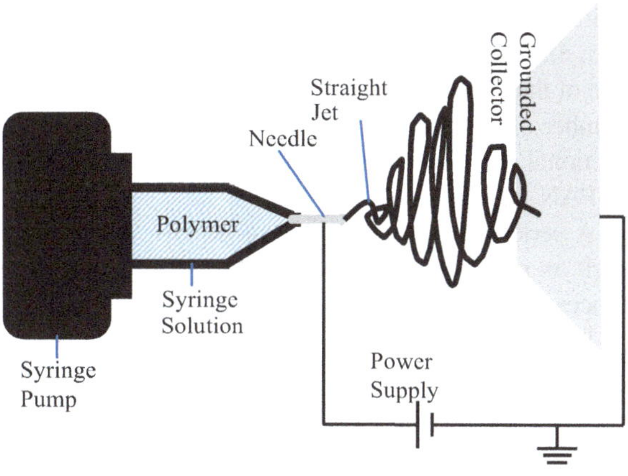

Fig. 9.8 Schematic of electrospinning apparatus and mechanism [36]

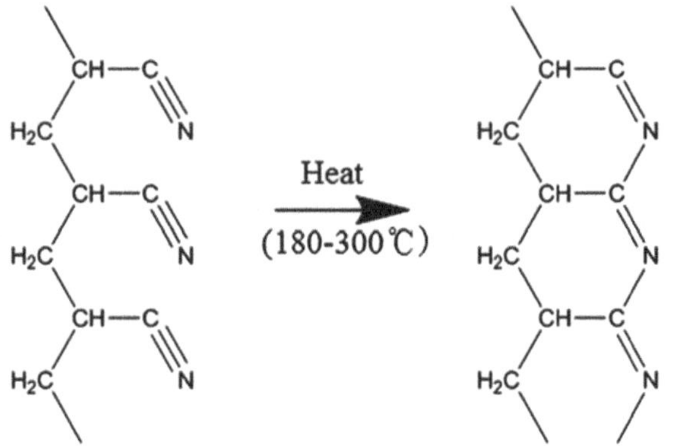

Fig. 9.9 Schematic diagram of cyclization reaction of PAN [44]

Fibers and yarns have a long history, although remarkable progress has been achieved in the technology in the last few decades. The applications of fibers and yarns have move beyond traditional textiles, and the materials are increasingly used in high value-added applications, as in composites, filtration media, gas separation, sensors, and biomedical engineering. However, according to process requirements, single nanofibers cannot be used directly in weaving, knitting, and embroidery partly because of their fragile nature. It is more convenient to handle linear nanofiber assemblies such as nanofiber bundles or yarns instead of a single nanofiber [47].

Though electrospun carbon nanofibers show many merits for easy commercialization, several bottlenecks still remain. Among others, electrospun nanofibers are in weblike form and a postprocess is required for the production of continuous carbon nanofibers. In the traditional textile engineering, a nonwoven web can be transformed into continuous fibers by the processing steps consisting of the orientation of the web, slitting of the web, and then twisting of the slit web. Yarn twist reduces the strength variability by increasing the interfiber lateral interaction and friction [48]. An examination of the mechanical properties of dried and twisted yarns of electrospun PAN nanofibers has shown that the initial modulus and the ultimate strength increase with increase in the twist angle. Properties, including yarn diameter, fiber count, and twist per unit length, can be adjusted by the rotation rate of the collector disk. Twist also lowers the strength, which can be attributed to a number of factors, including the increased fiber angle relative to the yarn axis. The frictional and normal forces arising from twist may damage or roughen the surface of PAN nanofiber in yarns.

A series processing steps can be also accomplished with other types of fibers such as polymer and ceramic fibers acting as core fibers. Figure 9.10 shows the process of transforming electrospun nanofiber webs into continuous nanofibers.

In order to avoid complex transform manufacturing , the development of a direct long nanofiber electrospinning has been researched. These processes are basically based on the collecting techniques, which result in highly oriented continuous nanofibers. Figure 9.11 shows a modified centrifugal electrospinning device [49].

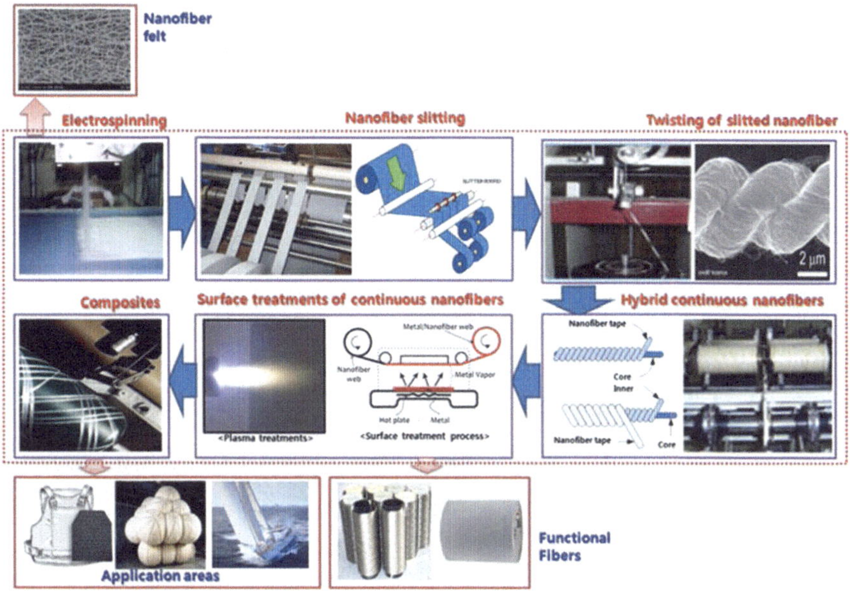

Fig. 9.10 Schematic of continuous nanofiber manufacturing process by electrospun felt

Fig. 9.11 Schematic of modified centrifugal electrospinning [49]

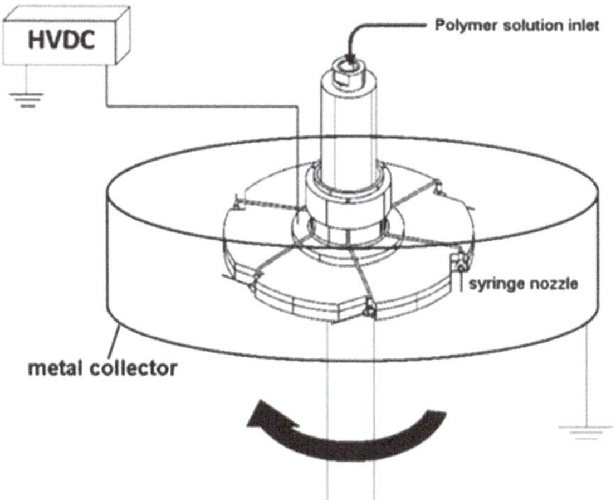

This electrospinning device is composed of a high-speed rotatable polymer feeder and a round metal collector. The conventional electrospinning device can produce nonoriented polymer nanofiber webs. However, this device can produce unidirectionally oriented nanofiber webs (or bundle) because of the high rotation speed of the polymer feeder during the spinning process.

Figure 9.12 shows the SEM and transmission electron microscopy (TEM) images of the nanofibers electrospun by the modified centrifugal electrospinning device. The nanofibers and CNTs in the fibers were observed to be well aligned along the axis of the electrospun nanofibers. Because of the elongation of the fluid jet, the nanotubes were oriented along the streamlines of the electrospinning solution [49].

Figure 9.13a illustrates the setup used for the electrospinning experiments. The configuration is essentially the same as that used in a conventional setup except for the use of a collector containing a gap in its middle. Such a collector could be simply fabricated by putting two stripes of electrical conductors together or by cutting a piece of aluminum foil. The width of the gap could be varied from hundreds of micrometers to several centimeters. Figure 9.13b shows a cross-sectional view of the electric field strength vectors between the needle and the grounded collector

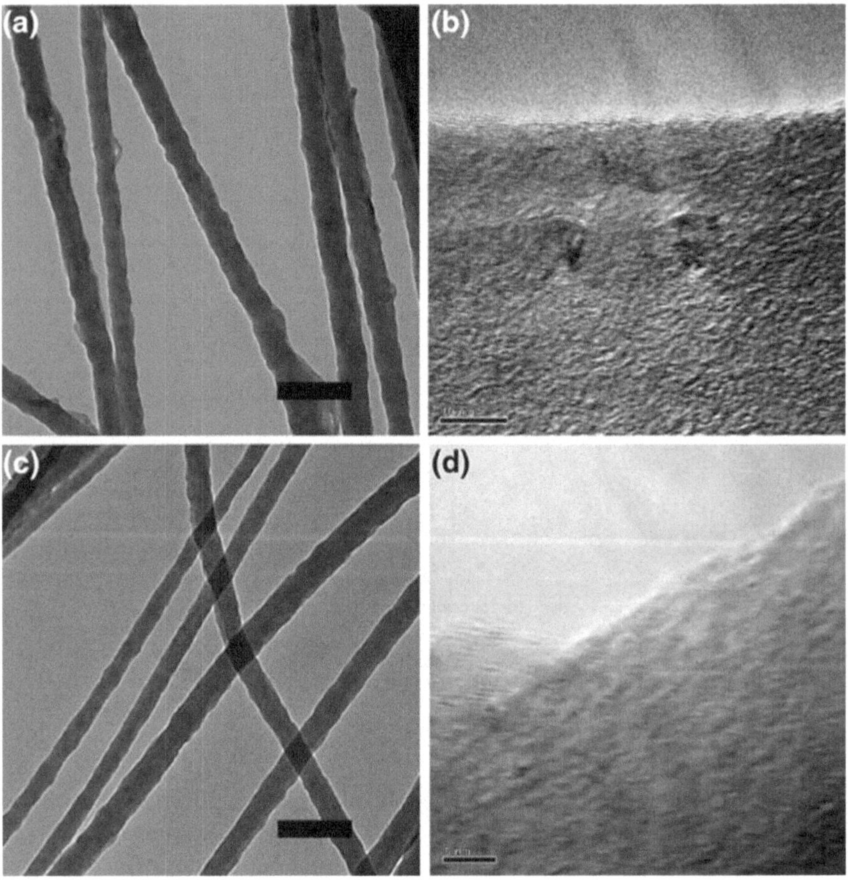

Fig. 9.12 TEM images **a** of CNT/PAN nanofibers with 2.0 wt% CNTs (scale bar = 200 nm), **b** of high-resolution image of CNT/carbon nanofibers (scale bar = 10 nm), **c** of PAN nanofibers (scale bar = 200 nm), and **d** of high-resolution image of carbon nanofibers (scale bar = 5 nm) [49]

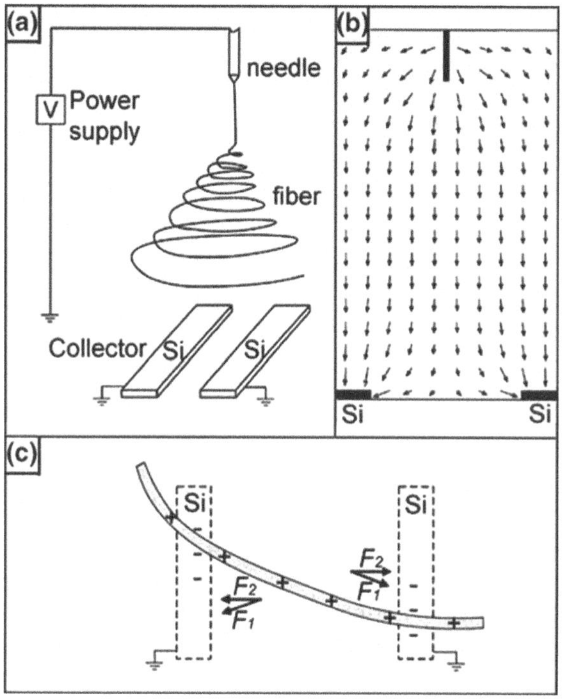

Fig. 9.13 **a** Schematic of setup used for electrospinning uniaxially aligned nanofibers. **b** Calculated electric field strength vectors in the region between the needle and the collector. **c** Electrostatic force analysis of a charged nanofiber spanning across the gap. Electrostatic force (*F1*) resulted from the electric field and the Coulomb interactions (*F2*) between the positive charges on the nanofiber and the negative image charges on two grounded electrodes [50]

calculated using the Student's Quick Field program. Unlike the conventional system, the electric field lines in the vicinity of the collector were split into two fractions pointing toward the opposite edges of the gap. Figure 9.13c illustrates the electrostatic forces acting on a charged nanofiber spanning the gap. The as-spun fiber can be considered to be a string of positively charged elements connected through a viscoelastic medium [50].

In general, the charged nanofiber should experience two sets of electrostatic forces, the first set (F1) originating from the splitting electric field and the second one between the charged fiber and image charges induced on the surfaces of the two grounded electrodes (F2).

Meanwhile, two separate collectors can produce uniaxially aligned polymer nanofibers. Figure 9.14a shows a typical optical micrograph of polyvinylpyrrolidone (PVP) nanofibers electrospun from a solution containing 0.5 g of PVP and 4.2 mg of tetramethylammonium chloride in a mixture of 8 mL of ethanol and 4 mL of water and collected over a gap formed between two stripes of silicon substrates. Figure 9.14b shows the SEM images of the same sample, which displays nanofibers

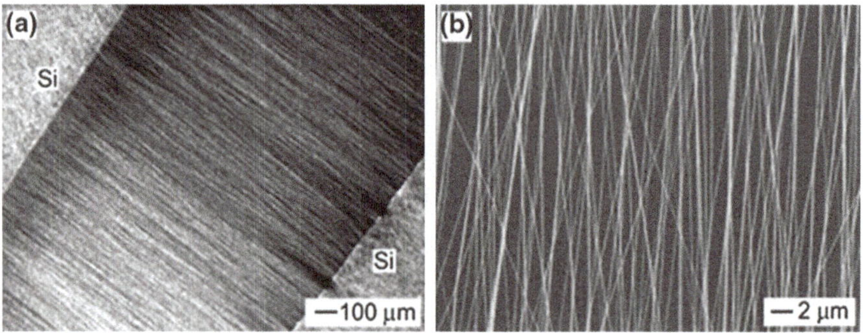

Fig. 9.14 Images showing orientation of PVP nanofibers on a collector, containing a gap in its middle. **a** *Dark field* optical micrograph of PVP nanofibers collected on top of a gap formed between two silicon stripes [50]

deposited across the gap and on top of the silicon stripe. These images clearly indicate that the polymer nanofibers were uniaxially aligned across the gap with their longitudinal axes oriented perpendicular to the edges of the gap [50].

Figure 9.15 shows a rotatable net cylindrical collector. This collector combines each merit of other collectors mentioned above. This collector can lead to uniaxially aligned and highly oriented nanofibers [51].

As shown in the SEM images (Fig. 9.16), good alignment could be achieved up to approximately 15 min of electrospinning. After 15 min, the mat of fibers begins to thicken and entanglements form. However, decent alignment could still be achieved present up to 40 min, as shown in Fig. 9.16g–h. After longer periods of time, the fibers spin in random patterns and the alignment is gradually lost as the thickness of the fiber mat increases, as shown in Fig. 9.16i after 2.5 h of electrospinning. Figure 9.16a–c show that excellent alignment is observed after only five minutes of electrospinning. Figure 9.16g shows an example of how the fibers can entangle or be disturbed if not carefully collected, vastly changing the properties of the aligned fibers [51].

Fig. 9.15 Plexiglas disk with copper wires. Electrospun nylon nanofibers are collected on copper wires. Nanofiber mat shows stratified layering in the magnified image [51]

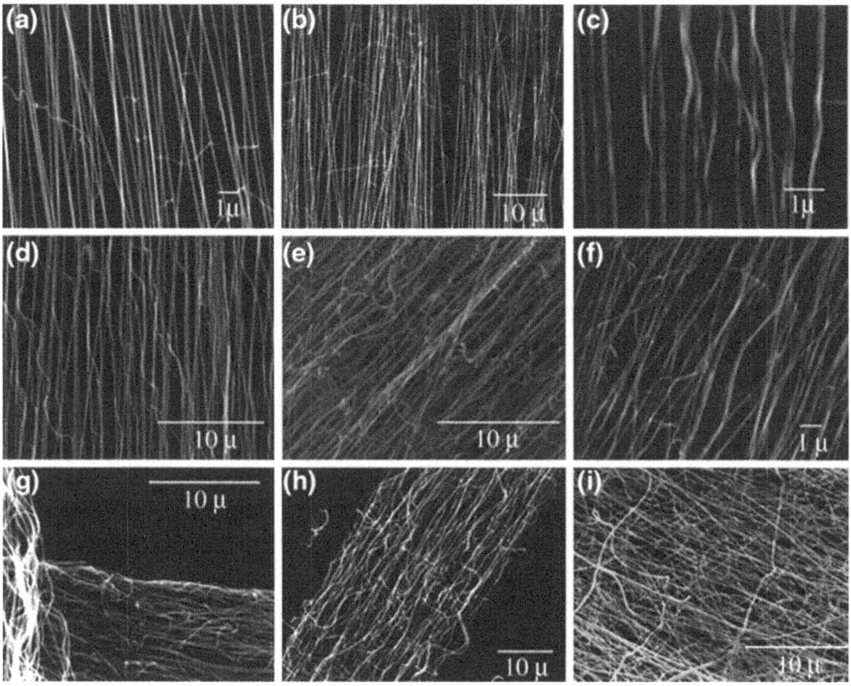

Fig. 9.16 SEM images of axially aligned polymer nanofibers collected on conductive copper wire drum. Images (**a**, **b**, **c**) show the alignment after 5 min of spinning. Images (**d**, **e**) show the alignment after 15 min. Images (**f**, **g**, **h**) show the alignment after 40 min and image (**i**) after 2.5 h of spinning [51]

Recently, novel electrospinning processes have been researched. However, the mechanical performances of continuous carbon nanofibers obtained from electrospun continuous nanofibers are insufficient for these materials to be used as fillers in composites. Many problems such as full orientation, advanced twisting, or hybridization with other fibers still remain.

Mass production of continuous electrospun nanofibers is now possible, and hence, obtaining super carbon nanofibers by electrospinning techniques with very high mechanical properties and optimized carbonization process is foreseeable.

The investigation of fabrics of nanofiber yarns is very much in its infancy, given that nanofiber yarns have not yet been produced on a commercial scale. Nanofiber woven fabrics can be used in artificial leather, filters, diapers, sanitary pads, sutures, antisetting agents, wiping cloths, artificial vessels, bone-fixing devices, and so on. Nanofiber fabrics with simple structures have been woven in recent years for a new generation of textiles such as scaffold fabrics, electronic fabrics, and responsive fabrics.

Nanofiber yarns with excellent mechanical properties have at least sufficient strength to allow them to be adopted by the mainstream textile industry. It has been generally recognized that both the inner structures of the fibers and the alignment between fibers play a key role in the mechanical properties of yarns.

9.2.2 Novel Carbon Nanotube (CNT) Yarns

In addition, an aspiring manufacturing method for super carbon fibers is using CNTs to form long fiber yarns. The first generation of CNT yarns was produced by wet-spinning process using polymer binder and CNT fillers. After the development of vertically grown CNTs, the CNTs were processed into fibers by the dry method, which is very much similar to pulling cotton yarns. Finally, the dry CNT yarning technique has evolved such that the CNT synthesis and yarning are concerted. The mechanical properties of CNTs vary depending on the arrangement of hexagonal rings constituting them and the number of walls, and the properties of individual CNTs are superior to commercialized high-performance fibers [52–60].

9.2.2.1 Carbon Nanofibers from CNT Yarns by Wet Methods

The first critical challenge in the development of this method is the difficulty of processing CNTs in liquid state. CNTs are inert in pristine state and tend to bundle together because of the strong van der Waals interactions, making them difficult to disperse uniformly in aqueous or any organic solvents. Some special methods involving oxidation and grafting with different functionalities [61–65] have been used to overcome this issue; however, these methods normally destroy the intrinsic structure and properties of the CNTs. Thus, these methods are unfavorable for fiber spinning. Shaffer and Windle [66] have previously suggested that CNTs can be viewed analogously with high-aspect-ratio rigid-rod polymers. According to this analog, CNTs can be used in two types of solution-based spinning methods, i.e., coagulation spinning and liquid crystal solution spinning. Figure 9.17 shows how CNT yarns can be produced by using the wet method.

Generally, the principle of the coagulation spinning used for synthetic fiber processing involves the extrusion of a polymer solution through a thin capillary tube and injection of the extruded material into a bath containing a second liquid, which removes the solvent, leaving behind the polymer, which condenses and forms a fiber because of phase separation. To employing the coagulation spinning method for CNT fiber fabrication, the CNTs need to be dispersed into a liquid solution at an almost molecular level so that they could be manipulated and aligned well. Surfactants are widely used for this purpose because of their ability to absorb/wrap at the surface of individual CNTs and prevent them from rebundling. This spinning approach was initially adopted by Vigolo et al. [67]. Their processes consisted of dispersing the CNTs in solution and then recondensing them in a stream of another solution, which served as the coagulant [67].

SWNTs produced by arc discharge were first dispersed in an aqueous solution using sodium dodecyl sulfate (SDS) as surfactant. The dispersion was then injected into a rotating bath of aqueous polyvinyl alcohol (PVA) solution, which served as the coagulant. During this process, PVA displaced the surfactant, causing CNTs to collapse and form ribbon-like elastomeric gel fibers. These fragile fibers were pulled

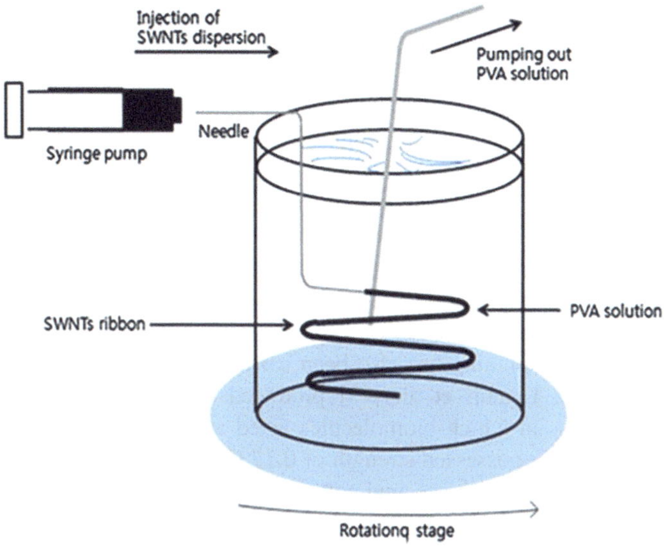

Fig. 9.17 Schematic of experimental setups used to make CNT yarns by wet method [67]

from the coagulation bath at a rate of approximately 0.01 m/min to form solid fibers. These fibers were washed by immersing in a water container to remove excess PVA and surfactant residues and were dried by pulling them out of water bath. The amount of SDS was one critical parameter determining the making of a good dispersion of CNTs . When the concentration of SDS was excessively low, large and dense clusters of CNTs remained in solution even after sonication, indicating that the amount of surfactant was excessively low to produce an efficient coating and induce sufficient electrostatic repulsion, which could counterbalance van der Waals attractions. On the other hand, when the concentration of SDS was excessively high, the osmotic pressure of the excess micelles caused depletion-induced aggregation. An optically homogeneous solution could be formed with 0.35 wt% of CNTs of a particular diameter and length and 1 wt% of SDS. In addition, flow-induced alignment could lead to the preferential orientation of CNTs in fibers [67–69]. The coagulant is required to flow faster than the gel fiber to stretch the fiber along the axial direction and promote alignment of the CNTs in the fiber. This could be accomplished by rotating the coagulant container. This coagulation-based fiber spinning technique is exciting because of its simplicity and ability to produce fibers with very high CNT loadings (60 wt%). The final CNT/PVA composite fibers exhibited a tensile strength in the order of 0.1 GPa, and the Young's modulus varied between 9 and 15 GPa. In contrast to most ordinary carbon fibers, the CNT fibers could be heavily bent and even tightly tied without breaking. The main challenges existing in this method include the dispersion of SWNTs at high concentrations, low processing rate, and the poor fiber performance. To improve the mechanical performance of the as-spun CNT fibers, various modified methods have been developed. By drying CNT fibers under load, improved mechanical properties were obtained and a tensile strength of 230 MPa and a Young's modulus of 45 GPa was achieved [70, 71].

By hot drawing the fibers, Miaudet et al. [71] drew such PVA fibers at elevated temperatures and the fibers yielded strength of 1.8 GPa, modulus of 45 GPa, and toughness of 55 J/g at 11% strain. Dalton et al. [72] further advanced the spinning apparatus to spin fibers continuously by injecting CNT dispersions into a cylinder with the coagulant flowing in the same direction. They were able to spin a reel of CNT gel fibers and then converted it into 100 m of solid CNT composite fibers at a rate of more than 0.70 m/min. The final fibers exhibited an increased mechanical performance with a tensile strength of 1.8 GPa and a Young's modulus of up to 80 GPa. Coagulation spinning has also been carried out with solutions other than PVA. For example, Lynam et al. [73] produced CNT biofibers based on a wet-spinning process in which biomolecules acted as both the dispersant and the coagulant. These fibers possessed strength of 0.17 GPa and modulus of 0.146 GPa.

Because the existence of the second component polymer can be expected to add complicity of processing and this second component is usually an insulator, which will compromise the conductive property of the as-spun fibers, pure CNT fibers are favorable in some circumstances. Kozlov et al. [74] developed a polymer-free solution spinning method. Pure CNT fibers could be produced from CNT/surfactant/water solutions. However, the mechanical properties of the as-spun fibers with specific strength of 65 MPa/g cm^3, specific modulus of 12 GPa/g cm^3, and electrical conductivity of 140 S/cm were not impressive .

9.2.2.2 Carbon Nanofibers from CNT Yarns by Dry Methods

To eliminate the challenges encountered in making dispersions in the case of solution-based spinning methods, spinning CNT fibers directly from as-grown CNT materials seems to be a more convenient way. A breakthrough was made by Jiang et al. [75] in 2002 when a neat CNT yarn could be simply drawn from a vertically superaligned CNT array. The CNTs could be self-assembled into yarns of up to 30 cm in length. Subsequently, Zhang et al. [76] produced highly orientated free-standing CNT transparent sheets using a similar method and further assembled CNTs into fibers by using a draw-twisting spin method [77]. The typical SEM images of the fiber spinning processes are shown in Fig. 9.6. Many applications of these fibers were proposed and demonstrated [78–82], and different postspinning methods were developed to improve their performances. Jiang et al. [75] found that the strength and conductivity of their yarns could be improved after heating at high temperatures.

By introducing twist during spinning, multiple torque-stabilized yarns could be obtained by Zhang et al. [77] who achieved yarn strengths greater than 460 MPa. They emphasized that the load could be transferred effectively between CNTs because of the twisting. In a twisted fiber, individual CNTs are inclined at an angle α with respect to the fiber axis, generating transverse forces, which lock the fibers together as a coherent structure. They also found that these twisted yarns deformed hysterically over large strain ranges from 0.5 to 8%, providing up to 48% energy damping. These yarns could also retain their strength and flexibility even after being heated in air at 450 C for an hour or being immersed in liquid nitrogen. In addition to using postspinning methods, Zhang et al. [83–85] observed that mechanical properties could be significantly improved by using longer CNT arrays. The tensile strength and stiffness of their fibers spun from a 1-mm-long CNT array were measured to be in the range 1.35–3.3 and 100–263 GPa, respectively. These values are many times higher per weight than the values shown by the best existing engineering fibers and CNT fibers reported previously. It is clear that the strength of the CNT fibers increased with increase in the CNT array length, which yields a much larger friction between CNTs. In the case of longer CNTs, there are fewer mechanical defects (like the ends of CNTs) per unit fiber length [86–89]. Other factors such as structure, purity, density, alignment, and straightness of CNTs [90] have all been investigated. For instance, to obtain densely packed CNT fibers, surface tension-driven densification [91, 92] was employed during fiber spinning. Zhang et al. found that after the CNT yarn was pulled through droplets of ethanol, the several centimeter-wide yarn shrank into a tight fiber, typically 20–30 μm in diameter, and the strength of the CNT yarn dramatically improved. The CNT alignment was found to be crucial for fiber properties [93, 94] and could be measured by Raman and X-ray diffraction (XRD) [95, 96]. Zheng et al. [97] observed a strong correlation between the array morphologies (the straightness of CNTs) and the fiber properties and well-aligned arrays yielded high performance, while wavy arrays led to poor performance.

Because CNTs in a yarn are nearly parallel aligned, the CNT yarn is intrinsically an anisotropic material and has a special axis along the drawing direction, which demonstrates many fascinating properties and applications. However, some key issues need to be solved in advance to realize their practical applications. Currently, the growth of CNT arrays is easy, but not all CNT arrays could be spun into yarns or fibers. Zhang et al. found that strong van der Waals interactions exist between individual CNTs within superaligned arrays, and this van der Waals force makes the CNTs join end to end, resulting in the formation of a continuous yarn during pulling. Meanwhile, Zhang et al. claimed that the formation of yarn was due to the disordered regions at the top and bottom of the CNT arrays, which entangled together to form a loop. Further investigation is required to understand the underlying spinning mechanism [98].

9.2.3 Novel Graphene Fibers

CNT and graphene are one- and two-dimensional independent carbon crystals, respectively, with long-range order up to the millimeter scale. Because of the intrinsic covalent connection of sp2-hybridized carbon atoms, these two carbon crystals have extremely high mechanical strength, up to 130 GPa tensile strength and 1.1 TPa elastic modulus in the case of graphene, together with the valuable flexibility to accommodate deformation and transcendental conductivity for electrical current and thermal flow [99–101]. Both CNT and graphene are possibly perfect building blocks for new carbonaceous fibers. By contrast to the regeneration of the graphitic structure in the case of conventional CFs, the fabrication of new CNT and graphene fibers should rely on the "bottom-up assembly" concept to fashion individual building blocks into an ordered state with atomic precision (Fig. 9.18). More importantly, the superior properties of CNT and graphene create high expectations for the combined performances of their fibers to catch up and even exceed those of conventional CFs.

CNT fibers can be fabricated by two methods: wet spinning of CNT fluids or dry spinning of CNT forests/cotton. Dry-spun CNT fibers generally have superior properties to wet-spun ones. To date, considerable efforts have yielded materials with combined properties comparable to those of CFs [102–104]. More impressive are their rich functionalities, which already exceed conventional CFs. These advances have been summarized in many reviews that readers can refer to [105], so it is unnecessary to repeat them further here.

Since its initial development in 2011, graphene fiber (GF) has become another promising carbonaceous fiber [106]. In this chapter, we mainly focus on newly

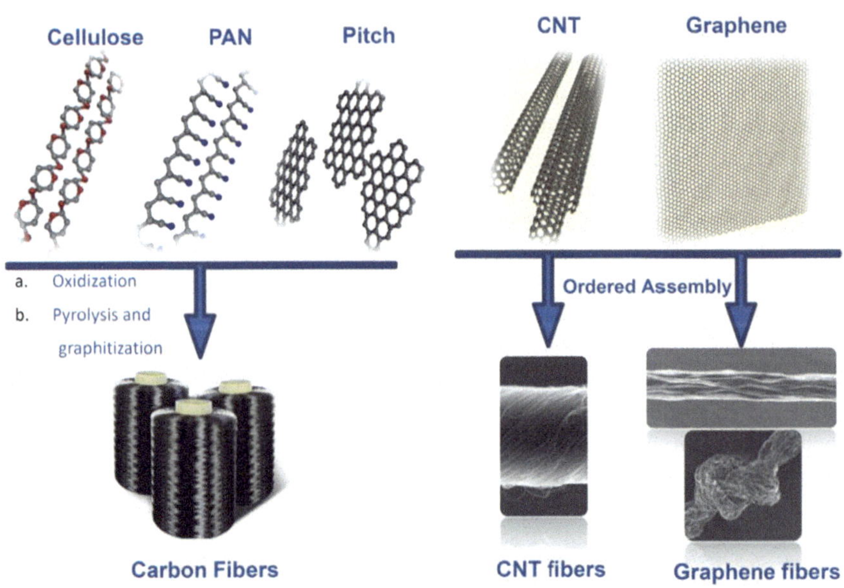

Fig. 9.18 Precursor system of conventional carbon fibers and the new species including CNT and graphene fibers. Oxidization cross-linking and pyrolysis (and/or graphitization) are employed to fabricate conventional CFs from cellulose, PAN, and mesophase pitch precursors. The polycrystalline graphitic structure of CFs is generated from these precursor macromolecules. By contrast, CNT fibers and graphene fibers are fabricated from individual carbon nanotubes and graphene sheets, by a well-controlled macroscopic assembly method [99]

sprouted GFs, describing their concept, strategies of fabrication, structural attributes, mechanical properties and functionalities. A tentative prospect in the vicinity of GFs is also presented.

As discussed above, the outstanding performances of conventional CFs are primarily because of the interior graphite crystallites consisting of turbostratically interconnected graphene. In their established structural model, the size (width La and thickness Lc) of graphite crystallites together with their alignment and interaction plays a key role in their performance (Fig. 9.19). However, further improvement in their combined performance is confined by their inherent polycrystalline nature of limited size and faulty alignment. Graphene sheets are actually 2D carbon monocrystals with a lateral size extending from dozens of microns, even to submillimeters, far larger than the size (smaller than 20 nm) of the graphene substructure in the turbostratic graphite of CFs. Principally, the scaling of graphite crystallite size allows enhancement of the combined performance of fibers. Provided that graphene is engineered into macroscopic fibers, the extremely long-range covalent crystalline nature is able to eliminate grain boundaries in the lateral direction to the utmost extent, which may bring about a breakthrough in the limitations of CFs.

9.2.3.1 Strategies to Fabricate GFs from Graphite

The building block of 2D topological graphene has extreme asymmetry in its structure and properties, and its attractive mechanical and electronic attributes are in the planar direction (note that, although we acknowledge the existence of graphene ribbons and quantum dots, in this review, graphene means graphene sheets unless specifically stated otherwise). Making GFs from graphene building blocks is actually a process of regularly aligning graphene sheets continuously in a uniaxial

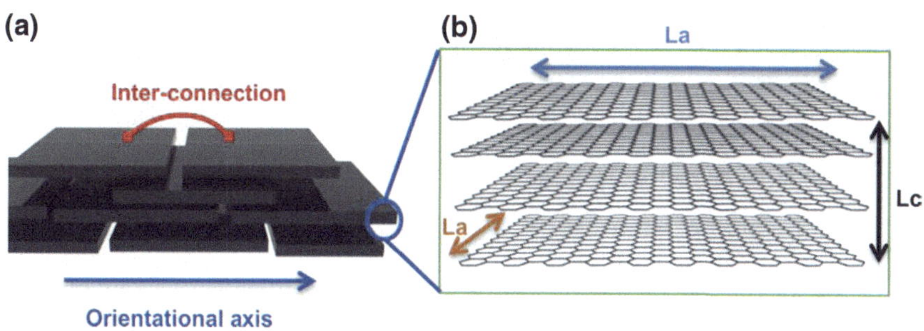

Fig. 9.19 Lamellar crystalline structural model of graphitic materials. The whole structure encompasses a two-level hierarchy: **a** lamellar graphite crystal platelets at the nano or sub-microscale. The orientation degree and interconnection among these crystals platelets are critical to the performance of macroscopic materials, including fibers, films, and bulks; **b** lamellar graphene sheets consist of graphite crystals at the atomic scale, with characteristic parameters of lateral width (La) and perpendicular thickness (Lc) [102]

direction along which the asymmetrical merits of graphene are translated into macroscopic properties [107]. Because of the unmeltable nature of graphene, it is inconceivable to process graphene into neat fibers by the melt-spinning processes for some polymeric fibers such as nylon. In addition, the lack of entanglement of 2D graphene also excludes the throwing method employed in ancient spinning for cotton threads and dry spinning used to make CNT fibers from CNT aligned forests or cottons. Thus, the production of GF must utilize the strategy of solution assembly, which resembles the wet spinning of polymeric fibers such as Kevlar and CNT fibers.

To fabricate GF from graphite, the "graphene LC-wet-spinning" system has been established in recent years and is becoming a reliable method to obtain continuous GFs as monofilaments, tows, and yarns (Fig. 9.20). A series of GFs with high performances and multifunctionalities has been developed. Importantly, many attempts have been made to understand the basic relationship between structure and properties. We now summarize these advances in detail and describe the deduction of a perspective for these new carbonaceous fibers.

9.2.3.2 Wet-Spinning Process for GFs

The transition from LCs to fibers is a familiar concept in polymer science and technology. Even pitch-based CFs are melt-spun from pitch mesophases. In this context, the discovery of graphene-based LCs points to the possibility of fabricating continuous GFs by the wet-spinning method. Our first attempt in 2011

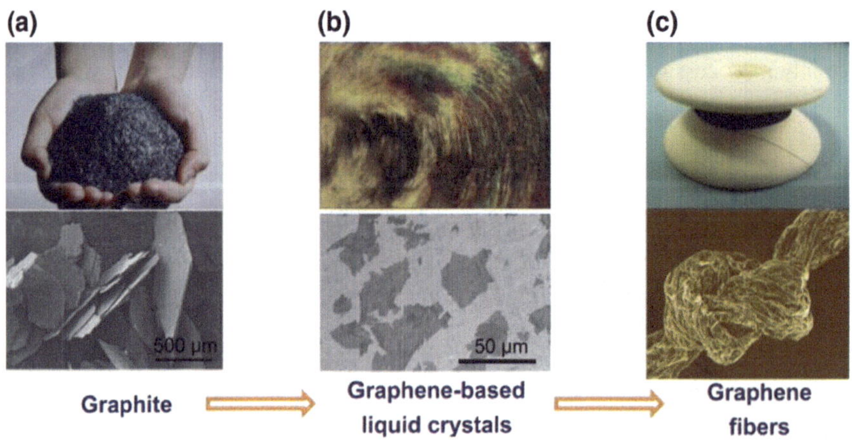

Fig. 9.20 The road from graphite to GFs. In the first step, graphite crystals (**a**) are exfoliated to individual graphene sheets, usually by chemical modification. The modified graphene sheets form liquid crystals in solvents with orientational or positional order (**b**). In the second step, wet-spinning assembly is employed to make continuous graphene fibers (**c**) from these graphene-based liquid crystals, which transform orders from the fluid state to orders in the solid state [103]

demonstrated the practicability of wet-spinning GF from GO LCs [8]. The experiment outlined the following basic processes in obtaining wet-spun GFs: (1) extruding LCs into coagulation baths with uniaxial flow; (2) phase transition by coagulation with solvents exchange; (3) solidification by evaporating solvent; and (4) chemical and/or thermal reduction. The whole process is depicted in Fig. 9.21. In the first section, extrusion forces GO LC dope through spinning pipes with uniaxial flow. Importantly, this common process unifies the random alignment of GO LCs in 3D space along the flow direction and fashions GO sheets oriented along the resultant fiber axis. SEM characterizations showed distinct differences in GO LCs between the still and flowing states. The flow-induced alignment of GO dispersions revealed their rheological attribute as typical shear flow. In the second section, coagulation baths using solutions compatible with solvents of GO LCs but

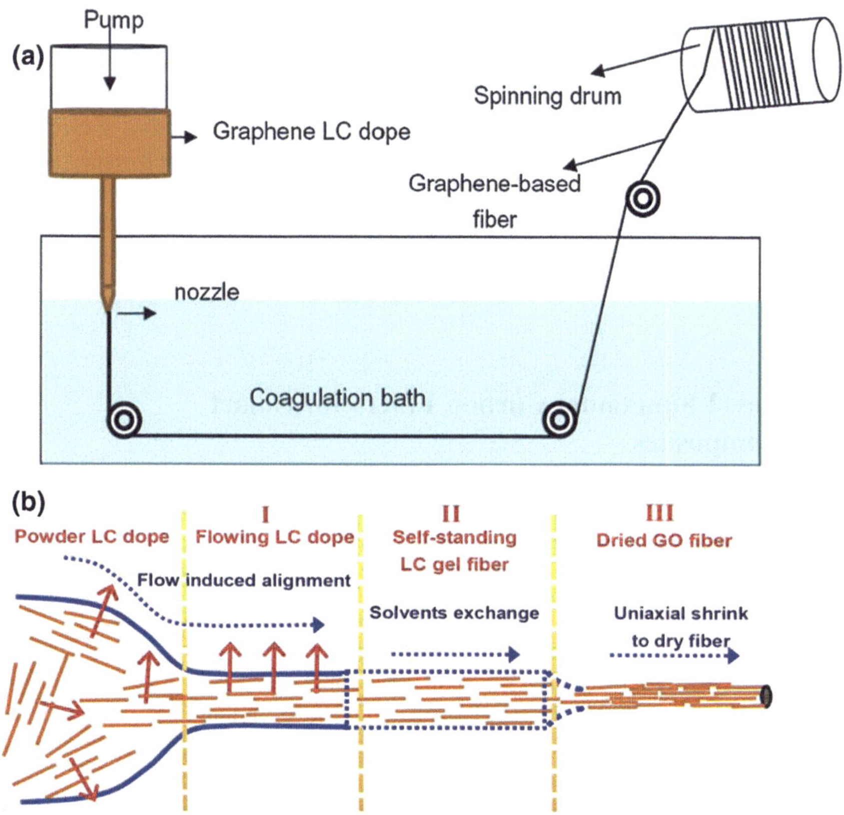

Fig. 9.21 Liquid crystal-based wet spinning for GF. **a** The schematic wet-spinning process from GO liquid crystals to fibers in a continuous manner. **b** The structural evolution in the spinning process, which indicates the enhanced alignment process under uniaxial flow during the spinning (Section I), from the powder GO LC with different orientation ordering to the regular alignment of the flowing GO LC dope, GO LC gel fibers (Section II), and final dried GO fibers (Section III). The red arrows denote direct vectors [106]

incompatible with GO sheets, such as aqueous solutions with acids, bases, diverse salts that screen the electrostatic repulsion of GO sheets, and some organic solvents such as ethyl acetate with a poor capability to disperse GO sheets were chosen. After solvent exchange, GO liquid crystalline gel fibers formed with maintaining the structural integrity by hydrogen bonding between GO sheets or ions cross-linking. During the drying section, GO gel fibers underwent shrinkage in the radial direction. The interior GO sheets folded in the same direction but remained aligned along the fiber axis. In the last section, chemical or thermal reduction was employed to transform GO fibers to GFs, which allowed the optimal expression of the outstanding mechanical, electronic, and thermal merits of graphene in the macroscopic performances and functionalities of GFs.

Graphene fiber is a new high-performance carbonaceous fiber with rich functionalities. Its creation significantly demonstrates a novel protocol to fabricate carbonaceous fibers just from mineral graphite. The preparative system of GF has been established in recent years, which encompasses reliable synthesis of solvated graphenes, formation of liquid crystals and wet spinning for neat or composite GFs. A basic understanding of the structure and properties of GFs has been established preliminarily, and the critical control parameters have also been concluded, including the lateral size of graphene building blocks, the alignment thereof, and the interaction between them. Guided by these concepts, the combined properties of GFs have been rapidly promoted to a GPa level of tensile strength and electrical conductivity comparable to conventional CFs, promising a thrilling trend toward higher levels on par with or even exceeding CFs.

9.3 Novel Functional Carbon Fibers for Smart Composites

9.3.1 Novel Metal-Coated Carbon Fibers and Their Applications

Traditionally, carbon fibers were used as good reinforcements for metals [108], ceramics [109], and polymers [110]. To increase the physical properties of carbon fiber-reinforced metal (CFRM), carbon fiber-reinforced ceramic (CFRC), and carbon fiber-reinforced polymer (CFRP) matrix composites, various surface treatments such as thermal oxidation, electric oxidation, ozone treatments, plasma treatments, and even metal coating were developed [111–119]. Among these, metal-coating methods lead to very interesting interfacial and electrical properties between carbon fibers and matrices. The adhesion of metal/carbon fiber composites has been shown to be excellent when the metal coating is grown from the carbon fiber surfaces. In particular, polymer/metal interaction is strong, mainly because of the high metal surface energetics, which allows extensive wetting with the polymer. According to Park's report [120], metal deposition on carbon fibers can lead to high surface free energy, resulting in good mechanical and interfacial properties of CFRPs. Figure 9.22 shows the surface free energy of Ni-plated carbon fibers [120]. The Ni plating on carbon fiber surfaces at a current density range 5–60 A/m^2 leads to better wetting levels than that of untreated carbon fibers. To improve the wettability, the surface free energy of the treated carbon fibers needs to be made larger than that of the matrix in a real composite system. When a current density of 10 A/m^2 was used, the surface free energy, γ_S, reached the highest value of approximately 67 mJ/m^2. Thus, it is expected that the degree of adhesion between metallized carbon fiber surfaces and epoxy resin can be enhanced, because the surface free energy of the metallized carbon fibers does allow extensive wetting with the epoxy resin (45 mJ/m^2).

It is well known that the resulting mechanical properties of the composites strongly depend on the fiber–matrix interfacial adhesion, which affects the ease of load transfer from one matrix to the other via the fibers. Figure 9.23 shows the interlaminar shear stress (ILSS) of the Ni-plated CFRPs as a function of the current density. All metallized CFRPs showed higher ILSS values than those of the untreated CFRPs. Further, the maximum strength was observed at a current density of 10 A/m^2 [120].

9.3.2 Novel Electroless Metal Coating for Electric Devices

All electric devices and systems such as television transmitters, police radios, and cellular phones can cause electromagnetic interference (EMI) because of the frequencies of their switching rates [121]. It is a well-known fact that EMI can cause monitors and other medical devices such as pacemakers to malfunction. To shield EMI, traditionally, the use of antistatic or electric materials has been strongly pursued [122]. In practical application, some composite materials, which have

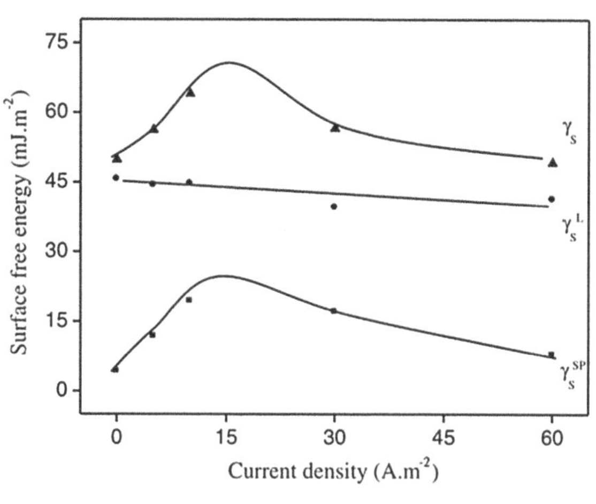

Fig. 9.22 Surface free energy and their components of carbon fibers [120]

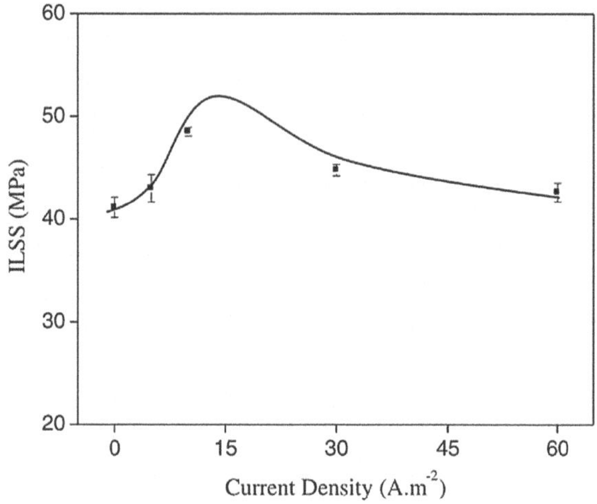

Fig. 9.23 ILSS of CRFPs as a function of electrolytic Ni-plating current density [120]

fillers with low electric resistance, can partially shield EMI in the case of electric devices. This is the reason for the extensive efforts of many researchers toward the development of highly conductive fibers or powders [123, 124].

Composites reinforced by conductive fillers can be actively applied in PTC materials, stealth aircrafts, sensors, and other many electric devices. The method to reduce the electric resistance of conductive fillers at room temperature is regarded as one of the key technologies in the composites.

Carbon fibers are widely used as reinforcements in polymer matrix composites. Because of good mechanical properties and the wide areas in which the carbon fiber-reinforced composites can be applied, carbon fibers are still attractive materials to researchers. If carbon fibers have a very high electric conductivity, the application of carbon fibers is bound to increase. Highly conductive carbon fiber-reinforced composites can be easily applied in EMI shielding or stealth materials.

To decrease the specific electric resistivity of carbon fibers, various metal plating techniques have been applied using Ni, Cu, Zn, Pt, and Ag [125–130]. Among these techniques, electroless Ni plating is a viable method to produce low resistive fillers for EMI shielding. Electroless Ni plating offers advantages of good metal deposition, very low electric resistivity and shielding effectiveness, and applicability to nonconducting materials with complicated shapes [131, 132].

The nickel-plating process is based on a redox reaction in which the reducing agent is oxidized and Ni^{2+} ions are reduced on the substrate surface. When the first layer of nickel is deposited, it serves as a catalyst for the process. As a result, a linear relationship between the coating thickness and time normally occurs. During electroless plating, carbon fiber surfaces are metalized when exposed to plating solutions, coating molten salt in the presence of a reductant.

Electroless Ni plating, in contrast to nickel electroplating, does not require an external electric current to produce a deposit. Deposition occurs in an aqueous solution containing metal ions, a reducing agent, and catalysts. Chemical reactions on the surface of the catalytic part being plated lead to deposition of the metal or alloy. The most common reducing agent used (also used here) is sodium hypophosphite, which leads to two types of simultaneous reactions.

The first reaction involves the cathodic reaction of Ni^{2+}, H^+, and $H_2PO_2^-$ or the deposition of Ni–P alloy and the production of hydrogen.

$$Ni^{2+} + 2e^- \rightarrow Ni \tag{9.1}$$

$$2H^+ + 2e^- \rightarrow H^2 \tag{9.2}$$

$$H_2PO_2^- + 2H^+ + e^- \rightarrow P + 2H_2O \tag{9.3}$$

The second reaction is the anodic oxidation of $H_2PO_2^-$.

$$H_2PO_2^- + H_2O \rightarrow H_2PO_3^- + H + H^+ + e^- \tag{9.4}$$

The Ni^{2+} ions adsorbed in the presence of the activated fiber surfaces are reduced by the adsorbed H^+ ions, and the plating proceeds as long as there is sufficient availability of metal ions and reducing agent. The reduction of nickel is confirmed by the formation of hydrogen from the fiber surfaces. In this case, a very vigorous reaction is observed during the plating initially.

Figure 9.24 shows the SEM micrographs of the Ni/carbon fibers as a function of plating time. Figure 9.24a, b, c shows the side view of the as-received and carbon fibers plated with Ni for 5 and 10 min, respectively. Figure 9.24d, e shows the cross-sectional images of the fibers plated for 10 and 30 min. The as-received

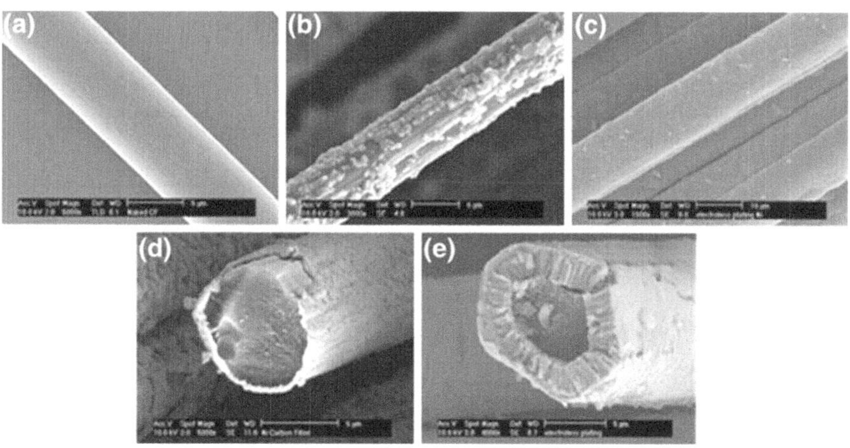

Fig. 9.24 SEM micrographs of carbon fibers. **a** As-received, side view. **b** After 5-min plating, side view. **c** After 10-min plating, side view. **d** After 10-min plating, cross section. **e** After 30-min plating, cross section [132]

carbon fibers (a) showed a smooth surface and uniform diameters. In case of the sample plated for 5 min (b), Ni particles were deposited irregularly. Perfect Ni layers were coated on carbon fibers in the sample shown in Fig. 9.24c, and the thickness of the Ni layer increased in the sample shown in (e) in comparison with the sample shown in (d). These results are in good agreement with the data shown in Fig. 9.24. In Fig. 9.24, the Ni content rapidly increased up to 10 min and then gently increased to 30 min. Abraham et al. [133] reported that metal plating on carbon fibers began at scratches and fine cracks existed on the surface of carbon fibers were connected to each other. Ni plating in this system is understandable through a similar mechanism of Ni loading, which can be applied on carbon fibers. In the sample plated for 5 min, an island-like Ni layer was observed, as shown Fig. 9.24b. On the other hand, in the sample plated for 10 min (c), a very smooth surface similar to that observed in the case of the as-received samples was observed. This result indicates that the island-like Ni clusters merged together and formed a layer. In the sample plated for 30 min (e), the diameter of the fibers is approximately 6–7 μm and the thickness of the nickel layer is 2–3 μm. As a result, perfect Ni layers formed and fully covered the carbon fiber surfaces.

Wide-scan spectra in the binding energy range 0–1, 200 eV were obtained to identify the surface elements present and to conduct a quantitative analysis. The XPS wide-scan spectra of the two carbon fiber samples (as-received and samples plated for 30 min) are shown in Fig. 9.25. The spectra distinctly show the carbon, oxygen, and nickel peaks of the samples. The peak intensity of the O_{1s} (BE = 532.8 eV) and the Ni_{2p} (BE = 857.6 eV) peaks of the Ni-plated carbon fibers was higher than that of the as-received carbon fibers. The enhancement of the O_{1s} peaks on the carbon fiber surface can be explained on the basis that severe plating conditions such as high pH and temperature may affect the surface structures and chemical compositions, resulting in an introduction of oxygen-containing groups on the carbon surfaces. Figure 9.26 shows the expanded scale of the

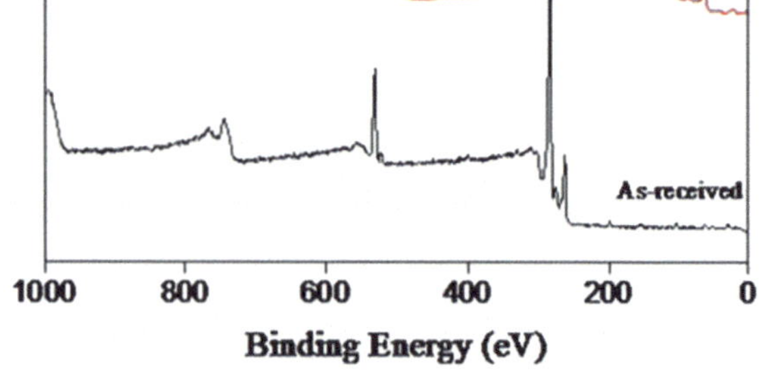

Fig. 9.25 XPS profiles of electroless Ni-plated carbon fibers as a function of plating time [132]

O_{1s} XPS profiles of the (a) as-received and (b) electroless Ni-plated carbon fibers after 30 min of plating. The O_{1s} profile of the as-received sample reveals the presence of two peaks. Peak 1 (binding energy, BE = 530.6 eV) corresponds to the C=O group (ketone, lactone, and carbonyl) or the –OH group, and peak 2 (BE = 532.6 eV) corresponds to the O–C–O group. The spectra of the Ni-plated carbon fibers reveal the presence of three peaks corresponding to the NiO group (BE = 527.8 eV), C=O or –OH groups (BE = 531.6 eV), and O–C–O groups (BE = 532.6 eV). The center of the C=O or –OH peak was shifted to the high BE region after Ni plating [133].

This indicates that the BE between elements oxygen and carbon is enhanced because of the introduction of Ni particles on the carbon surfaces. As is shown in Fig. 9.26b, the NiO peak was newly formed and it can be understood that the source of elemental oxygen is the O–C–O, C=O, or –OH of the as-received carbon fibers. From Fig. 9.26, the O–C–O peak decreased in intensity and the C=O or –OH peak was enhanced after the Ni plating. Hence, it can be concluded that the introduction of Ni particles helps the formation of C=O or NiO by reducing O–C–O. In addition, some of the C=O also contributes to the formation of NiO, resulting in an increase in the binding energy of the C=O peaks. The experimental XPS results show that the carbon content of the Ni-plated fibers decreased when they were plated with nickel, whereas the oxygen and nickel contents of the plated fibers were higher in comparison with those of the as-received fibers. These active groups on the carbon fiber surfaces produced through nickel plating may help change the polarity and the functionality of the fiber surfaces [133].

Figure 9.27 shows the specific electric resistivity of the electroless Ni-plated carbon fibers as a function of the plating time. The electrical conductivity increased proportionally with the Ni-electroplating time. It was also found that the specific electric resistivity was significantly reduced after 10 min of plating, after which no additional decrease occurred up to 30 min of plating. This can be ascribed to the formation of perfect Ni layers on the carbon fibers after a specific point of the process.

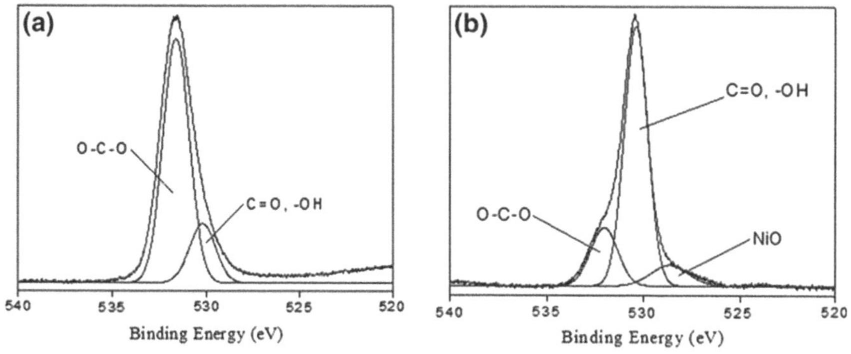

Fig. 9.26 High-resolution O_{1S} profiles of as-received and electroless Ni-plated carbon fiber surfaces **a** as-received and **b** after plating for 30 min [133]

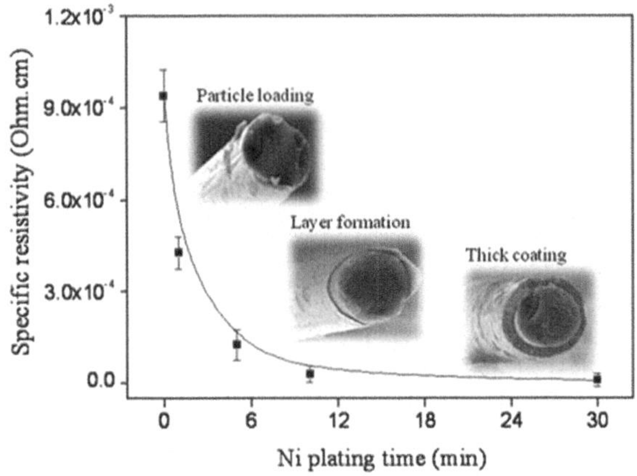

Fig. 9.27 Specific resistivity of Ni-plated carbon fibers as a function of Ni-plating time [133]

Normally, electroless plating is achieved through three steps : (1) the introduction of nuclei, (2) the formation of metallic islands (or particles), and (3) the formation of perfect layers. In Fig. 9.27, the specific resistance of the sample subjected to plating for 1 min was decreased in comparison with the value shown by the as-received sample, indicating that the electric network on the fiber surfaces was highly enhanced in the sample plated for 1 min. It can be understood that the island-like Ni loading in the sample plated for 1 min strongly leads to a reduction in the electric resistivity of the carbon fibers, though the metallic layer on the fiber surfaces has not yet formed.

From the results, it is concluded that the electric resistivity can be controlled by electroless Ni plating when a perfect layer is formed. Excessive plating conditions make the microstructure of the Ni layers and the crystal sizes of the nickel particles dense and small, resulting in superlow-specific electric resistivity and saturation state.

9.3.3 Novel Electrolytic Metal Coating for Electric Devices

All carbon fibers must be pretreated before metal coating to remove the polymeric sizing materials. The electric resistance of the heat-treated carbon fibers (350 °C, 10 min) and the chemically treated fibers (acetone, 30 min) is 0.75×10^{-3} and $0.81 \times 10^{-3}\ \Omega$ cm, respectively, while the value shown by the as-received carbon fibers is $0.94 \times 10^{-3}\ \Omega$ cm. Heat treatment was more effective for removing sizing polymers than the chemical treatment. However, the heat treatment caused many partial defects in the carbon fibers because of the occurrence of burning. In conclusion, to avoid defects in carbon fibers, we chose the chemical method as a pretreatment even though the electric resistivity was slightly higher than that of carbon fibers subjected to heat treatment.

As shown in Fig. 9.28, the nickel particles grew gradually and formed a metallic layer on the fiber surface. Nickel particles formed a perfect nickel layer after 10 s of plating, and then, the thickness of the nickel layers increased with increase in the plating time. From the SEM analyses, a mechanism of formation of nickel-coated carbon fibers can be proposed. Normally, there are many gaps or cracks on the top surface of the as-received carbon fibers. In the initial state of plating, fine nickel particles are introduced at these points, which form an island-like state, which constitutes the filling step. These metal particles combine and become big clusters during the "large-cluster formation" step. Each cluster grows, and the clusters meet each other and form a metallic layer, during the layer formation step. Then, the

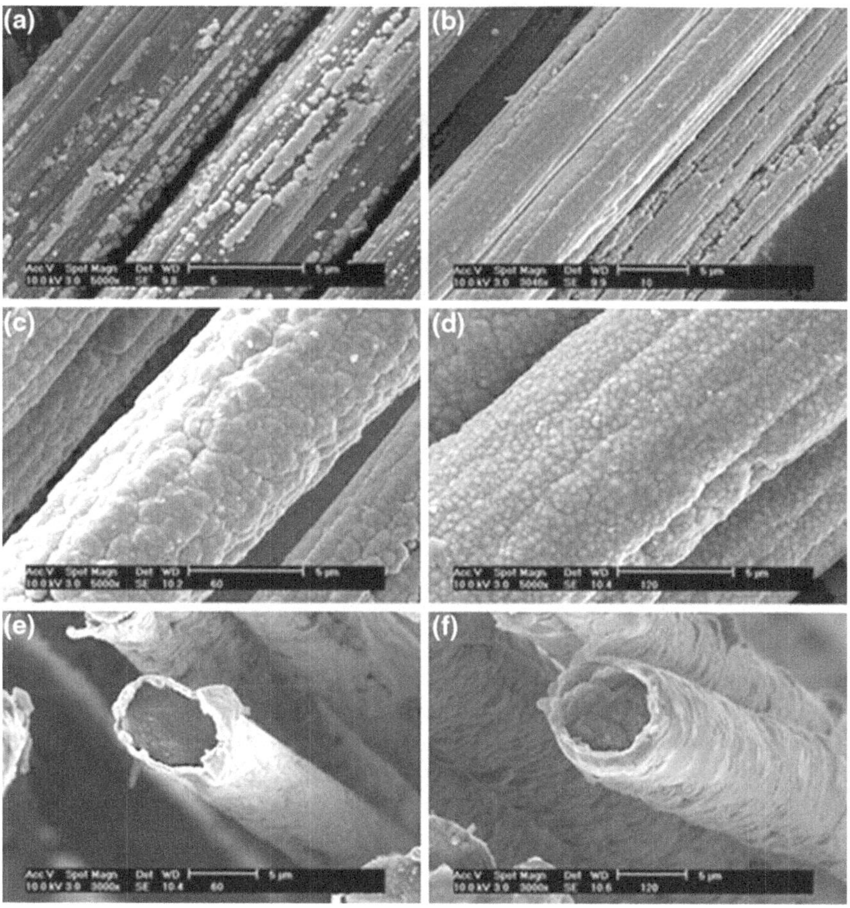

Fig. 9.28 SEM images of Ni-plated carbon fibers after **a** 5 min, **b** 10 min, **c** 60 min, and **d** 120 min of plating. Cross-sectional images of samples plated for 60 min **e** and 120 min **f** are also shown [134]

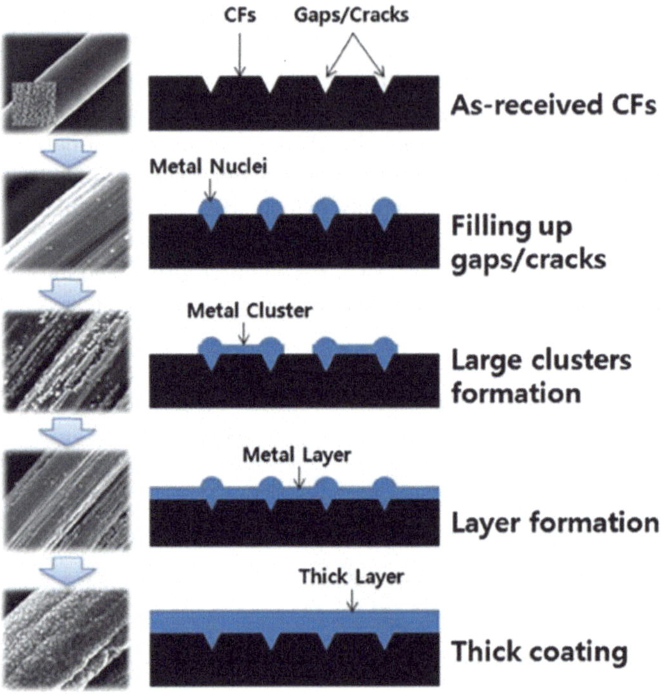

Fig. 9.29 Schematic of nickel layer plating behaviors on carbon fibers [134]

thickness of the nickel layer grows with plating time, which is called the thick coating step. In this study, the large-cluster formation and layer formation steps were observed at 10–30 s of plating (Fig. 9.29) [134].

XRD analysis was employed to observe the microstructures of the nickel/carbon hybrid fibers, and the results are shown in Fig. 9.30. The characteristic peaks of nickel, such as Ni(111) at 44° (2θ), Ni(200) at 52° (2θ), and Ni(220) at 76° (θ), were found in all the Ni-plated carbon fibers. It was also found that all the Ni peaks increased in intensity with increase in plating time. In case of C(002) at 25° (2θ), the characteristic peak significantly decreased in intensity with increase in plating time, indicating that metallic Ni formed perfect layers and covered the outer surface of carbon fibers. Figure 9.30b, c, d shows the magnified peaks of Ni(111), Ni(200), and Ni(220), respectively. It is interesting to note that each characteristic peak was shifted to the right side with increasing plating time, indicating that the high amount of nickel loading affected the microstructures of the nickel layers.

In the case of Ni(111), the peak position altered from 44.39° to 44.50°. Normally, the peak position in a material indicates the distance between standard crystals, and the change in the peak position indicates a change in the distance between the crystals, resulting in a change in the material density. In all the Ni peaks, the peak position shifted to a high value, indicating that the distance between the crystals became narrow, and the nickel layer formed on the fibers became dense.

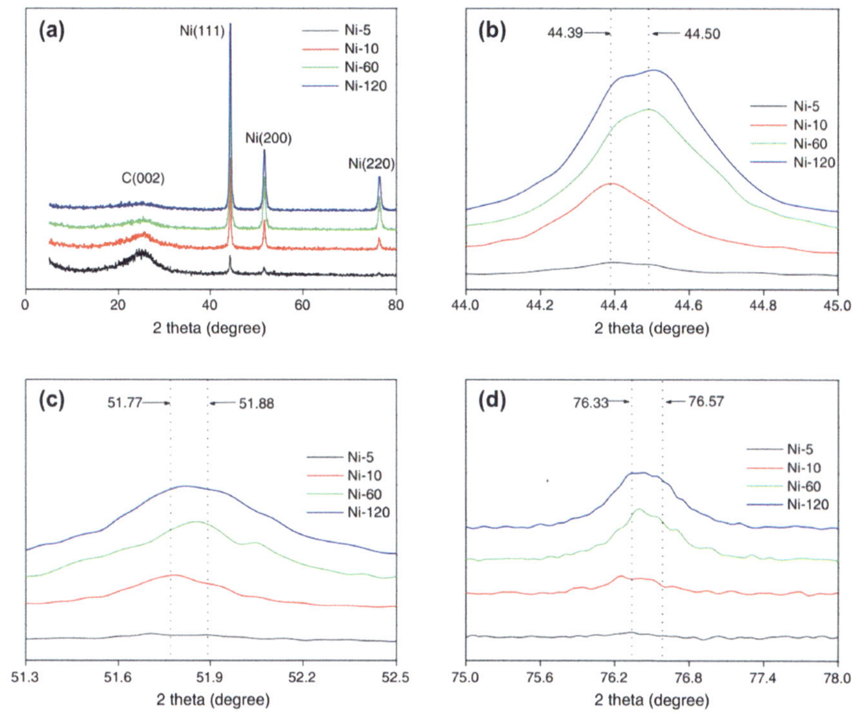

Fig. 9.30 XRD patterns of Ni-plated carbon fibers as a function of electroplating time. **a** Wide-angle XRD spectra of all samples. **b** High-resolution peaks of Ni(111). **c** High-resolution peaks of Ni(200). **d** High-resolution peaks of Ni(220) [134]

The nickel layer is composed of nickel particles. Hence, the gaps between the nickel particles are present. The results of the XRD analysis prove the occurrence of changes in the packing density with plating time. The average particle size decreased with increase in plating time, resulting in the enhancement of electric conductivity of the fibers. Figure 9.31 is a schematic of the nickel particle deposition behaviors on carbon fibers. The change in the particle size and packing density can indicate a decrease of inner pores, which results in the reduction of contact loss between nickel particles.

Figure 9.32 shows the electric resistance of the nickel-coated carbon fibers. The specific resistance decreased with increase in plating time. Also, the specific resistivity of the as-received sample exceeded 0.80×10^{-3} Ω cm, although this value decreased dramatically with increase in plating time up to 60 s. From the good linearity of results in the range 5–30 s, a strong correlation was observed between the electrical conductivity and plating time. In the case of samples subjected to plating for more than 60 s, the conductivity did not significantly enhance. Hence, a perfect Ni layer can be formed in the range 30–60 s, resulting in the saturation of electrical conductivity due to the minimization of inner pores (or contact loss between nickel particles) [134].

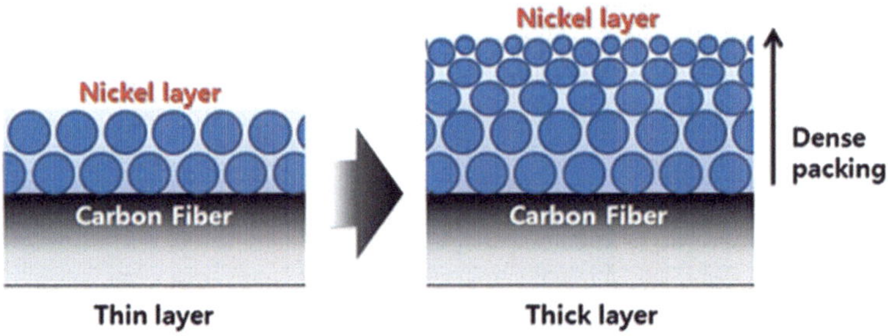

Fig. 9.31 Schematic of deposition behaviors of nickel particles on carbon fibers [134]

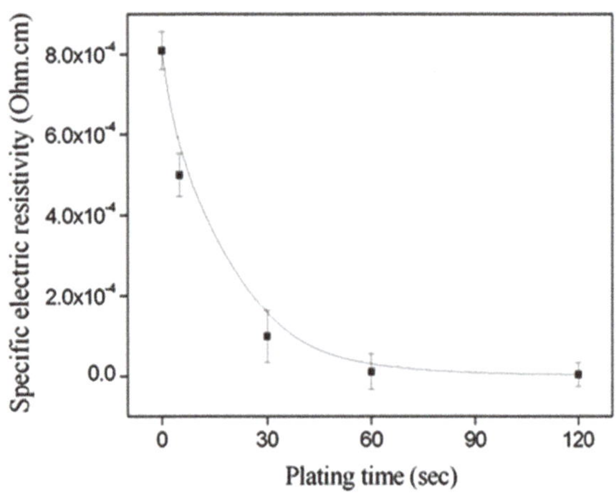

Fig. 9.32 Specific electric resistivity of Ni-plated carbon fibers as a function of plating time [134]

9.3.4 Novel Nanocarbon Coating on Carbon Fibers

Recently, carbon/carbon composites (CCCs) have stimulated intense scientific research on account of their exceptional physical and chemical properties such as excellent high-temperature resistance, mechanical properties, durability, and electrical conductivity [135–147]. Examples of CCCs include carbon nanofibers/activated carbon fibers (CNFs/ACFs), carbon nanotubes/carbon fibers (CNTs/CFs), carbon nanotubes/carbon black (CNTs/CB), carbon nanotubes/carbon papers (CNTs/CPs), and graphene/CNTs [143–147]. These CCCs have been used in a wide range of applications, i.e., in polymer composites, in electromagnetic interference shielding materials, in electrode materials, and in thermal conductivity materials.

The growth of CNTs or CNFs on the surface of CFs has specifically attracted enormous attention as model systems for nanoscience and nanotechnology [140–156]. CNTs or CNFs are a highly important class of carbon materials with novel properties, particularly high aspect ratios, small radius of curvature, high mechanical strength, unique electrical properties, and chemical stability [157, 158]. By far, some studies have been reported about the growth of CNFs on the surface of CFs. Kim et al. [151] reported a simple method for the preparation of ACFs coated with graphite nanofibers to enhance the electrical conductivity of ACFs. Because of the unique electrical properties of the CNFs, the volume resistivity of CNF-coated ACFs was twofold greater than pristine ACFs. Mathur et al. [152] reported that the mechanical properties of CNT/CF–resin composites increase with the increase in the amount of deposited CNTs. Hsieh et al. [156] fabricated superhydrophobic carbon fabric with micro/nanoscaled two-tier roughness by decorating CNTs onto microsized CFs, using catalytic chemical vapor deposition and a subsequent fluorination surface treatment.

While the growth of CNTs or CNFs on the CF substrate can further improve the properties of the CFs, there are some challenges encountered in the synthesis of CNFs or CNTs on the CF substrates by catalytic chemical vapor deposition (CCVD). For example, the metal catalysts (e.g., Ni, Fe, and Co) easily aggregate into large particles and form mixed phases of carbon materials (CNFs, CNTs, and amorphous carbon), and the formation of CNTs or CNFs is not uniform on the CFs, requiring purification processes. Previous researches have reported that the particle size of the catalyst is the factor determining the diameter of the CNFs grown on the substrates and the key is to control the morphology of the obtained nanoparticles [159–162].

To synthesize CNFs with a given diameter, it is necessary to stabilize their sizes on the supporting substrates. Therefore, the preparation of catalyst nanoparticles has played an important role in the growth of CNFs on the CF substrate. Several catalyst preparation methods have been reported in the literature [148, 150, 161]. Sun et al. reported the growth of CNTs on carbon paper by ohmically heating silane-dispersed catalytic sites [161]. This method helped to prevent catalyst precursors from diffusing into the carbon matrix and allowed them to play a similar role of the catalysts used in the growth of CNTs on the CFs. As many research studies have reported, ordered mesoporous silica materials are excellent catalyst supports and have been used in the growth of carbon nanomaterials [162–164]. Zheng et al. reported that mesoporous silica films can direct the growth of well-aligned CNTs perpendicular to parallel to the substrate surface through the control of the distribution of catalysts [164]. Therefore, the purpose of this study was to use different methods of uniformly dispersing Ni catalysts on the CF substrate so that dense carbon nanomaterials can grow on the CF surfaces. Thermal CVD was used for the subsequent growth of carbon nanomaterials.

1. Preparation of nanocarbon coated on CFs

In order to prepare nanocarbon-coated CFs, three steps are required, which include (1) surface treatments of CFs, (2) catalyst coating on CFs, and (3) direct nanocarbon synthesis on CFs. CFs (Tairyfil, TC36-3 K) used in our experiments consist of filaments with a diameter of 3–7 lm. Before being used, the CFs were desized by ultrasonically cleaning in acetone and ethanol and drying at room temperature. Then, the CFs were subjected to thermal debinding at 723 K for 30 min in air to burn the organic binder. Finally, they were immersed in a methyl ethyl ketone solution (butanone) to dissolve the burnt carbon and clean the CF surfaces. The desized CFs are immersed in a solution of HNO_3/H_2SO_4 (with a volume ratio of 1:3) for 12 h at room temperature and then washed with water to obtain a pH level of 7 [165].

The Ni-containing catalysts supported on CFs were prepared via three methods. In the first method, the acid-treated CFs were impregnated with a 5 wt% nickel nitrate solution and dried in air. The CF samples collected before the CNFs were grown on the CF surfaces are designated as CF-1, while the CF samples collected after the CNFs were grown on the CF surfaces are designated as CF-1-CNF.

In the second method, the acid-treated CFs were impregnated with mesoporous silica sol and dried at 353 K. Mesoporous silica precursor solutions were prepared by mixing tetraethyl orthosilicate (TEOS, Aldrich), surfactant (Pluronic P123, EO20PO70EO20, $M_{av} = 5800$, Aldrich), distilled water, ethyl alcohol, and hydrochloride at a molar ratio of 1:0.0103:9.36:21.4:0.04. The CFs were then calcined at 723 K for 2 h. Finally, the CFs coated with mesoporous silica film were impregnated with a 5 wt% nickel nitrate solution for 12 h and dried in air. The samples collected before the CNFs were grown on the CF surfaces are designated CF-2, while the CF samples collected after the CNFs were grown on the CF surfaces are designated as CF-2-CNF.

In the third method, the acid-treated CFs were impregnated with a Ni-doped mesoporous silica precursor solution. The solution was prepared by mixing TEOS, P123, distilled water, ethyl alcohol, hydrochloride, and nickel nitrate in a molar ratio of 1:0.0103:9.36:21.4:0.04:0.47. The CFs were then calcined at 723 K for 2 h. The samples collected before the CNFs were grown on the CF surfaces are designated as CF-3, while the CF samples collected after the CNFs were grown on the CF surfaces are designated as CF-3-CNF.

The CFs were placed in an alumina boat located in the middle of a flow through a tube furnace and heated to a certain temperature (1023 K) in 200 ccm of flowing Ar. At the same temperature, they were then subjected to a reduced flow of 5% H_2/Ar (150 ccm) for 30 min to obtain the reduced Ni nanoparticles. Acetylene (30 ccm) was then introduced into the furnace for 30 min to help the CNFs grow on the surface of the CFs. Figure 9.33 illustrates the process of growth of CNFs on the surface of the CFs the three methods .

2. Characteristics of nanocarbon coated on CFs

Figure 9.34 shows the SEM images of CFs treated with the three methods. As shown in Fig. 9.34a, the surface of the CF is very smooth. However, after the CFs are immersed in a solution of $Ni(NO_3)_2$, the surfaces of the CFs contain Ni $(NO_3)_2$ particles with sizes ranging from approximately 10–100 nm (Figure 9.34b). As shown in Fig. 9.34c, some $Ni(NO_3)_2$ particles with different sizes are adsorbed on the CF surfaces coated with the mesoporous silica film; the particles are bigger because some of the $Ni(NO_3)_2$ particles are supplemented in the mesoporous silica layer. In Fig. 9.34d, the CF surfaces are coated with a smooth Ni-doped mesoporous silica thin film. It can be seen that the surface of CF-3 is covered by a smooth thin film, which can be attributed to the catalysts incorporated into

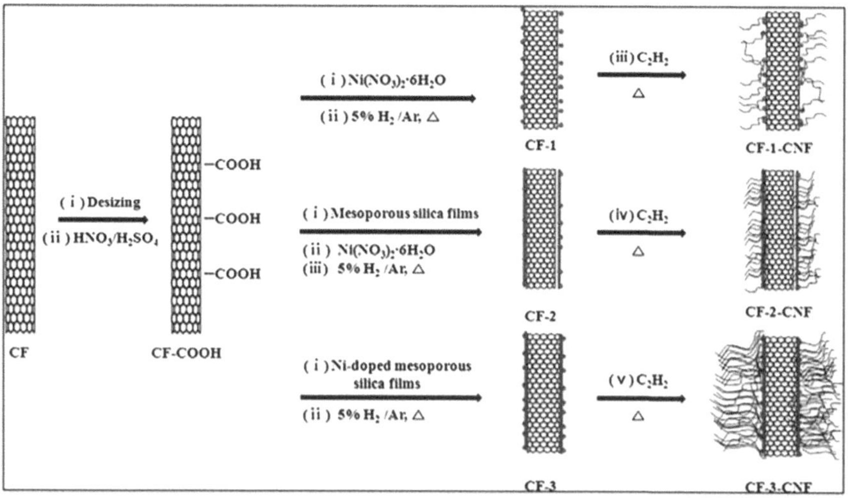

Fig. 9.33 Preparation process of CFs coated with CNFs by various methods [142]

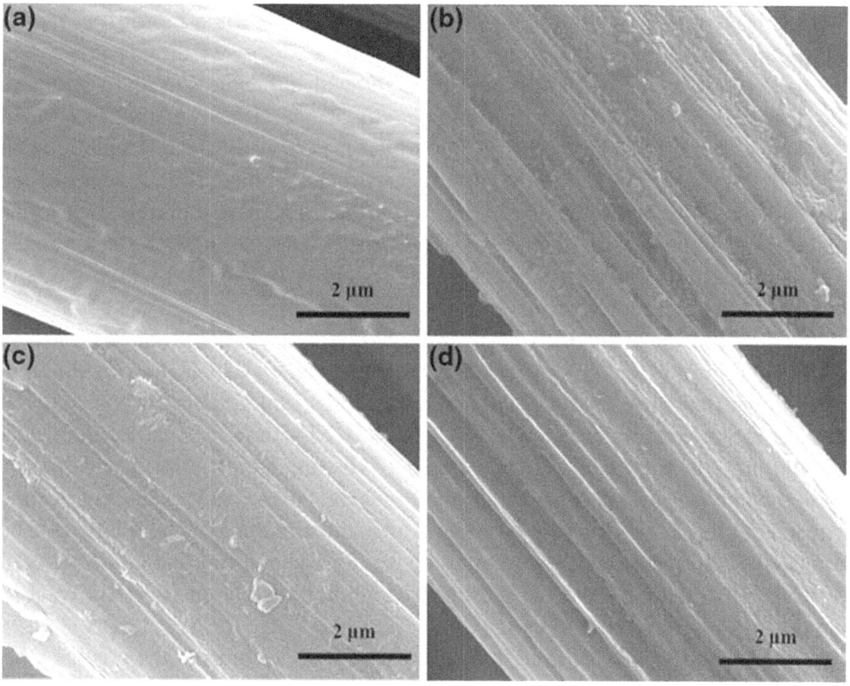

Fig. 9.34 SEM images of pristine **a** CFs, **b** CF-1, **c** CF-2, and **d** CF-3 [142]

mesopores with a molecularly well-defined structure [164, 166]. The successful coating of this film on the CF surfaces can directly facilitate the uniformity and density of the CNFs that grow on the CFs through strong interactions with the Ni catalyst [167–169]. A comparison of the CNFs grown on the CF surfaces via the various methods leads to the conclusion that the content and size of Ni particles distributed on the CF surfaces can be correlated with the density and structure of the CNFs [159].

To understand how the CF pretreatment affects the growth of CNFs, we used SEM to analyze the morphology and microstructure of the obtained CNFs. For the growth of CNFs, the CFs were first reduced by hydrogen gas. Because of the growth of carbon materials on the CFs without hydrogen, $Ni(NO_3)_2$ did not play the role of catalyst [156, 169]. Figure 9.35 shows the SEM images of CNFs grown on the CF surfaces with various types of pretreatment. As shown in Fig. 9.35a, b, the heterogeneous anchoring of CNFs on the CF-1 surfaces produces CFs with a thickness of 10 lm and a diameter in the range 60–120 nm. This outcome can be attributed to a weak interaction of the active components with the CFs, in which case the Ni particles easily aggregate to bulk particles on the CF surface. In Fig. 9.35c, d, the growth of the CNFs on the CF surfaces is nonuniform. The diameters and thicknesses of the CNFs are in the range 20–200 nm and 12–20 μm because the active component of Ni particles is more uniformly distributed in the mesopores and the mesopores limit the catalysts from aggregating into a bulk particle [33]. As shown in Fig. 9.35e, f, after the Ni-doped mesoporous silica thin films have been wash-coated, the CF surfaces are covered entirely by the uniform growth of CNFs with a thickness of up to 52.5 μm. The diameters of the CNFs are in the range 20–50 nm. The formation of straight CNFs may be related to the pore structure of the mesoporous silica. Because the Ni nanoparticles are distributed in the mesopores structures, the Ni catalyst incites the growth of the CNFs on the top of the CF surfaces coated with mesoporous silica films. Hence, the content and size of Ni particles adsorbed on the CF surfaces are correlated with the density of CNFs, and Ni nanoparticles are well dispersed within the pore channels of the mesoporous silica film for the growth of CNFs.

The morphology of the CNFs was also analyzed using TEM and is shown in Fig. 9.36. Figure 9.36a, b shows the entangled bundles of CNFs with different sizes. Impurities such as carbon nanotubes, CNFs, amorphous carbon, and Ni particles are present in the sample. The grown CNFs show a range of diameters (20–400 nm). In Fig. 9.36c, d, the grown CNFs are more uniform and have a herringbone-type alignment in which the graphene planes are at an angle to the fiber axis [170, 171]. Our results confirm that the structure of the CNFs can be correlated with the Ni catalyst size, and the results also confirm that interactions of the CNFs with the CFs and the mesoporous silica films are an excellent catalyst support for the growth of CNFs on CF surfaces.

N_2/77 K isotherms were used to measure the pore structures of the pristine CFs, CFs coated with Ni-doped mesoporous TiO_2 film, CF–GNF, and CF–GNF–FP, as shown in Fig. 9.37. It was found that CF–GNF shows a typical type IV isotherm. There is a measurable increase in the specific surface area and the total pore volume of CF–Ti because the mesoporous TiO_2 film is calcined at 723 K in air to remove

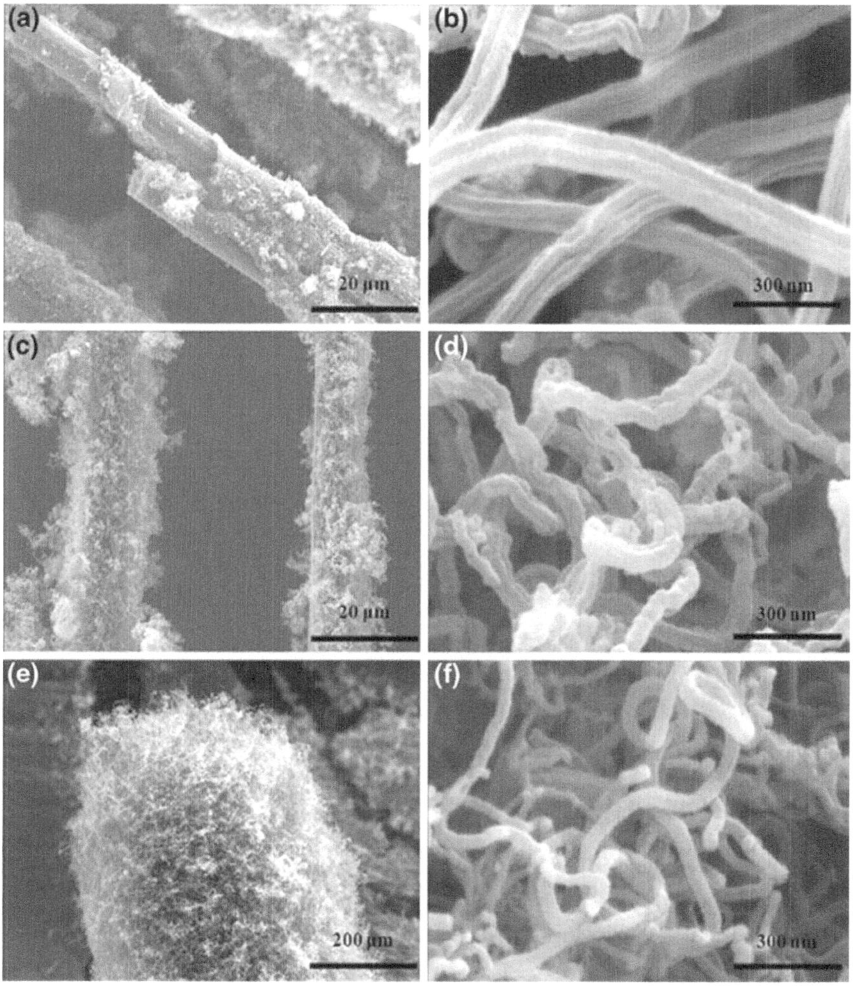

Fig. 9.35 SEM images of **a**, **b** CF-1-CNF, **c**, **d** CF-2-CNF, and **e**, **f** CF-3-CNF [142]

any surfactant (Pluronic P123), and this process produces a mesoscopically porous layer. It is noteworthy that the Ni particles are uniformly dispersed in the TiO_2 framework in the case of CF–Ti [169].

The specific surface area and total pore volume of CF–GNF are 248 m^2/g and 0.293 cm^3/g, respectively. The result demonstrates that the Ni-doped mesoporous TiO_2 film leads to the growth of 3D GNF network on the CF surfaces. It also shows that the adsorbed N_2 volume of CF–GNF–FP is less than that of CF–GNF. The decrease in the adsorbed volume for the GNF-coated CFs was approximately 32%. The surface area and total pore volume of the CF–GNF–FP are 169 and 0.240 cm^3/g, respectively. This indicates that the fluoropolymer coating did not seriously diminish the pore structures of CF–GNF.

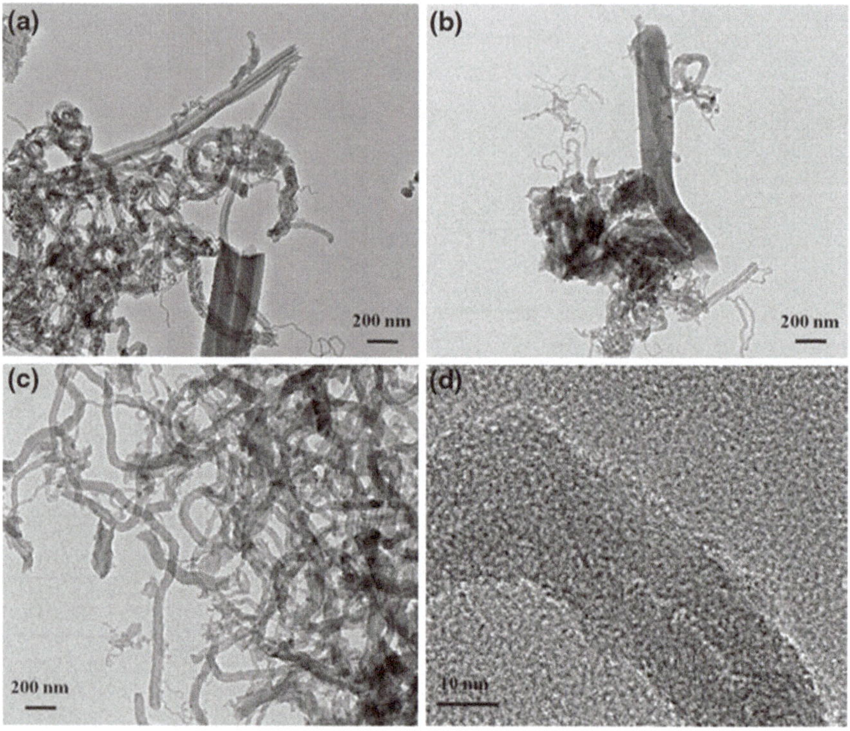

Fig. 9.36 TEM images of **a** CF-1-CNF, **b** CF-2-CNF, and **c**, **d** CF-3-CNF [142]

Fig. 9.37 N_2 full isotherms of pristine CFs, CFs coated with Ni-doped mesoporous TiO_2 films, CFs coated with GNFs, and fluoropolymer-coated CF–GNF [143]

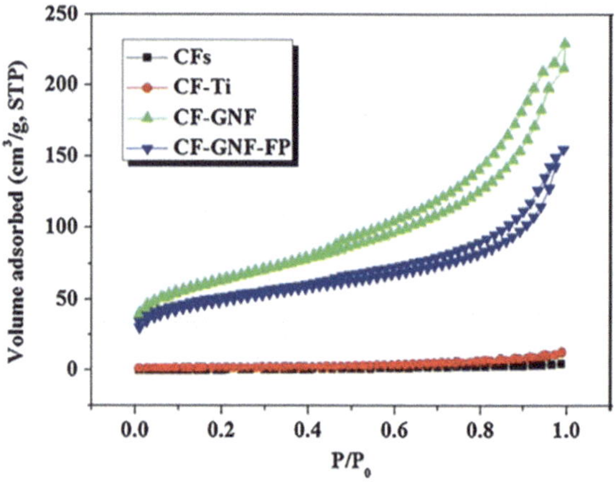

The surface morphology of the CF surfaces was examined by SEM. Figure 9.38a, c shows the diameter of a single filament of CFs measured from this micrograph, and the filament diameter is 6 μm. As shown in Fig. 9.38d, the CF surfaces are coated with a smooth Ni-doped mesoporous TiO_2 thin film. The

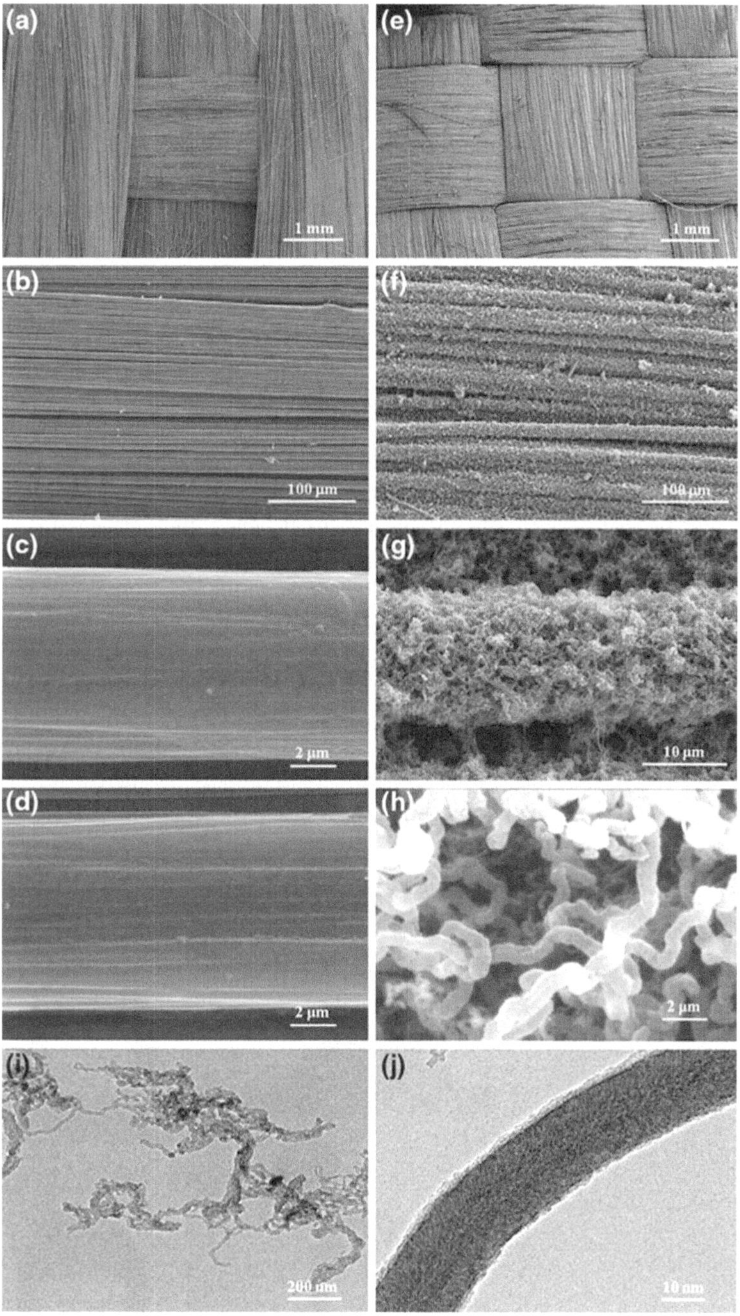

Fig. 9.38 SEM (**a–h**) and TEM (**i**, **j**) images of CFs. **a–c** Pristine CFs. **d** CFs coated with Ni-doped mesoporous TiO_2 films. **e–j** CFs coated with GNFs [134]

successful coating of this film on the CF surfaces can directly facilitate the uniformity and density of the GNFs grown on the CFs through strong interactions with the Ni catalysts [142]. After catalytic C_2H_2 decomposition at lower temperature, the CF surfaces are covered entirely by the uniform growth of GNFs with a thickness of up to 15.4 μm, as shown in Fig. 9.38e–h. Figure 9.38e, f shows entangled bundles of GNFs, and the diameters of the GNFs range from 60 to 80 nm. The morphologies of the grown GNFs have also been analyzed using TEM and are shown in Fig. 9.34i, j. The GNFs are uniform and show a herringbone-type alignment in which the graphene planes are at an angle to the fiber axis [157].

Figure 9.39 shows the changes in the contact angle (CA) and the surface properties of the CFs. As shown in this Fig. 9.39, without any treatment, the water droplet retains an ellipsoidal shape on the CFs, with a CA of 27.2°, suggesting that the raw CF surfaces exhibit a hydrophilic character. Interestingly, the water drop rapidly passes through the CFs and then completely wets the CFs, as illustrated in Fig. 9.39a. The CA of CF–FP increases to 121.5° after being coated with a fluoropolymer because of the hydrophobic CF_3 groups, which mimic the wax epicuticular on the surface of a lotus leaf to decrease the surface free energy (Fig. 9.35b). Previous studies have reported that F-coated CF surfaces are hydrophobic (CA > 90°) [176]. The sharp increase of the CA from 27.2° to 145.3° in the case of CF–GNF (Fig. 9.39c) originates from the 3D GNF networks. As many works have revealed, surface roughness plays an important role in determining the wetting behavior of solid surfaces [142, 172, 173]. Generally, fluorinated compounds and silanes are typically low surface energy compounds and can endow various substrates with high hydrophobicity [174, 175]. In accordance with this, the CA of CF–GNF–FP after treatment with fluoropolymer increases to 153.5°, and the surface shows superhydrophobicity (Fig. 9.39). The falling water droplet could bounce up from the CF–GNF–FP surfaces, indicating the low contact angle

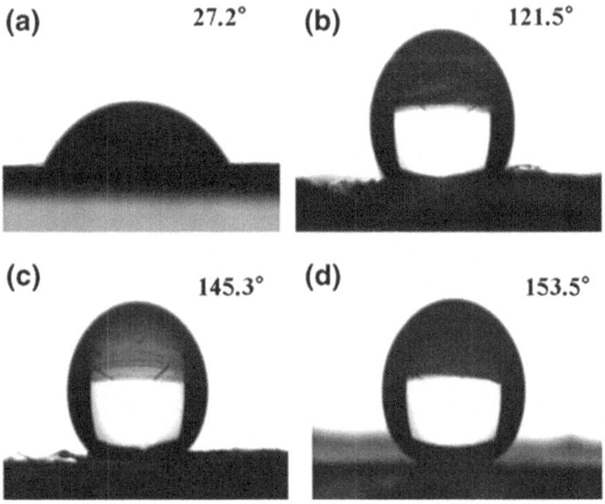

Fig. 9.39 Changes in water-contact angle of CFs. **a** Pristine CFs. **b** Fluoropolymer-coated CFs. **c** CFs coated with GNFs. **d** Fluoropolymer-coated CF–GNF [143]

hysteresis. The superhydrophobic CF–GNF–FP surface is believed to be caused by the presence of synergistic binary geometric roughness at the micro/nanometer scale and the low surface energy functional groups, which reduce the surface energy.

The cyclic voltammetry and galvanostatical charge/discharge test in a KOH electrolyte is considered as a suitable tool for estimating the electrochemical properties of CFs before and after the chemical treatment. Figure 9.40a shows the cyclic voltammetry curves of pristine CFs and CFs coated with GNFs at a voltage scan rate of 20 mV/s. It shows the voltammetry profiles of all the carbon samples. CF–GNF and CF–GNF–FP show marked deviation from the ideal rectangular shape, which is typical and is ascribed to the formation of many micro- or mesopores. The GNFs show the graphite-like structure, in which the stacking of nano-sized graphene is quite uniform along the whole fiber axis [172]. An evident increase in the current density of CF–GNF composites is clearly observed because of the higher specific surface area and higher electrical conductivity enhanced by the growth of GNFs [174]. The galvanostatical charge/discharge curves of CFs and CFs coated with GNFs tested at a current of 2 A/g are also shown in Fig. 9.40b. It can be seen that the charge/discharge curves of CFs were very symmetrical and triangular. After the growth of GNFs on the CF surfaces, the times required for both charge/discharge processes are longer than those observed for pristine CFs. The specific capacitance of CF–GNF composites was 19 F/g, which is more than that of the pristine CFs (2 F/g) and CF–GNF–FP (15 F/g). After chemical treatment with a fluoropolymer, the specific capacitance of CF–GNF–FP is lower than that of CF–GNF because of the lower specific surface area and higher ionic transfer resistance. The results indicate that the more nanoporous structure of GNFs can effectively reduce the diffusion length and the diffusion time of ions and thus lead to rapid charge–discharge characteristics at the same scanning rate.

The future of carbon fibers is bright. Traditionally, the application of carbon fibers has been limited to very special fields such as aerospace and military because of their high cost. However, various techniques for low-cost carbon fibers are under development in terms of using low-cost precursors, low-cost manufacturing processes, and even functional coating methods. Moreover, future applications of carbon fibers can widen not only as structural reinforcements but also in information technology-based applications such as housings for electric devices, smart cloths, and healthcare items. In particular, EMI shielding can be a new field because EMI shielding materials normally need high mechanical strength, good electric conductivity, and good EMI absorbability. In this respect, carbon fibers coated with functional materials can be optimal materials.

However, the bottleneck inhibiting the real application of carbon fibers is its price. The price of carbon fibers is approximately $25–80/kg. For automobile applications, the price of carbon fibers must be under $15/kg. The price in the case of application in electric devices must be under $30/kg.

Hence, continued research on carbon fibers is necessary to achieve the goal of producing low-cost and highly functional future carbon fibers .

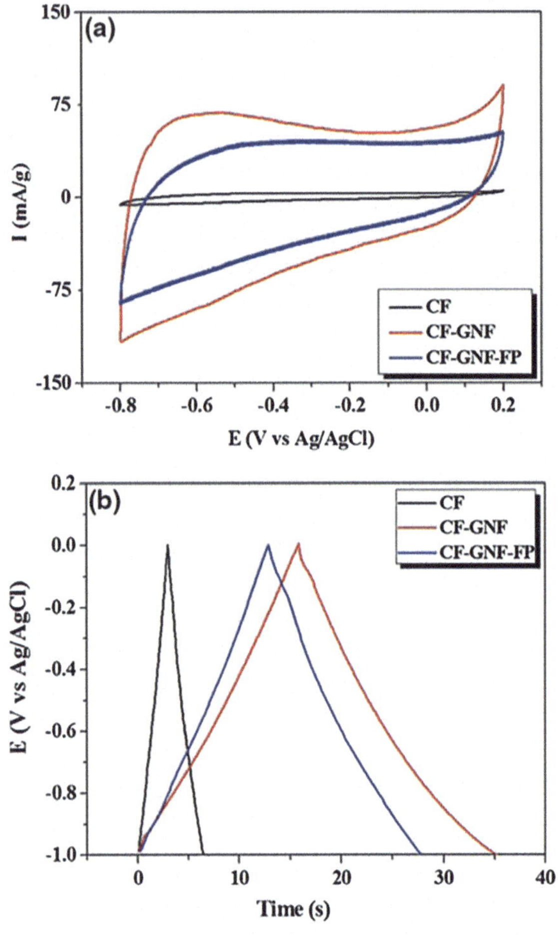

Fig. 9.40 **a** Cyclic voltammogram. **b** Galvanostatic charge/discharge curves of pristine CFs, GNFs coated CFs, and fluoropolymer-coated CF–GNF in 6-M KOH electrolyte [143]

References

1. C.D. Warren, *Future Lower Cost Carbon Fiber for Autos: International Scale-up & What is Needed* (Oak Ridge Nation Laboratory, USA)
2. E. Sjostrom, *Wood Chemistry* (Academic Press, San Diego, 1993)
3. J.F. Kadla, S. Kudo, R.A. Venditti, R.D. Gilbert, A.L. Compere, W. Griffith, Carbon **40**, 2913 (2002)
4. W. Qin, J.F. Kadla, Ind. Eng. Chem. Res. **50**, 12548 (2011)
5. S.P. Maradur, C.H. Kim, S.Y. Kim, B.H. Kim, W.C. Kim, K.S. Yang, Synth. Met. **162**, 453 (2012)
6. O. Faruk, A.K. Bledzki, H.P. Fink, M. Sain, Prog. Polym. Sci. **37**, 1552 (2012)
7. S.W. Beckwith, SAMPE J. **44**, 64 (2008)
8. D. Cho, J.M. Kim, I.S. Song, I. Hong, M1ater. Lett. **65**, 1492 (2011)

9. D.A. Baker, T.G. Rials, J. Appl. Polym. Sci. **130**, 713 (2013)
10. J. Lin, K. Koda, S. Kubo, T. Yamada, M. Enoki, Y. Uraki, J. Wood Chem. Technol. **34**, 111 (2014)
11. M. Thunga, K. Chen, D. Grewell, M.R. Kessler, Carbon **68**, 159 (2014)
12. S. Chatterjee, A. Clingenpeel, A. Mckena, O. Rios, A. Johs, RSC Adv. **4**, 4743 (2014)
13. M. Zhang, A.A. Ogale, Carbon **69**, 626 (2014)
14. D.D.L. Chung, *Carbon Fiber Composites* (Butterworth-Heinemann, Boston, 1994)
15. J. Bijwe, J. Indumathi, J.J. Rajesh, M. Fahim, Wear **249**, 715 (2001)
16. B.S. Gupta, W. George, Abs. Pap. Am. Chem. Soc. **37** (1972)
17. B.S. Gupta, W. George, Textile Res. J. **45**, 338 (1975)
18. H.S. Kim, H.H. Cho, H. Ito, T. Kikutani, N. Okui, Seikei Kako. **9**, 449 (1997)
19. K. Twarowska-Schmidt, D. Grochowska-Lapienis, Fib. Tex. East. Eur. **8**, 37 (2000)
20. B. Marcher, Tappi J. **74**, 103 (1991)
21. J.H. Southern, D.H. Martin, D.G. Baird, Textile Res. J. **50**, 411 (1980)
22. D. Godshall, C. White, G.L. Wilkes, J. Appl. Polymer. Sci. **80**, 130 (2001)
23. Z.M. Huang, Comp. Sci. Techol. **63**, 2223 (2003)
24. L.C. Wadsworth, S.R. Malkan, Rev. Melt Blowing Technol. INB Nonwovens **2** (1991)
25. W.J.G. McCulloch, Int. Nonwovens J. **8**, 66 (1999)
26. B. Vishal, R.L. Shambaugh, Ind. Eng. Chem. Res. **37**, 1799 (1998)
27. T. Kaino, Kobunshi Ronbunshu **46**, 257 (1985)
28. J. Doshi, D.H. Reneker, J. Electrostatics **35**, 151 (1995)
29. M. Bognitzki, W. Czado, T. Frese, Adv. Mater. **13**, 70 (2001)
30. E. Zussman, D. Rittel, A.L. Yarin, Appl. Phy. Lett. **82**, 3958 (2003)
31. A. Elzatahry, Int. J. Electrochem. Sci. **9**, 22 (2014)
32. A.A. Taha, A.A. Hriez, Y.N. Wu, H. Wang, F. Li, J. Colloid Interface Sci. **417**, 199 (2014)
33. M. Endo, Chem. Tech. **18**, 568 (1988)
34. C. Journet, W.K. Maser, P. Bernier, A. Loiseau, M. Lamy dela Chapelle, S. Lefrant, P. Deniard, R. Lee, J.E. Fischer, Nature **388**, 756 (1997)
35. A. Thess, R. Lee, P. Nikolaev, H. Dai, P. Petit, J. Robert, C. Xu, Y.H. Lee, S.G. Kim, A.G. Rinzler, D.T. Colber, G.E. Scuseria, D. Tomanek, J.E. Fischer, R.E. Smalley, Science **273**, 483 (1996)
36. S. Maity, L.N. Downen, J.R. Bochinski, L.I. Clarke, Polymer **52**, 1674 (2011)
37. Y. Zhang, M. Park, H. Kim, B. Ding, S.J. Park, Appl. Surf. Sci. **384**, 192 (2016)
38. Y. Zhang, M. Park, H. Kim, S.J. Park, J. Colloid Interf. Sci. **500**, 155 (2017)
39. M.S.P. Shaffer, A.H. Windle, Adv. Mater. **11**, 937 (1999)
40. D. Matatagui, M.J. Fernández, J. Fontecha, I. Sayago, I. Grácia, C. Cané, M.C. Horrillo, J. P. Santos, Talanta **120**, 408 (2014)
41. W.J. Jin, H.K. Lee, E.H. Jeong, W.H. Park, J.H. Youk, Macr. Rap. Commun. **26**, 1903 (2005)
42. C. Yin, J. Dong, Z. Li, Z. Zhang, Q. Zhang, Comp. Part B Eng. **58**, 430 (2014)
43. M. Peng, D. Li, L. Shen, Y. Chen, Q. Zheng, H. Wang, Langmuir **22**, 9368 (2006)
44. N. Yusof, A.F. Ismail, J. Anal. Appl. Pyrol. **93**, 1 (2012)
45. V.I. Kostikov, *Fibre Science and Technology* (Chapman and Hall, London, 1995)
46. P. Rajalingam, G. Radhakrishnan, JMS-Rev. Macromol. Chem. Phys. **31**, 301 (1991)
47. S. Moon, R.J. Farris, Carbon **47**, 2829 (2009)
48. C.K. Liu, R.J. Sun, K. Lai, C.Q. Sun, Y.W. Wang, Mater. Lett. **62**, 4467 (2008)
49. I.H. Chen, C.C. Wang, C.Y. Chen, J. Phys. Chem. C **114**, 13532 (2010)
50. R.A. Mrozek, B.S. Kim, V.C. Holmberg, T.A. Taton, Nano Lett. **3**, 1665 (2003)
51. P. Katta, M. Alessandro, R.D. Ramsier, G.G. Chase, Nano Lett. **4**, 2214 (2004)
52. M.F. Yu, O. Lourie, M.J. Dyer, K. Moloni, T.F. Kelly, R.S. Ruoff, Science **287**, 637 (2000)
53. S. Frank, P. Poncharal, Z.L. Wang, W.A. de Heer, Science **280**, 1744 (1998)
54. J. Hone, M. Whitney, A. Zettl, Synth. Met. **103**, 2498 (1999)
55. S. Berber, Y.K. Kwon, D. Tomanek, Phys. Rev. Lett. **84**, 4613 (2000)

56. Q.W. Li, Y. Li, X.F. Zhang, S.B. Chikkannanavar, Y.H. Zhao, A.M. Dangelewicz, L.X. Zheng, S.K. Doorn, Q.X. Jia, D.E. Peterson, P.N. Arendt, Y.T. Zhu, Adv. Mater. **19**, 3358 (2007)
57. T. Qian, X. Zhou, C. Yu, S. Wu, J. Shen, J. Mater. Chem. A **1**, 15230 (2013)
58. F. Tardani, P. Strobbia, A. Scipioni, C. La Mesa, RSC Adv. **3**, 25917 (2013)
59. J. Marchewski, J.B. In, D. Poulikakos, C.P. Grigoropoulos, Carbon **68**, 308 (2014)
60. R. Hamzaoui, S. Guessasma, B. Mecheri, A.M. Eshtiaghi, A. Bennabi, Mater. Design **56**, 60 (2014)
61. Y.H. Zhang, K.Y. Rhee, S.J. Park, Compos. B **114**, 111 (2017)
62. Y.N. Zhang, L.X. Zheng, Nanoscale **2**, 1919 (2010)
63. J.B. Xu, K.F. Hua, G.Z. Sun, X.Y. Lv, Y.J. Wang, Electrochem. Comm. **8**, 982 (2006)
64. Y. Zhang, K. Rhee, S.J. Park, Compos. B Eng. **114**, 111 (2017)
65. A. Star, Y. Liu, K. Grant, L. Ridvan, J. Fraser Stoddart, D.W. Steuerman, M.R. Diehl, A. Boukai, J.R. Heath, Macromolecules **36**, 553 (2003)
66. M.S.P. Shaffer, A.H. Windle, Macromolecules **32**, 6864 (1999)
67. B. Vigolo, A. Penicaud, C. Coulon, C. Sauder, R. Pailler, C. Jonrnet, P. Bernier, P. Poulin, Science **290**, 1331 (2000)
68. C. Mercader, A. Lucas, A. Derré, C. Zakri, S. Moisan, M. Maugey, P. Poulin, Proc. Natl. Acad. Sci. U.S.A. **107**, 18331 (2000)
69. B. Vigolo, P. Poulin, M. Lucas, P. Launois, P. Bernier, Appl. Phys. Lett. **81**, 1210 (2002)
70. E. Muñoz, A.B. Dalton, S. Collins, M. Kozlov, J. Razal, J.N. Coleman, B.G. Kim, V.H. Ebron, M. Selvidge, J.P. Ferraris, Adv. Eng. Mater. **6**, 801 (2004)
71. P. Miaudet, S. Badaire, M. Maugey, A. Derré, V. Pichot, P. Launois, P. Poulin, C. Zakri, Nano Lett. **5**, 2212 (2005)
72. A.B. Dalton, S. Collins, E. Munoz, J.M. Razal, V.H. Ebron, J.P. Ferraris, J.N. Coleman, B. G. Kim, R.H. Baughman, Nature **423**, 703 (2003)
73. C. Lynam, S.E. Moulton, G.G. Wallace, Adv. Mater. **19**, 1244 (2007)
74. M.E. Kozlov, R.C. Capps, W.M. Sampson, V.H. Ebron, J.P. Ferrais, R.H. Baughuman, Adv. Mater. **17**, 614 (2005)
75. K.L. Jiang, Q.Q. Li, S.S. Fan, Nature **419**, 801 (2002)
76. M. Zhang, S. Fang, A.A. Zakhidov, S.B. Lee, A.E. Aliev, C.D. Williams, K.R. Atkinson, R. H. Baughman, Science **309**, 1215 (2005)
77. M. Zhang, K.R. Atkinson, R.H. Baughman, Science **306**, 1358 (2004)
78. Y. Zhang, S.J. Park, Carbon **122**, 287 (2017)
79. C. Feng, K. Liu, J.S. Wu, L. Liu, J.S. Cheng, Y.Y. Zhang, Y.H. Sun, Q.Q. Li, S.S. Fan, K.L. Jiang, Adv. Funct. Mater. **20**, 885 (2010)
80. Y.H. Sun, K. Liu, J. Miao, Z. Wang, B. Tian, L. Zhang, Q. Li, S. Fan, K. Jiang, Nano Lett. **10**, 1747 (2010)
81. Y. Zhang, S.J. Park, J. Catal. **355**, 1 (2017)
82. L.X. Zheng, M.J. O'Connell, S.K. Doorn, X.Z. Liao, Y.H. Zhao, E.A. Akhadov, M.A. Hoffbauer, B. Roop, Q.X. Jia, R.C. Dye, D.E. Peterson, S.M. Huang, J. Liu, Y.T. Zhu, Nat. Mater. **3**, 673 (2004)
83. X.F. Zhang, Q.W. Li, T.G. Holesinger, P.N. Arendt, J. Huang, P.D. Kirven, T.G. Clapp, R.F. DePaula, X. Liao, Y. Zhao, L. Zheng, D.E. Peterson, Y.T. Zhu, Adv. Mater. **19**, 4198 (2007)
84. X.F. Zhang, Q.W. Li, Y. Tu, Y. Li, J.Y. Coulter, L.X. Zheng, Y.H. Zhao, Q.X. Jia, D.E. Peterson, Y.T. Zhu, Small **3**, 244 (2007)
85. Q.W. Li, X.F. Zhang, R.F. DePaula, L.X. Zheng, Y.H. Zhao, L. Stan, T.G. Holesinger, P.N. Arendt, D.E. Peterson, Y.T. Zhu, Adv. Mater. **18**, 3160 (2006)
86. N.K. Mehra, V. Mishra, N.K. Jain, Biomaterials **35**, 1267 (2014)
87. Z.Y. Zhan, Y.N. Zhang, G.Z. Sun, L.X. Zheng, K. Liao, Appl. Surf. Sci. **257**, 7704 (2011)
88. L. Zheng, B.C. Satishkumar, P.Q. Gao, Q. Zhang, J. Phys. Chem. C **113**, 10896 (2009)
89. S.K. Doorn, L.X. Zheng, M.J. O'Connell, Y.T. Zhu, S.M. Huang, J. Liu, J. Phys. Chem. B **109**, 3751 (2005)
90. J.J. Jia, J.N. Zhao, G. Xu, Carbon **49**, 1333 (2011)

91. X.B. Zhang, K.L. Jiang, C. Feng, Adv. Mater. **18**, 1505 (2006)
92. Y. Wei, K.L. Jiang, X.F. Feng, Phys. Rev. B. **76**, 7 (2007)
93. V.K. Rangari, M. Yousuf, S. Jeelani, M.X. Pulikkathara, V.N. Khabashesku, Nanotechnology **19**, 245703 (2008)
94. L.J. Lanticse, Y. Tanabe, K. Matsui, Carbon **44**, 3078 (2006)
95. H.H. Gommans, J.W. Alldredge, H. Tashiro, J. Park, J. Magnuson, A.G. Rinzler, J. Appl. Phys. **88**, 2509 (2000)
96. V. Pichot, M. Burghammer, S. Badaire, Europhys. Lett. **79**, 46002 (2007)
97. L.X. Zheng, G.Z. Sun, Z.Y. Zhan, Small **6**, 132 (2010)
98. T.W. Chou, A. Kelly, A. Okura, Composites **16**, 187 (1985)
99. K.S. Novoselov et al., Science **306**, 666 (2004)
100. A.K. Geim, Science **324**, 1530 (2009)
101. C.N.R. Rao et al., Angew. Chem. Int. Ed. **48**, 7752 (2009)
102. L.Q. Liu, W.J. Ma, Z. Zhang, Small **7**, 1504 (2011)
103. W. Lu et al., Adv. Mater. **24**, 1805 (2012)
104. N. Behabtu et al., Science **339**, 182 (2013)
105. N. Behabtu, M.J. Green, M. Pasquali, Nano Today **3**, 24 (2008)
106. Z. Xu, C. Gao, Nat. Commun. **2**, 571 (2011)
107. Z. Xu, C. Gao, Acc. Chem. Res. **47**, 1267 (2014)
108. Y. Zhang, K. Rhee, D. Hui, S.J. Park, Compos. B Eng. **143**, 19 (2017)
109. A.J. Rodriguez, M.E. Guzman, C.S. Lim, B. Minaie, Carbon **49**, 937 (2011)
110. Y. Zhang, S.J. Park, J. Polym. Sci. Pol. Phys. **55**, 1890 (2017)
111. K. Rhee, S.J. Park, D. Hui, Y. Qiu, Comp. Part B **43**, 2395 (2012)
112. S.J. Park, B.J. Kim, Mater. Sci. Eng., A **408**, 269 (2005)
113. G.Y. Heo, Y.J. Yoo, S.J. Park, J. Ind. Eng. Chem. **19**, 1040 (2013)
114. L.Y. Meng, S.J. Park, Curr. Appl. Phys. **13**, 640 (2013)
115. S.Y. Lee, S.J. Park, J. Colloid Interface Sci. **389**, 230 (2013)
116. L.Y. Meng, K.Y. Rhee, S.J. Park, J. Ind. Eng. Chem. **20**, 1672 (2014)
117. S.Y. Lee, S.J. Park, Carbon **68**, 112 (2014)
118. J.R. Choi, Y.S. Lee, S.J. Park, J. Ind. Eng. Chem. **20**, 3421 (2014)
119. M. Li, H. Liu, Y. Gu, Y. Li, Z. Zhang, Appl. Surface Sci. **288**, 666 (2014)
120. S.J. Park, Y.S. Jang, J. Colloid Interface Sci. **237**, 91 (2001)
121. Y.L. Yang, M.C. Gupta, K.L. Dudley, R.W. Lawrence, J. Nanosci. Nanotechnol. **5**, 927 (2005)
122. J. Joo, A.J. Epstein, Appl. Phys. Lett. **65**, 2278 (1994)
123. Y.L. Yang, M.C. Gupta, K.L. Dudley, R.W. Lawrence, Adv. Mater. **17**, 1999 (2005)
124. S. Kim, S.J. Park, J. Electrochim, Acta **53**, 4082 (2008)
125. B.J. Kim, W.K. Choi, H.S. Song, J.K. Park, J.Y. Lee, S.J. Park, Carbon Lett. **9**, 105 (2008)
126. S.J. Park, B.J. Kim, Y.S. Lee, M.J. Cho, Int. J. Hydrogen Energy **33**, 1706 (2008)
127. S. Lee, J. Kim, B.C. Ku, Y. Chung, Carbon Lett. **12**, 26 (2011)
128. C.S. Ramesh, H. Adarsha, S. Pramod, Z. Khan, Mater. Design **50**, 597 (2013)
129. S.M. Goushegir, J.F. dos Santos, S.T. Amancio-Filho, Mater. Design **54**, 196 (2014)
130. S. Tengsuwan, M. Ohshima, J. Supercritical Fluids **85**, 123 (2014)
131. B.J. Kim, W.K. Choi, K.M. Bae, C.W. Moon, H.S. Song, J.K. Park, J.Y. Lee, S.S. Im, S. J. Park, Bull. Korean Chem. Soc. **32**, 1630 (2011)
132. S. Lee, J. Kim, B.C. Ku, J. Kim, Y. Chung, Carbon Lett. **12**, 26 (2011)
133. S. Abraham, B.C. Pai, K.G. Satyanarayana, V.K. Vaidyan, J. Mater. Sci. **25**, 2839 (1990)
134. B.J. Kim, W.K. Choi, M.K. Um, S.J. Park, Surf. Coat. Tech. **205**, 3416 (2011)
135. S. Shaikh, K. Lafdi, Carbon **48**, 813 (2010)
136. C.T. Hsieh, W.Y. Chen, Carbon **48**, 612 (2010)
137. T.Q. Li, Z.H. Xu, Z.J. Hu, X.G. Yang, Carbon **48**, 924 (2010)
138. K.S. Kim, K.Y. Rhee, K.H. Lee, J.H. Byun, S.J. Park, J. Ind. Eng. Chem. **16**, 572 (2010)
139. D.H. Lee, J.E. Kim, T.H. Han, J.W. Hwang, S. Jeon, S.Y. Choi, S.H. Hong, W.J. Lee, R.S. Ruoff, S.O. Kim, Adv. Mater. **22**, 1 (2010)

140. J.Y. Gu, K.X. Li, J. Wang, H.W. He, Micropor. Mesopor. Mater. **131**, 393 (2010)
141. P. Li, T. Li, J.H. Zhou, Z.J. Sui, Y.C. Dai, W.K. Yuan, D. Chen, Micropor. Mesopor. Mater. **95**, 1 (2006)
142. L.Y. Meng, C.W. Moon, S.S. Im, K.H. Lee, J.H. Byun, S.J. Park, Micropor. Mesopor. Mater. **142**, 26 (2011)
143. L.Y. Meng, S.J. Park, Mater. Chem. Phys. **132**, 324 (2012)
144. L.Y. Meng, S.J. Park, Curr. Appl. Phys. **13**, 640 (2013)
145. S. Rahmanian, A.R. Suraya, R. Zahari, E.S. Zainudin, Appl. Surf. Sci. **271**, 424 (2013)
146. C.A. Dyke, Z. Ophir, J. Appl. Phys. **113**, 024313 (2013)
147. X. Liu, X. Yin, L. Kong, Q. Li, Y. Liu, W. Duan, L. Zhang, L. Cheng, Carbon **68**, 501 (2014)
148. L. Zeng, W. Wang, D. Lei, J. Liang, Y. Xia, H. Zhao, X. Kong, Carbon **46**, 359 (2008)
149. S. Lim, S.H. Yoon, Y. Shimizu, H. Jung, I. Mochida, Langmuir **20**, 5559 (2004)
150. N. Sonoyama, M. Ohshita, A. Nijubu, H. Nishikawa, H. Yanase, H. Hayashi, T. Chiba, Carbon **44**, 1754 (2006)
151. B.J. Kim, S.J. Park, J. Colloid Interface Sci. **315**, 791 (2007)
152. R.B. Mathur, S. Chatterjee, B.P. Singh, Compos. Sci. Technol. **68**, 1608 (2008)
153. S.S. Tzeng, K.H. Hung, T.H. Ko, Carbon **44**, 859 (2006)
154. P.H. Maheshwari, R.B. Mathur, Electrochim. Acta **54**, 7476 (2009)
155. L. Zeng, W. Wang, D. Lei, J. Liang, H. Zhao, J. Zhao, X. Kong, Phys. B **403**, 2662 (2008)
156. C.T. Hsieh, W.Y. Chen, F.L. Wu, Carbon **46**, 1218 (2008)
157. D. Sebastian, I. Suelves, R. Moliner, M.J. Lazaro, Carbon **48**, 4421 (2010)
158. D. Sebastian, I. Suelves, M.J. Lazaro, R. Moliner, J. Power Sources **192**, 51 (2009)
159. S. Takenaka, S. Kobayashi, H. Ogihara, K. Otsuka, J. Catal. **217**, 79 (2003)
160. Y.F. Guan, R.C. Pearce, A.V. Melechko, D.K. Hensley, M.L. Simpson, P.D. Rack, Nanotechnology **19**, 235604 (2008)
161. X. Sun, B. Stansfield, J.P. Dodelet, S. Desilets, Chem. Phys. Lett. **363**, 415 (2002)
162. S. Morgan, R. Mokaya, J. Phys. Chem. C **112**, 15157 (2008)
163. D. Barreca, W.J. Blau, G.M. Croke, F.A. Deeney, F.C. Dillon, J.D. Holmes, C. Kufazvinei, M.A. Morris, T.R. Spalding, E. Tondello, Micropor. Mesopor. Mater. **103**, 142 (2007)
164. F. Zheng, L. Liang, Y. Gao, J.H. Sukamto, C.L. Aardahl, Nano Lett. **2**, 729 (2002)
165. S.J. Park, J.S. Shin, J.W. Shim, S.K. Ryu, J. Colloid Interface Sci. **275**, 342 (2004)
166. Q. Luo, H. Zhu, Y. Zhou, G. Zheng, D. Zhao, J. Mater. Chem. **11**, 2934 (2001)
167. J.T. Feng, Y.J. Lin, D.G. Evans, X. Duan, D.Q. Li, J. Catal. **266**, 351 (2009)
168. B.J. Kim, Y.S. Lee, S.J. Park, Int. J. Hydrogen Energy **33**, 4112 (2008)
169. A.B. Gacia, I. Camean, I. Suelves, J.L. Pinilla, M.L. Lazaro, J.M. Palacios, R. Moliner, Carbon **47**, 2563 (2009)
170. B.J. Kim, S.J. Park, Nanotechnology **17**, 4395 (2006)
171. J.L. Pinilla, I. Suelves, M.J. Lazaro, R. Moliner, J.M. Palacios, Int. J. Hydrogen Energy **35**, 9801 (2010)
172. R. Blossey, Nat. Mater. **2**, 301 (2003)
173. L.Y. Meng, S.J. Park, J. Colloid Interface Sci. **342**, 559 (2010)
174. L.Y. Meng, S.J. Park, Macromol. Res. **19**, 209 (2011)
175. J. Qu, L. He, Carbonhyd. Polym. **98**, 1056 (2013)
176. W.K. Cho, S.J. Park, S.Y. Jon, I.S. Choi, Nanotechnology **18**, 395602 (2007)

Chapter 10
Prospective and Outlook of Carbon Fibers

Carbon materials have been used by human beings in their living environment for a long time. In modern times, they have been widely applied as high-tech materials through various processes of research and development. Among the diverse carbon materials, CFs have been researched because of attractive characteristics such as lightweight, superior mechanical, electrical, thermal, and chemical properties. However, despite the fact that CFs have existed for quite some time, the synthesis of CFs with custom tensile strength and high modulus fibers from a precursor has yet to be completed. This is due in large part to lack of understanding of the polymer orientation, crystal size, crystallinity, and mechanism of how the polymer structure of chosen precursor converts to the structure of the resulting CFs. In addition, the effect of the CF structure as mentioned above on mechanical properties such as tensile strength and Young's modulus, thermal and electrical conductivity is far from being comprehended. To reduce currently high-production costs and produce CFs with desired properties, it is essential to understand the structure and structure formation of CFs on a molecular scale.

The CFs-based composites were prepared by blending CFs and polymer such as epoxy resin. The composites primarily designed for high-strength structural applications, for instance, aircraft and automobiles. The surface incompatibility between CFs and resin causes negative effect on the physical and mechanical performance of resulting composites. Improving compatibility between CFs and resin will be a key factor in the use of CFs-based composites. Unique surface treatment technologies, such as oxidation, electrochemical and plasma treatments, electron-beam irradiation pretreatment, will be a key technology to achieve this. If the precursor, synthesis conditions, and surface treatment techniques are optimized, not only the strong exoskeleton but also the energy problem will be solved (Fig. 10.1).

With an increasing consumption of CFs, the demand for CFs in the coming two years is expected to at least double as shown in Table 10.1. As soon as possible, alternative raw precursors for CFs must be identified and performed. Not only the

S.-J. Park, *Carbon Fibers*, Springer Series in Materials Science 210,
https://doi.org/10.1007/978-981-13-0538-2_10

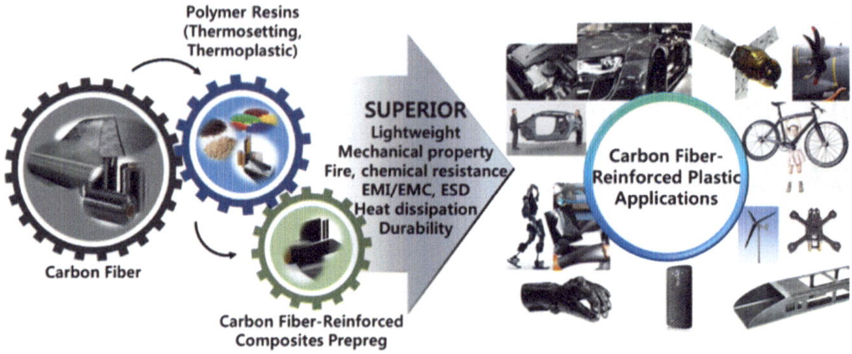

Fig. 10.1 Carbon fiber-reinforced composites for modern and future applications

Table 10.1 Global annual carbon fiber use and prospects

Field	2006	2008	2010	2011	2016	2020
Consumer goods	5900	6700	7000	9000	10,800	13,300
Industrial	12,800	15,600	17,500	29,800	68,500	120,000
Aerospace	6500	7500	9800	7000	13,100	19,700
Total	25,200	29,800	34,200	45,800	92,400	153,000

aerospace and automotive industry but also the construction, power, and consumer goods industry will require such novel, low-cost fibers.

Recently, in addition to CFs, the development of large-area sheets and fibers using graphene-based carbon materials has been actively pursued. In particular, research into a form applicable to portable smart devices and wearable electronic devices is in full swing. With the increasing interest in such wearable devices, there is an increasing demand for energy storage devices designed for them. Especially, the electrodes to be mounted on electronic devices are required not only to be inexpensive, but also to have nature such as flexibility, lightweight, and large size. To satisfy the requirements, electrode materials that can be produced in various forms, including films, fibers, and yarns, and maintain energy storage performance even when bent or twisted, have been developed. Many studies using graphene or CNTs as electrode materials have been carried out steadily. To make graphene films or CNT fibers directly, CVD methods have been applied using various transition metal catalysts such as Ni, Cu, and hydrocarbon as carbon sources. However, CVD methods require very high temperature and are not suitable for manufacturing large-area films or long fiber; hence, the CVD methods are inefficient for commercial use. To produce graphene or CNT films or fibers in a large scale, there are the approaches of preparing in a solution. The dispersions in which graphene-based carbon materials are dispersed can be made into fibers or films by methods such as vacuum filtration, drop casting, wet-spinning, and electrospinning. However, fibers or films prepared from their dispersions have some drawbacks such as relatively low mechanical strength and still high cost. Therefore, recently, in order to develop cost-effective, enhanced flexibility, and mechanical performance, several methods including carbon material coating or deposit on a desired substrate (e.g., cellulose-based materials or polymers) have been developed. If the developed techniques, including simplification of the manufacturing process, stability, and uniformity for commercialization, allow carbon materials to be stabilized and supplied smoothly, they will play a key role in solving the energy problem in the future.

Index

S.-J. Park, *Carbon Fibers*, Springer Series in Materials Science 210,
https://doi.org/10.1007/978-981-13-0538-2

R

The manufacturer's authorised representative in the EU is Springer Nature Customer Service Centre GmbH, Europaplatz 3, 69115 Heidelberg, Germany. If you have any concerns regarding our products, please contact ProductSafety@springernature.com

Printed and bound by CPI Group (UK) Ltd, Croydon, CR0 4YY
22/09/2026
02220948-0018